Preserving Wildlife

Flashpoints

Series Editor: Roger S. Gottlieb
             Worcester Polytechnic Institute

PRESERVING WILDLIFE

An International Perspective

EDITED BY
Mark A. Michael

Humanity Books

an imprint of Prometheus Books
59 John Glenn Drive, Amherst, New York 14228-2197

Published 2000 by Humanity Books, an imprint of Prometheus Books

Preserving Wildlife: An International Perspective. Copyright © 2000 Mark A. Michael. All rights reserved. No part of this publication may be reproduced, stored in a retrieval system, or transmitted in any form or by any means, electronic, mechanical, photocopying, recording, or otherwise, without prior written permission of the publisher, except in the case of brief quotations embodied in critical articles and reviews. Inquiries should be addressed to Humanity Books, 59 John Glenn Drive, Amherst, New York 14228–2197. VOICE: 716–691–0133, ext. 207. FAX: 716–564–2711.

04 03 02 01 00   5 4 3 2 1

Library of Congress Cataloging-in-Publication Data

Preserving wildlife : an international perspective / edited by Mark A. Michael.
    p.   cm.
  Includes bibliographical references.
  ISBN 1-57392-727-9 (pbk. : alk. paper)
  1. Wildlife conservation. I. Michael, Mark A., 1954–
QL82.P77 1999
333.95'416—dc21   99–38312
                                                                                           CIP

Printed in the United States of America on acid-free paper

Contents

Introduction 7

PART I: INDIVIDUALS AND WILDLIFE PRESERVATION

1. Environmental Ethics and the Case Against Hunting
   Roger J. H. King 21
2. The Medical Treatment of Wild Animals
   Robert W. Loftin 47
3. Ethical Considerations and Animal Welfare in Ecological Field Studies
   R. J. Putman 57

PART II: STRATEGIES FOR CONSERVATION AND MANAGEMENT

4. Launching the Natural Ark
   James R. Udall 77
5. Weeding the Garden
   Andrew Neal Cohen 84
6. The Olympic Goat Controversy: A Perspective
   Victor B. Scheffer 93
   Response to Scheffer
   Cathy Sue Anunsen and Roger Anunsen 96

Reply to the Anunsens
 Victor B. Scheffer   101

7. Captive Breeding of Endangered Species
 Robert W. Loftin   106

8. Helping a Species Go Extinct: The Sumatran Rhino in Borneo
 Alan Rabinowitz   122

PART III: PEOPLE, POLITICS, AND WILDLIFE

9. Approaches to Conserving Vulnerable Wildlife in China: Does the Color of Cat Matter—If It Catches Mice?
 Richard B. Harris   139

10. Local Challenges to Global Agendas: Conservation, Economic Liberalization and the Pastoralists' Rights Movement in Tanzania
 Roderick P. Neumann   167

11. Is This the Way to Save Africa's Wildlife?
 Victoria Butler   189

12. Wildlife Conservation Outside Protected Areas: Lessons from an Experiment in Zambia
 Dale Lewis, Gilson B. Kaweche, and Ackim Mwenya   194

13. Wildlife Conservation in Tribal Societies
 Raymond Hames   210

PART IV: UTILIZING WILDLIFE

14. Economics: Theory Versus Practice in Wildlife Management
 Raymond Rasker, Michael V. Martin, and Rebecca L. Johnson   239

15. Elephants and Economics
 Graeme Caughley   263

16. Kangaroo Harvesting and the Conservation of Arid and Semi-Arid Rangelands
 Gordon Grigg   268

17. Tourism as a Sustained Use of Wildlife: A Case Study of Madre de Dios, Southeastern Peru
 Martha J. Groom, Robert D. Podolosky, and Charles A. Munn   274

18. Going, Going, Gone
 Jim Mason   296

Contributors   305

Introduction

On the surface, the claim that we should preserve wildlife is unobjectionable. Few people would argue that driving elephants or pandas or whales to extinction would result in an insignificant loss. There is a consensus that humanity's interactions with wildlife in the past were based on ignorance or self-interest at best and maliciousness at worst, that the decimation of wildlife throughout the world that stemmed from these attitudes is tragic and reprehensible, and that we need to act to reverse current trends and to ensure the continued existence of wildlife for the sake of ourselves, future generations, and perhaps even the animals themselves. But once we move beyond decrying the state of elephants, pandas, whales, and other charismatic species, and the conversation turns to hard policy questions such as how we should go about preserving wildlife, what species should be preserved and at what cost, who is to bear those costs, and what other values and priorities should be sacrificed for the sake of wildlife preservation, the consensus breaks down and serious disagreement takes its place.

Furthermore, while people generally support the concept of wildlife preservation when it is someone else who must do the preserving, a NIMBY ("not in my backyard")-like reaction often surfaces when people are asked to bear the costs of preserving wildlife in their backyards. Wildlife preservation is a marvelous idea when it occurs somewhere else, and at no cost to us. It is exciting to travel and see wildlife in exotic locales. But no one wants to come home and have to worry about tigers attacking her children or elephants destroying her garden. The preservation of wildlife does have costs, however, and it is intellectually irresponsible to ignore this fact. Herbivores can wreak havoc on crops, carnivores can cause extensive losses to ranchers. Opportunity costs must also be factored into calculations of the cost of preserving wildlife. Many species, particularly large carnivores, have extensive range requirements; the land they need could be devoted to more profitable and perhaps even more socially beneficial uses. Additionally, when people refrain from killing wildlife, they forego whatever profit they could make from the sale of dead wildlife parts such as leopard skins and elephant tusks.

This anthology highlights some of the hard issues that must be addressed and resolved in the quest to preserve wildlife. We want to know which methods of preserving wildlife are most effective, whether there are any moral limits to how we should carry out our preservation efforts, and what should be done when preservation efforts run counter to the interests of humans. These questions are broadly philosophical in that their solutions will either implicitly or explicitly invoke some theory of value. But the solutions must also incorporate findings concerning animals and their behavior, the ecosystems that contribute to their flourishing or foundering, human behavior and political feasibility, and the likely outcomes of various conservation strategies. For example, if we think we have a duty to preserve wolves, we need to know something about their range requirements, their reproductive rates and the sort of population level which is needed to ensure a healthy population, and the likely responses of people who live in proximity to wolves, especially when wolves cause economic losses. Thus in discussing the issues that will concern us here, we ignore the findings of biology, economics, sociology, and political science at our own risk. The issue of wildlife preservation demonstrates the inadequacy of the once prevalent dogma that questions of fact and questions of value can and should be investigated independently of one another.

THE SELECTIONS

While all of the papers in this anthology start from the assumption that wildlife should be preserved, a variety of visions concerning exactly what we are trying to preserve, and why we should preserve it, are represented here. These disparate visions give rise to a range of strategies for translating the principle of wildlife preservation into a workable policy, and for determining whether the policy is morally permissible. It is these concerns that are highlighted in the present collection.

Part I: Individuals and Wildlife Preservation

The papers in the first section focus on the moral permissibility of human activities that directly affect individual wild animals. They are perhaps the most explicitly philosophical of all the papers in this collection. Roger King's paper looks at the ethics of sport hunting. Sport hunting is a unique form of hunting because there is no goal to the activity; hunters engage in the hunt for its own sake. While the hunted animal may in fact be eaten, securing food is not its primary purpose; sport hunters usually have full refrigerators. Sport hunting differs in this respect from subsistence hunting, in which an individual hunts in order to feed himself or his family, and hunting for markets, which involves hunting with the intention of selling some product derived from the animal on the open market. These other forms of hunting will be touched upon in subsequent sections of the anthology.

Sport hunting typically takes place under controlled conditions. It usually requires the purchase of a license, and these are issued on the basis of biological findings concerning the population size and reproductive rates of the prey species. Bag limits are set to ensure that there will be no detrimental effects on the overall population. In these controlled conditions, sport hunting may serve the same functions once performed by natural predators and so may be used by wildlife managers as a means of managing populations. There is, consequently, little chance that a successful hunting season will have any serious effects on the overall well-being of the prey species. Inasmuch as sport hunting only affects individual wild animals, what, if any, moral considerations are at work with respect to this activity? King's paper addresses this issue.

The selection by Robert Loftin asks whether we are under a moral obligation to try to cure and rehabilitate animals that are injured in the wild. Loftin suggests that if we are concerned with environmental wholes such as ecosystems, then the fate of individual animals has no moral dimension. Loftin defends a view known as holism, arguing that while there is nothing wrong with engaging in the medical treatment of wildlife, we have no moral obligation to do so. The failure to aid individual wild animals is not wrong.

Finally, we might wonder whether there are any moral constraints on biologists and ecologists when they are engaged in field research. What moral limits are there on our search for information about animals and ecosystems? While knowledge is surely a good thing, there are some moral constraints on what we may do to obtain knowledge when, for example, we are trying to learn about human psychology or physiology. Putnam's paper explores this issue with respect to animals in the wild. What kind of acquisition of knowledge would justify harming individual animals in the course of conducting experiments or observing their behavior? May we harm an individual animal if the knowledge we gain as a result will improve the well-being of other members of its species, or other animals generally?

Part II: Strategies for Conservation and Management

In an ideal world, in which people went out of their way to keep environmentally harmful activities to a bare minimum, a laissez-faire approach to wildlife preservation might be the most effective and rational policy. In such a world the whole concept of managing wildlife would be misguided; the best way to preserve wildlife would be to leave it alone. But in our actual, far from perfect world, that sort of attitude is counterproductive. We have not gone out of our way to preserve either the environment or individual species. Human encroachment on the habitat of almost all wild species has taken a variety of forms: overhunting; clear-cutting land for farming; altering landscapes and ecosystems in order to exploit natural resources at minimal expense; and polluting rivers, fields, and the air in the process of industrial and agricultural expansion. Human activity has driven species to extinction and damaged ecosystems, many beyond repair, and this makes a hands-off approach to wildlife preservation a nonstarter. We must repair damage that has already occurred in order to create the conditions that will make it possible for threatened species to rebound, and we need to pursue strategies and policies that will give wild species a reasonable chance of flourishing in the future.

The question of what sorts of policies will be most effective in achieving these goals involves problems of management. These problems are largely biological, and many of the issues here are highly technical and esoteric. Ecologists and biologists need to gather data and then perform extrapolations based on various models to determine the effects of factors such as population growth, predator-prey ratios, and genetic isolation on species health. They must then make predictions about what conditions must exist at present in order for a given species to have a reasonable chance of persisting for a given number of years into the future. I have not included that sort of material here. Rather, the papers included in this section confront the reader with more general issues that arise from attempts to ensure the health of wild species. The question that is the main focus of this section is this: What sort of steps must be taken to achieve and maintain healthy populations of wild species, both endangered and nonendangered?

It is probably no exaggeration to suggest that the most pressing problem confronting those involved in efforts to preserve endangered species and wildlife in general is the loss of habitat. What seems a rather obvious point has too often been ignored or forgotten: Wildlife cannot be preserved in a way that incorporates saving what makes it valuable, as opposed to saving wild animals as museum curiosities, without simultaneously saving the habitat and ecosystems in which it thrives. Any effort to save and protect endangered species without a parallel effort to protect habitat will be futile. But how is this to be achieved, given the tremendous pressures being exerted on wildlife habitat by humans? How much habitat is required, and how free of human activity does it need to be?

In the opening paper of this section, James Udall discusses some of the changes in thinking among conservation biologists and wildlife managers concerning the need for habitat protection. Most now believe that preservation efforts should be focused on habitats and ecosystems rather than on specific endangered or key species, since neither of these is as effective a method of preserving true biological diversity as the protection of habitats. But what form should habitat protection take? In the past, the drive to maintain habitats led to the establishment of national parks and nature preserves. The problem with this, as pointed out by Udall, is that the amount of land needed to maintain a healthy population of grizzly bears or mountain lions, for example, is staggering. The amount of land currently protected as part of the national park system is inadequate if the goal is to maintain healthy populations of all native species, including these large carnivores. But it is unrealistic to think that the large amount of additional land needed for these species could be incorporated into the park system without significant political opposition. So the national parks can safeguard only relatively small and isolated pockets of habitat, and these cannot sustain sufficiently large populations of many species. The end result will be inbred populations, and the genetic weakening that results from this process will make these populations susceptible to diseases that could ultimately decimate each population. These small populations would also be vulnerable to climatic changes and other natural perturbations. Thus, the effect of isolating populations in national parks is likely to be the eventual elimination of each separate population, and, with the death of the final population, the end of the species.

Udall stresses the need to develop a new model of ecosystem and species management which eliminates this fragmentation of habitats. This new model would involve the creation of multiple-use buffer zones around highly protected core areas, with the expectation that wildlife and humans could coexist in these areas. Wildlife would be protected from human predation in these buffer zones, but human activities that did not severely degrade or permanently damage the ecosystem would be permitted. Examples of these uses include recreation, sustainable agriculture, and perhaps even light industry. The crucial element of this model is the provision of corridors through which wild animals could travel safely from one core reserve to another, thereby ensuring a healthy, interbreeding species. It is absolutely critical to design these corridors in such a way that they can guarantee safe movement for animals. Unless such models are adopted, the prognosis for many species in the United States, especially large carnivores, is not good.

The next two articles address the more active and intrusive management strategies that are sometimes thought necessary. Species can become overabundant within whatever area is left to them after human encroachment, and when this occurs the population must be actively controlled. This kind of situation usually occurs when natural predators have been removed from an ecosystem. Overabundant species can have seriously harmful effects, both on other species within the ecosystem and on the ecosystem itself. In these situations, decisions must be made about whether or not culling is wise, likely to succeed, and morally justifiable. Of course, we might wonder whether we can give objective, biological criteria to define the concept of overabundance, or whether a species will be said to be overabundant whenever it stands in the way of human projects and plans. Andrew Cohen's article offers a personal look at the sorts of dilemmas faced by wildlife managers when they must resort to the destruction of overabundant animals.

Victor Scheffer's article focuses on the problem of alien species. Once introduced into ecosystems—usually by humans, often inadvertently—these species either outcompete or directly prey on native species. The result is the destruction of the native species, which upsets the ecological balance of that ecosystem. Scheffer argues that mountain goats are not native to Olympic National Park, and because their numbers have steadily increased since their introduction in the 1920s, they are currently having a deleterious effect on some plant species in the park. Hence they should be removed, destroyed if necessary. In their response to Scheffer's article, Cathy Sue and Roger Anunsen argue that there is some evidence that the goats are, in fact, native to the park and so should be protected rather than destroyed. They also argue that the damage caused by the goats is not as extensive as claimed by Scheffer, and that much of it is attributable to actions of park biologists in the 1960s. This discussion is of interest not only because it offers insights into the specific problem of goats in Olympic National Park, but also because it illustrates that problems of this sort are neither wholly scientific nor wholly moral, but involve elements of both, and so cannot be resolved without addressing both.

The final two articles in this section look at the management of species under extremely critical conditions. One method for saving a species on the brink of extinction is to initiate a captive breeding program. Robert Loftin's paper raises questions about the effectiveness of captive breeding and reintroduction programs. After presenting a number of case studies, he concludes that these programs are justified only under fairly narrow conditions. One question raised by Loftin concerns the expense of these programs. Is this strategy really cost-effective? Could not the money and effort devoted to these programs be spent more efficiently by directing them to in situ conservation activities?

The latter question comes to the fore in Alan Rabinowitz's article on the Sumatran rhino. In spite of all sorts of financial expenditures and international efforts to save this species through captive breeding programs, the plight of the Sumatran rhino has become more rather than less dire. Rabinowitz argues that what is needed is in situ preservation. This will require both an increase in the amount of land devoted to rhino habitat, and consequently not available for economic development, and a serious effort to end poaching. But policies that advance these goals will never be put in place if we focus solely on the biological dimensions of the problem, since it is human activities and desires, largely for financial gain, that drive both poaching and shrinking rhino habitat, thereby pushing the rhino inexorably toward extinction. This section of the anthology shows that while we have a pretty good idea about what sort of changes will have to be made if we are to preserve wildlife, we still need to know how to get people to accept and implement those changes. That problem is the focus of the next two sections of the anthology.

Part III: People, Politics, and Wildlife

No matter which management strategies turn out to be most effective, they will not be implemented in a social or economic vacuum, and so must be not only biologically efficacious but politically and socially feasible as well. It is one thing, for example, to point out that large carnivores need vast tracts of habitat if their population levels are to remain healthy; it is something else to convince people to accept policies that provide the habitat when this appears contrary to their own well-being and self-interest. The papers in the third section all explore the social and political ramifications of wildlife preservation.

For better or worse, there are now so many humans that almost all wildlife habitat is caught in a competition between those who want to develop the land and those who favor leaving it to wildlife. The tension between these two outlooks is especially acute in many Third World countries as a result of the confluence of two factors. First, ecosystems in Third World countries have suffered much less intense exploitation than those in industrialized nations, so there are more possibilities and opportunities to preserve wildlife and biotic richness in these countries. Second, however, because of relatively steep increases in the human population of many Third World nations, there is tremendous pressure to develop resources that are still in the ground. This demand is exacerbated by the fact that many people in these nations are impoverished and view development as a means of eliminating their poverty. If the West had a right to develop by exploiting their own natural resources and driving some species to extinction, why shouldn't Third World nations have that right as well? People in the Third World may even view wildlife and ecosystem preservation as one more gambit by the "haves" of the world for impeding development, progress, and ultimately economic independence for the "have nots." Historically, formal conservation was often instituted by the colonial powers, who maintained all sorts of prerogatives for their own people at the expense of the indigenous local people. For example, while local people were usually prevented from hunting on what had once been tribal lands, the colonial governments often allowed settlers and trophy hunters to hunt in these areas. Thus, in the minds of many, conservation is associated with colonial rule.

In the first paper in this section, Richard Harris examines attitudes toward wildlife in China. The upshot of this paper is that the ideal of wildlife preservation that dominates Western thinking, in which wildlife is preserved and protected for its own sake and in which any mention of the utilization of wildlife is considered heretical, simply will not work in all societies. Forcing this model on people will only cause a backlash; the result will be greater hostility toward wildlife and, consequently, its destruction. Therefore, if our goal is to preserve wildlife, some compromises will have to be made in our dealings with other societies that will enable them to derive benefits from wildlife, at least for the immediate future. If we insist on maintaining a rigidly preservationist stance, then no accord will be reached, and the species we want to save are likely to become extinct.

Issues of justice must be addressed in tandem with issues of political feasibility. Significant human needs and aspirations often come into conflict with the goals of wildlife preservation, and these must be weighed alongside the interests of wildlife. Relatively wealthy tourists or hunters from industrialized nations will pay to exploit wildlife in a Third World country. They get what they want from wildlife. The central government and tour operators located in the capital of a foreign country benefit economically from the fees they charge and the services they provide. Local people, however, frequently do not share in the benefits of the presence of wildlife on their lands. Yet they are expected to bear the burden of living in close proximity to wild animals that destroy crops, that use resources such as vegetation and water that might be consumed by their domesticated stock, and that may injure and even kill them. Is it really fair or just to expect local people who are already impoverished to willingly bear these costs? After all, why should the burden of saving wildlife be placed on the rural poor in Third World countries when wealthy ranchers in the United States often are unwilling to bear less imposing costs, and when the government appears unwilling to force them to bear such costs?

Of course, hard-nosed environmentalists may argue that while this is unfortunate, wildlife must be protected, regardless of its effects on local people. But if local attitudes toward wildlife deteriorate when it appears that wildlife is being protected at the behest of environmental organizations representing a constituency based in industrialized nations, we need to ask whether even draconian measures will ensure wildlife's continued existence. Roderick P. Neumann's paper presents a case study in some of the problems that develop when wildlife preservation is perceived as a policy foisted on local populations by a central government as a result of pressure from foreign conservation organizations and because of the potential economic gains that will be enjoyed by the government and those in positions of power.

Many governments, and some conservation organizations, have decided that this unfair distribution of burdens and benefits needs to be rectified if conservation is to be successful, and experimental programs have been implemented that seek to devolve both the decision making and the economic benefits of wildlife preservation to the people who live with it. People in Third World societies have, in many cases, existed in close proximity to wildlife for hundreds and thousands of years, so it may be possible to devise solutions that encourage coexistence, rather than a sharp separation between people and wildlife. It has been argued that the whole concept of national parks is a thoroughly Western idea and an inappropriate model for Third World countries. These experimental programs reject the Western paradigm of parks and nature preserves that completely exclude economic activities by local people inside the boundaries of the preserve. Instead, they allow local people to benefit economically from the presence of wildlife. The question here is this: Do these programs really work? Do they promote conservation goals while simultaneously benefiting local people?

The article by Victoria Butler offers a personalized account of how well CAMPFIRE, which is perhaps the best known of these experimental programs, is working. Dale Lewis, Gilson B. Kaweche, and Ackim Mwenya discuss a project that seeks to protect elephants and rhinos in the Luangwa Valley of Zambia by giving local people greater control over wildlife. The most controversial aspect of these programs is that they allow for limited hunting of elephants and other wildlife as long as the hunting will have no significant effect on population levels. These programs seem to be working, but we need to ask whether such projects will continue to work and whether they can be implemented successfully on a large scale and in other societies. We also need to raise the moral issue of whether this kind of utilization, which involves the death of highly evolved mammals, is morally justifiable.

Where the interests of wildlife and local populations are opposed to one another, central governments will have either to coerce people not to destroy wildlife, which is likely to work only if harsh punishments are introduced for crimes such as poaching, or to readjust the distribution of burdens and benefits through political action, so that the interests cease to be in conflict. But must the interactions of local people and wildlife otherwise always be adversarial? This may be an overly pessimistic assessment. Perhaps technologically primitive people are somehow naturally more attuned to their own ecosystems and know how to live in harmony with them. If that were the case, large wildlife preserves might be created which would be protected from the depredations of outsiders, but which could be used in traditional ways by the aboriginal inhabitants of these areas. Ray Hames's paper looks at whether this idea of an ecologically noble savage who has a natural sense of how to maintain a healthy balance in his own ecosystem has any basis in reality in the Amazon.

Part IV: Utilizing Wildlife

Advocates of free-market economics believe that wildlife can best be preserved by utilizing it, and this view has gained ground among some conservationists as well. The articles in the final section look at various aspects of this claim. The central question here is whether allowing wildlife to take on economic value will contribute to its preservation. Those who answer in the affirmative argue that people only voluntarily preserve something when it is in their own best interest to do so, and the only way preserving wildlife can be in an individual's or group's interest is by allowing wildlife or products derived from wildlife to be bought and sold in markets. Of course this will require turning wildlife into a commodity. But if what is valuable is not the individual animal but rather the species in the ecosystem, there is no reason not to sacrifice individuals if it will help to preserve the species. The fact that individual animals will die is of no moral relevance to the holist, as long as the level of killing required by the demands of the market is not so high that it outstrips the species' ability to reproduce itself.

The paper by Raymond Rasker, Michael Martin, and Rebecca Johnson, which opens this section, is a wide-ranging overview of the pros and cons of markets in wildlife products. They examine possible sources of market failure that are especially pertinent to wildlife, and offer some suggestions concerning how these might be overcome through governmental intervention. They conclude that there is no general answer to the question of whether or not markets in wildlife will contribute to conservation; rather, each case must be looked at individually to determine the unique features that make it more or less amenable to market solutions. Graeme Caughley's paper offers a response, showing just how unique each situation is by comparing the status of elephants in Zimbabwe and Kenya.

One of the arguments for opening up markets and utilizing wildlife is that in many cases the animals that are killed and used would have to be killed anyway in managing for conservation, and it seems better to kill something and then use it than to kill it and let it rot. Gordon Grigg invokes this consideration, among others, in the course of arguing for a marketing drive to encourage markets in kangaroo meat in Australia. Kangaroos are not endangered there and often must be culled because of the destructive impact they have on grasslands. He claims that markets in kangaroo meat would have a salubrious effect on the Australian grasslands and the kangaroos themselves, as well as economically benefiting ranchers with wild kangaroos on their lands.

Ecotourism is one use of wildlife that is nonconsumptive. Many see it as a win/win situation that is compatible with any moral outlook. Local people will have a motive to protect wildlife because of its economic benefits, and yet this activity usually will not result in harm to the wildlife. Some conservationists have expressed concern about the impact that additional human traffic will have both on the habitat of the animals being observed and on their behavior. Others are concerned that ecotourism will work as a conservation tool only where enough of its financial benefits remain in the local economy. The paper by Martha Groom, Robert Podolsky, and Charles Munn looks at some of these issues in the course of discussing the efficacy of ecotourism in an area of the Peruvian Amazon.

The final paper is a cautionary tale. Jim Mason's discussion of the trade in exotics and hybrids shows what can happen when market ideology is pushed to its logical conclusion. The commerce in freaks, novelty animals, and wild animals that should never be kept as pets is illustrative of how perverse human attitudes can be toward wildlife. It shows how far we have yet to go in trying to change the attitudes that have led to the current abysmal state of wildlife on our planet.

FINAL NOTES ON METHOD

Three important decisions had to be made in compiling this anthology. One of them was substantive, the other stylistic, and the third a question of format. The topic of this anthology is the preservation of wildlife, but that term is vague; exactly what is to count as wildlife? I have adopted the mean between two extremes. The broadest account would include all nondomesticated animals, possibly even insects, under the rubric of wildlife. While from a biological perspective there is no reason not to include insects and while in some contexts doing so makes perfectly good sense, people do not expect a discussion of bugs when they pick up a book on wildlife. This does not mean that I think insects are unimportant or unworthy of study. It is just that they appear to be outside our everyday notion of what constitutes wildlife. On the other hand, wildlife might be construed very narrowly so that only fairly large, nondomesticated mammals would count. This, I think, is overly narrow. When people talk about and express concern for the future of wildlife, I think they are generally worried about the continued existence of nondomesticated reptiles, birds, and fish, as well as mammals. And while much of what is said in this anthology picks out large mammals as exemplars, most of it can be applied with equal appropriateness to these other forms of wildlife.

The stylistic issue involved whether to include highly technical articles or to opt instead for selections in which technical and biological terms were absent, more akin to anecdotal narratives in popular publications than to what one might find in a professional journal. Again, I have sought a middle ground here. A few articles from popular magazines have been included to introduce a topic or to give the reader a feel for the lay of the land. I thought it important to include selections which personalized these issues. While this may detract from the seamless (and possibly dry) tone of the anthology, it is important to put a human face—or in some cases, an animal face—on these issues. The bulk of the selections, however, were taken from professional journals in fields such as biology, sociology, and philosophy. I tried to use articles that kept the use of technical terms to a minimum and could be understood by a patient and inquisitive reader. Most readers are quite content to read with a dictionary within arm's reach, and one of the pleasures of reading in a new field is learning at least some of the terminology that is employed in that field. The present collection, which resulted from this selection process, is stylistically eclectic, but that is by design.

The third question was whether or not to present these issues in a pro and con format. This anthology was conceived as an introduction to the problems surrounding wildlife preservation, so no attempt has been made to lay out the full range of solutions to each of the problems highlighted here. It would have been difficult to do that in a single volume anyway. Thus, almost everyone who reads this book will find one or another of the solutions included to be ethically unpalatable, politically impractical, or morally oversensitive. I think that is all to the good. If this anthology provokes people to think more deeply about wildlife preservation, it will have achieved its purpose.

Part I
Individuals and Wildlife Preservation

1

Environmental Ethics and the Case Against Hunting

Roger J. H. King

Introduction

Much has been written in recent years on the welfare of animals. Both utilitarians and rights theorists have attacked the mainstream tradition in which "moral considerability" is denied to nonhuman species.[1] According to these critics, we are inconsistent when we restrict membership in the moral community to human beings alone because some animals possess the traits which are commonly thought to be relevant to membership, while some human beings do not.[2] This animal liberationist critique has focused on human use of domesticated animals by questioning the morality of animal experimentation, factory farming, and meat eating in general. In contrast, however, proponents of Aldo Leopold's land ethic have argued that the individualistic bias of animal liberation theories misses an important dimension of the ecological situation.[3] An ecosystem is a community of interrelated beings and processes that has an integrity and stability of its own. Moral attention, they argue, should be paid primarily to the ecosystem as a whole, rather than to the individual members of that system. A holistic ethic geared to respecting the integrity of the whole is more in tune with the ecological reality of biotic communities than a liberationist individualism.

When we look at the conflict between animal liberation and the land ethic from the standpoint of ecofeminist thinking, however, the disagreement takes on the appearance of a fraternal rivalry. Neither position includes a theoretical accounting of the social and political contexts within which human relations to

---

This article originally appeared as "Environmental Ethics and the Case for Hunting" in *Environmental Ethics* 13 (spring 1991): 59–85. Reprinted by permission. Copyright © 1991 Roger J. H. King. All rights reserved.

nature are embedded. According to ecofeminists, such as Mary Daly and Andree Collard, ignoring the social and political context of an individual's action significantly restricts our understanding of the depth and seriousness of the environmental crisis, and hence minimizes the usefulness of such theories for guiding us in radical change.[4]

An examination of the morality of hunting presents an opportunity to evaluate the conflicts between these three perspectives on environmental ethics. In this paper, I argue that inquiry into the case for hunting reveals serious inadequacies in the perspectives of both the liberationist and the land ethic approaches to environmental ethics. Although the land ethic marks an advance over the animal liberationist critique of traditional ethics, it fails to address the question of the place human beings have in the biotic community successfully. As a result, the land ethic treatment of hunting is unconvincing. Moreover, I argue that an ecofeminist critique of hunting reveals an important and neglected perspective on the significance of this practice. In this discussion, I expand discussion of the land ethic to include what I call the primitivist positions advocated by José Ortega y Gasset and Paul Shepard because their views are consistent with the land ethic and deal more explicitly with the problem of the human's place in the biotic community.[5]

Because hunting is such a complex issue, it is important to state at the outset why it warrants philosophical attention. First, hunting presupposes the deliberate and intentional infliction of pain and death on individual animals. This killing affects not only the individual animal, but the species and the biotic community of which the individual is a member. Because environmental ethics has emerged in contemporary philosophy as a response to the destructive consequences of human interventions in the natural world, hunting stands out prima facie in need of moral justification.

Second, to hunt means to kill wild, not tame or domesticated, animals. It is an activity which, unlike farming or animal experimentation, directly interferes with undomesticated biotic communities. When people hunt, therefore, they step out of the human-animal community established by human cultures.[6] For Shepard and Ortega y Gasset, this is a natural activity of a predatory and carnivorous human species. It is an expression of our participation in nature and way of overcoming the alienation from nature of our industrialized Western culture. Hunting, therefore, poses specific moral problems of its own that should not be reduced to those of other forms of human interaction with animals.

Finally, hunting is an activity carried on primarily by males, and hunters often celebrate it for this very reason. Ecofeminism has drawn philosophical attention to the existence of parallels between the treatment of women and the treatment of nonhuman nature. Andree Collard, for example, attacks hunting as a symptom of the violence and aggression inherent in a patriarchal culture. On this view, we cannot fully separate our moral judgment of the legitimacy of hunting from our moral assessment of the broader cultural patterns of which it is a constitutive part. Moral inquiry must, therefore, address hunting from the standpoint of the role it plays in constituting a peculiarly male outlook on the natural world. A moral assessment of hunting ultimately plays a part in the moral assessment of a male-dominated culture.

These three features of hunting highlight different dimensions of a single human activity. They also indicate that different kinds of questions can be asked about the morality of this activity. In trying to answer these questions we are forced to seek the context within which hunting most truly reveals itself. It is disagreement about the proper place from which to evaluate hunting, as much as disagreement about the moral evaluation itself, which currently separates many discussions of the morality of the hunt.

To hunt is to perform an act, localized in time and space. Yet hunting is also at the same time a social practice embedded in a broader social and political context which constitutes its meaning, its implications for nature, and the modes of belief which surround it. The aboriginal hunter, the ancient Assyrian king, the medieval poacher, the Victorian trophy hunter, and the modern sports hunter all kill animals. It would be mistaken to suppose, however, that they all perform the same act. The weapons used, the game pursued, the reasons and justifications which they offer, the symbolic functions which their hunting performs, the legal restrictions which apply to it, and the impact on the ecological situations in which they hunt, are all different. A fully developed inquiry into the morality of sport hunting must, therefore, address both the act and the context which constitutes the act, and understand the unavoidable relationship which links the two.

In this essay, I examine four different attempts to understand exactly what hunting is, and how it should be assessed. In sections 2 and 3, I argue that both the animal liberationist critique of humanist ethics and the land ethic are unsatisfactory in part because they do not recognize the contextual nature of moral acts. In section 4, I argue, on the other hand, that while the primitivist position does situate hunting, it has incorrectly presented and defended its interpretation of the proper context in which hunting should be evaluated. In section 5, I argue that the ecofeminist position provides a fruitful standpoint from which to interpret the relationship between hunting and its broader social context. I conclude that a certain kind of hunting, namely sport hunting, is not morally justified. By sport hunting I mean the practice of intentionally killing wild animals for reasons other than the need for survival. This definition applies to virtually all hunting in North America. In evaluating the morality of the hunt, we necessarily end up embedding that critique in the analysis of the broader social, political, and human context of which hunting is merely a sign.

THE LIBERATIONIST CRITIQUE

Peter Singer is one of the most important of those philosophers currently drawing attention to the moral status of animals.[7] As a utilitarian, he claims that we have a moral obligation to minimize the pain and suffering that are the consequences of our actions. Singer offers strong prima facie grounds for taking the morality of sport hunting seriously. Since hunting necessarily involves the infliction of pain and death, it is incumbent upon us to query the arguments that are raised in its defense. In the case of sport hunting, the hunter does not depend upon killing the animal in order to survive. For this reason, such hunting appears to be gratuitous and unnecessary.

A hunter with utilitarian propensities, however, might argue that the hunter's pleasure could at times outweigh the animal's pain. In such cases, no absolute condemnation of hunting would apply. Nevertheless, the validity of the objection raised here is hard to assess. Although it is conceivable that the ecstasy felt by a hunter who successfully kills an animal might outweigh the pain felt by the animal—especially if the animal is one of those rare victims of the perfect shot—it will be difficult to decide whether it also outweighs the distress caused to other animals linked in some way to the one who was killed. If we admit that such distress, or at least confusion, could be a consequence of killing an animal, then the calculation of the pains and pleasures of hunting risks losing the determinacy necessary for it to render a serious moral judgment in utilitarian terms. The hunter's objection to Singer then does little to show that hunting is morally justified as a general practice, despite the theoretical loophole it finds in the utilitarian position. At the same time, however, it helps to reveal one of the difficulties with the utilitarian approach.

The indeterminacy of the utilitarian calculation arises from the multiplicity of the empirical elements to be weighed. The more that is relevant to evaluating the particular kill, the more specious our ultimate conclusions become. Because the same imponderables are involved in hunting as a general practice as are involved in individual acts of killing, rule utilitarianism can offer no assistance here.[8] The notion that we might reach a truly calculated judgment of the morality of hunting loses some of its plausibility if we consider a few of the elements that must be weighed in the moral balance.

Suppose, for example, that in addition to the pain of the wound, the distress and confusion of other animals linked in sociality with the prey, and the exultation at the hunter's prowess, we add other relevant consequences of hunting: the economic benefits of hunting to the state and to the commercial and tourist industries, the benefits to those game species which have overpopulated their habitats, the losses to the natural predators who are deprived of food and are themselves killed because they are competing with the sports hunter, and the distress felt by the antihunting public.

Utilitarianism requires that we calculate the balance of pleasure and pain, of values and disvalues, benefits and costs, which these additional consequences embody. Only then can the utilitarian make a clear moral judgment about hunting, as opposed to a speculative, intuitive guess. Calculating this balance, however, presupposes that all factors can be quantified and made commensurable with one another. Because quantitative weight cannot be obtained for some of the consequences, the utilitarian calculus fails to provide an unambiguous defense or critique of the morality of hunting. As a result, the hunter's objection that he may feel greater pleasure at killing than he causes pain to the prey or to others turns out to be specious and unsubstantiated.

Seeking the support of utilitarian arguments, defenders of hunting have also argued that hunting is morally justifiable because it acts as an escape valve for people with violent tendencies.[9] Because shooting animals is preferable to shooting people, they claim that the utilitarian ethic must condone hunting in these cases. But this conclusion does not follow. At best, hunting is justified only if no alternatives are available for identifying and neutralizing these violent and sociopathic dispositions. Even supposing no such alternatives to exist, the hunter gains utilitarianism's support for the act of shooting an animal only by identifying himself as a person with dangerous criminal tendencies—a person to be feared both because he is a physical threat and because he is morally vicious—a tactic which, therefore, does not provide a satisfactory way around the prima facie utilitarian condemnation of hunting either.

In contrast to Singer's utilitarian critique of speciesism, Tom Regan offers a deontological alterative.[10] Hunting animals, on this view, violates the inherent value which these animals possess. It threatens their welfare, their ability to pursue their natural ends, and their ability to determine, within the scope of their species' instincts, what is best for them.[11] Of course, the inherent value and rights of animals may conflict with the rights of human individuals, for example, if a person were starving and had no vegetable matter to eat. In such a moral conflict of absolute rights, the deontologist must defend an ordering of these rights to show the priority of one over the other. Nevertheless, sport hunting does not provide us with a conflict of absolute rights, and thus fails to find moral support from the deontological branch of the animal liberationist position.

Despite the differences between the consequentialist and the deontological approach to animal liberation, the arguments of Singer and Regan share one significant feature. They identify moral considerability with the possession of traits which attach to individuals. Does that particular animal feel pain or distress? Is that particular animal a being possessing consciousness of self? Neither of the two approaches leaves much room for a defense of hunting. Nevertheless, despite the fact that Singer and Regan identify considerations which are prima facie relevant to any deliberation on the morality of hunting animals for sport, there is something wrong with the scope of the arguments themselves. The individualistic orientation of animal liberation theories ignores the ecological reality of the biotic community. Animals regularly inflict pain on one another. Animals kill one another. Predation, disease, and extinction are integral moments in a natural existence.[12] It would be foolish to say that we have a moral obligation to eliminate predation and natural extinction. We perpetrate an anthropocentric fiction when we think of animals in terms of our moral categories of villain and victim. If the intent of environmental ethics is to encourage respect for the autonomous processes of nature and for the animals who are parts of these processes, then we must be careful not to extend to the animal world values and concerns which are primarily the concern of human beings.

The animal liberationist critique of hunting does not adequately acknowledge the context in which hunting takes place. First, it draws no distinction between the domestic and the wild animal. This distinction is essential. Domestic animals live by definition within the ambit of human use and desire; they are part of a "mixed community" of humans and animals that establishes a relationship between the species which is not present between humans and wild animals. Because humans actively breed and raise domestic animals, there is good reason for arguing that we have obligations to the individuals which we cause to be near us. Animal liberation ignores context in a second sense as well. It has nothing to say about the cultural place of hunting as opposed to farming, experimenting, or pet keeping. These practices vary through history and between societies, and the motivations, intentions, and rationales for them vary accordingly. In order to evaluate hunting successfully, therefore, we must ultimately incorporate these contexts into our very understanding of the hunt itself, moving us away from the analysis of animal welfare which Singer and Regan have offered.

In the next section, I examine the land ethic's holistic approach to the question of hunting. This approach emphasizes the domestic-wild distinction and thus offers a different perspective on the morality of the hunt.

Hunting and the Land Ethic

Aldo Leopold's *A Sand County Almanac* is one of the seminal texts in environmental thinking about human relations to nature. Leopold argues that we must rid ourselves of the economic, instrumental attitudes toward land which have guided mainstream American development. The land, he suggests, is a community of beings and processes of which humans are and should be members, rather than some extrinsic and inessential place in which we only contingently happen to live. Being a member, or "plain citizen," of the biotic community, however, carries with it responsibilities which our primarily Western, instrumental understanding of nature ignores.[13] In denying these responsibilities, humans have simplified and destabilized the land community, threatening the viability of many species, exterminating some, and ultimately degrading both the physical and spiritual space of human existence.

J. Baird Callicott has sought to uncover the philosophical grounding for Leopold s "land ethic" by attacking the narrowness of animal liberationist positions. In "Animal Liberation: A Triangular Affair," Callicott uses hunting to highlight the difference between the land ethic and what he calls "humane moralism," the ethics of animal rights proponents.[14] He notes that Leopold, the author of the land ethic, was an avid hunter and an articulate writer on the subject of hunting. Callicott is thus led to suppose that hunting can be defended within the framework of the land ethic, and that therefore the land ethic is grounded differently from animal rights approaches. Unlike Singer and Regan, neither Leopold nor Callicott think that respect for nature and natural processes entails a theory of animal rights or concern about the pain and suffering of individual animals.

Callicott is correct to suppose that the land ethic offers a genuine alternative to moral theories of animal rights. The land ethic is clearly not concerned primarily with the individual animal, but with the role that that animal plays in the broader ecosystem of which it is a part. In addition, the land ethic is designed to resolve the moral questions which arise from human encounters with the nonsentient natural world. Thus, the land ethic does not give special status to those entities which possess features shared by human beings. It does not make pain, or rationality, or language use the criterion of moral "considerability." Most fundamentally, the land ethic is holistic in its orientation, not individualistic.

From the standpoint of the land ethic, the question of whether or not hunting is morally permissible must be answered with reference to Leopold's well-known moral principle: "A thing is right when it tends to preserve the integrity, stability, and beauty of the biotic community. It is wrong when it tends otherwise."[15] Unfortunately, when we look for guidance on the question of hunting, the principle turns out to be seriously ambiguous.

Leopold's principle is ambiguous because it is not clear when the stability, integrity, and beauty of a particular biotic community have been threatened. When a stream is blocked by a landslide, for example, and a pond is created on what was previously forested land, have the stability, integrity, and beauty of the stream ecosystem been destroyed in ways which the principle condemns, or have natural processes simply transformed one biotic community into another? The problem, in other words, has to do with the individuation of the biotic communities to which the principle is intended to apply.

One plausible interpretation of the principle limits its application to alterations of a biotic community which are the product of human actions. In the case above, we need to ask whether the landslide was caused by clear-cutting a slope or by mining activities, for example. The purpose of the principle, it might be argued, is simply to guide our decisions; it is a principle to be used whenever we have choices to make about what sort of impact we will have on a particular ecosystem. If, on the other hand, a biotic community is altered by the autonomous processes of nature, then the principle has no relevance, because these processes themselves are forces of the broader biotic community which the principle enjoins us to respect.

Although this is a plausible reading of Leopold's principle, Callicott does not consistently adopt it when he discusses its application. Callicott suggests that the principle could be used to justify "trapping or otherwise removing" a beaver who is beginning to dam a previously free flowing stream.[16] Although the beaver's activity will result in changes in the flora and fauna of the stream's biotic community, undermining its integrity, stability, and beauty, this application clearly imports human preferences concerning what is desirable and beautiful in the way of biotic communities, and seems to condone fairly radical human interference in the autonomous processes of nature. If this is the correct interpretation of Leopold's principle, then the land ethic does not provide any very serious protection against human manipulation of the environment to suit human conceptions of what is pleasant to live with or look at.

Callicott's claim that the land ethic is compatible with hunting follows from the same interpretation used to condone removing the beaver. Callicott uses the principle to argue against the hunting of predators, but to justify the hunting of deer which have overpopulated their habitat.[17] Predators are essential to the health of the ecosystem, while an overpopulated deer herd becomes diseased and overgrazes the vegetation. But why does this justify hunting? Faced with too many deer, we might instead try to restore the predators which we have eliminated, thereby restoring the stability and integrity of the predator-prey relationship as part of that biotic community. If hunting is offered as a solution instead, then game managers will determine the optimum number of deer that should be permitted to live in the area and the surplus number that may be shot. This solution ensures that what was once a wild biotic community will henceforth be an artfully managed and supervised community in which the fate of wild animals is transferred to human hands from the autonomous forces of the environment. If, further, management is concerned to ensure a viable game herd, that is, deer that may be hunted, then management will necessarily preserve the overpopulated status of the deer in order to guarantee a huntable surplus, and will be wary of too many predators lest they interfere with this surplus. Certainly the biotic community would be healthier, more complex and beautiful, if both predators and prey were part of it. But, in practice, large predator populations are a liability to the sport hunter and to the economic web of relations within which modern sport hunting exists. There is then a practical contradiction between sport hunting and Leopold's ethic, for hunting offers no long-term solution to overpopulation that is consistent with Leopold's principle. At best, hunting could be sanctioned as a temporary measure while predators were reintroduced into the region.

Hunting might, nevertheless, be compatible with the land ethic if it could be argued that human beings are themselves a part of the wild biotic community, and thus that they, as hunters, are playing a constitutive role in sustaining a wild biotic community. In other words, hunting might be defensible from the standpoint of the land ethic if human beings who hunt could lay claim to the status of natural predator. As natural predators, exclusion from the hunt would amount to an injury to the biotic community comparable in kind to the injury caused by the extermination of predators such as wolves, cougars, foxes, bears, and coyotes.

This "primitivist" position is attractive to those who defend hunting and I address it in more detail in the next section. It is important to note, however, that Callicott himself cannot reconcile the land ethic with hunting in this way. In the same article in which he defends hunting, Callicott also argues against the animal liberation movement's support for vegetarianism. This argument prevents him from taking a primitivist approach.

According to Callicott, it is important to distinguish domestic from wild animals. The land ethic's holism is intended to apply to wild animals living and dying in wild biotic communities. Domestic animals, on the other hand, do not have a natural niche in wild biotic communities. They are human artifacts, the living results of human genetic engineering and training.[18] Because the land ethic is concerned with how human beings should relate to the wild biotic communities of nature, it can provide us with no guidance on the morality of our relations to domesticated animals.

From the standpoint of the land ethic, the fate of domestic animals rests on a different foundation from that of wild animals. We may choose to continue using animals which human cultures have created, or we may allow these species to become extinct. But there is no question of "liberating" them and returning them to their "natural" place in the biotic community.[19]

Callicott's treatment of domestic animals is crucial for evaluating his defense of hunting. Domestic animals cannot be returned to the wild for one of three reasons. First, some species have been so engineered by human beings that they cannot survive without constant human attention. Second, some species would simply be killed by natural predators better equipped to live in the wild environment to which the domesticated species are being returned. Third, some domestic species would interfere with the natural functioning of wild species, thereby undermining the stability of ongoing, naturally evolved ecosystems. Because the ethic gives moral priority to the healthy functioning of these autonomous wild communities, it makes no sense to suppose that domesticated animals can, or should, be reinserted into a "natural" place in the wild. As a result, there is no such place for them to go.

The first two reasons which Callicott gives for not liberating domesticated animals make reference to the welfare of the animals themselves. He claims that many could not in fact survive a return to the wild and that therefore their very existence depends upon them remaining of use to human beings. The last reason, however, makes reference to the harm done to the wild animals and plants who would have to compete with the incursion of new domesticated species. Within the context of these three reasons, no noninterventionist alternative is available. We either interfere with wild biotic communities by "liberating" domesticated animals or we continue to support and use the animals we have already domesticated. Letting go does not turn out to be a way of minimizing human impact on nonhuman beings.

If domestic animals cannot be released into the wild in part because they would seriously disrupt ongoing ecological relationships, the same must surely be said for human beings. Human beings have lost the particular wild niche which they once had, just as the animals which they took from the wild and domesticated have lost their place. Not only have most human beings lost the skills necessary to survive in the wild—they would die if "returned" there—but human populations are so large that any attempt to reinsert them directly into wild biotic communities would cause irreparable harm to those communities. In short, the notion that we can simply reinsert ourselves into the biotic community without destroying those communities makes no more sense than the idea of liberating domestic animals so that they can return to the wild. It is this separation of humans from the wild that makes the claim that hunters are natural predators so unconvincing.

The holistic orientation of the land ethic makes it possible for Leopold and Callicott to identify one aspect of the contextual character of hunting, namely, the distinction between the domestic and the wild animal community, but it does not provide us with a very detailed understanding of hunting. The hunter himself remains invisible in this analysis. He is there to "cull" the herd in the interests of the health of the biotic community. Or he is hunting in order to "liberate" himself from his domesticated state. But who is this hunter and how might hunting be "liberating"? The hunter is never just an anonymous cipher, but a member of a particular culture, living at a particular moment in that culture's history. The hunter brings certain technologies to bear on the hunt, together with distinct beliefs and attitudes. A theory of the morality of hunting must, therefore, locate the place of hunting in this broader analysis.

Those who defend hunting, however, are unlikely to be convinced by my brief rejoinder to the claim that human beings are themselves natural predators who have a natural place in the biotic community from which they should not be excluded. To many hunters, and to philosophers who defend hunting, hunting seems to be unlike any other human activity in its potential for returning human beings to nature and to what is essential and "primitive" in human nature. Hunting, it is argued, is not just the isolated killing of animals, but a constitutive part of a way of life that is morally important, even essential, to the well-being of the human species. Any adequate inquiry into the morality of hunting must therefore, confront this defense of the hunter's behavior.

THE PRIMITIVIST DEFENSE

Both animal liberation and the land ethic address the morality of hunting from a decontextualized point of view. Singer and Regan ask us to consider the individual animal who is killed and demand to know by what right human beings arrogate to themselves the license to inflict suffering and violate inherent value. Leopold and Callicott find a place for some kinds of hunting by arguing that some hunting does not undermine the integrity, stability, and beauty of the wild biotic communities which hunters enter. For all of them, the problem of hunting is the problem of the relationship between hunter and nonhuman beings. But this description of hunting is insufficient. Hunting is also a way of life for some people: a way of life that itself may be the object of moral inquiry. In addition, hunting is one practice among the many that constitute a particular culture at a particular time. As such, it may serve as a sign of a particular way of looking at the nonhuman world. It is this contextualized picture of hunting which I discuss in this and the next section.

Ortega y Gasset's *Meditations on Hunting* is perhaps the most influential philosophical defense of hunting.[20] Echoes of his argument can be heard in a number of works, of which Paul Shepard's *The Tender Carnivore and the Sacred Game* is the most fully articulated. Together, these two works offer a sustained attempt to demonstrate that hunting has a place in the fully human life. The roots of the argument are both biological and cultural. Human beings, according to Shepard, retain their prehistoric, paleolithic nature beneath a thin layer of culture. Hunter-gatherer societies can give us glimpses into that "golden age" of human existence which can then serve as the basis for a critique of contemporary industrial-agricultural life. Hunting offers the possibility of returning to our own biological nature.

Ortega's argument starts with the importance of hunting to those who have had the leisure and good fortune to be able to choose the life they wished to lead. Hunting, in this view, is an integral part of the happy life for a human being, because it is an activity which those who are able to choose, namely the nobility and aristocracy, consistently have chosen.[21] Work is not a part of this life; indeed, the contrast between the life of drudgery and the life which includes hunting is a recurring leitmotiv in both Ortega's and Shepard's discussions. Hunting is a part of a return to nature, an escape from the life of labor which both Ortega and Shepard hold in considerable contempt. Ortega concludes that "the most appreciated, enjoyable occupation for the normal man has always been hunting. This is what kings and nobles have preferred to do, they have hunted. But it happens that the other social classes have done or wanted to do the same thing."[22]

Hunting, according to Ortega, has value as an activity, not just as a fleeting pleasure. Hunting is not simply about killing or capturing the animal. Rather, the object of the hunter is the hunt itself, the process of hunting down an animal.

> To the sportsman the death of the game is not what interests him: that is not his purpose. What interests him is everything that he had to do to achieve that death—that is, the hunt. . . . Death is essential because without it there is no authentic hunting: the killing of the animal is the natural end of the hunt and that goal of hunting itself, not of the hunter . . . one does not hunt in order to kill: on the contrary, one kills in order to have hunted.[23]

As an essential part of the happy life, then, hunting is distinguished from extermination by the interest of the hunter in the process of pursuing the animal.[24] It is essential to hunting, in Ortega's view, that it be "problematic," that is, that the outcome not be guaranteed in advance.[25] Hunting, it is true, always relates a superior species to an inferior one, but this "essential inequality between the prey and the hunter does not keep the pursued animal from being able to surpass the pursuer in one endowment or another."[26] Nonetheless, the hunter necessarily retains the advantage.

It is important to understand Ortega's characterization of what hunting is in order to see clearly just what he is defending. If hunting is essentially the pursuit, the stalking of game animals in a context in which they may exercise their natural capacities of evasion so that the outcome is uncertain, then much of what passes for hunting in contemporary industrial societies is irrelevant to Ortega's discussion.[27] Those who go to game preserves where the animals are guaranteed to be plentiful and available will not be engaged in hunting, according to Ortega's definition. Nor are those who shoot birds from commercial blinds or at watering holes along seasonal migration routes engaged in hunting. At best, these "hunters" engage in target practice with live targets instead of clay ones, having reduced hunting to the kill, the easiest and most mechanical of the processes involved in what Ortega defines as hunting.[28] Such practices, therefore, cannot find their moral justification in Ortega's defense of hunting.

What then leads Ortega to define hunting as he does and to value it so highly? Paradoxically, hunting is both the preeminently human activity and the activity that returns the human to nature. As such, hunting is the only activity which returns human beings to themselves. Paul Shepard puts it this way: "The cynegetic [hunting] life is authentic because it is close to the philosophical center of human life. It constantly contrasts two central mysteries: the nature of the animal and of death. These are brought together in hunting. All other ways of life weakly confront the wild or are designed to avoid it altogether."[29]

Hunting restores the contemporary human to his essential place in the natural world. The hunt makes possible a return to "that early state in which, already human, he still lived within the orbit of animal existence."[30] This return is important for both Ortega and Shepard; we have not transcended or escaped this natural condition; the prehistoric is still a part of the contemporary. Ortega's statement of the thesis makes the point most clearly:

> It has always been at man's disposal to escape from the present to that pristine form of being a man, which, because it is the first form, has no historical suppositions. Hunting begins with that form. Before it, there is only that which never changes, that which is permanent, Nature. 'Natural' man is always there, under the changeable historical man.... 'Natural' man is first 'prehistoric' man—the hunter.[31]

According to Ortega, this predicament underlies everything that human beings do.[32] We are constantly faced with the duality of life as it is lived now and life as it was lived ten thousand years ago. We are fundamentally and inescapably predators.

Ortega's position thus is that hunting, properly defined, can return us to our paleolithic selves and serve as an escape from the banality of our contemporary urban or agricultural lives. By hunting we recapture our animal selves and can once again become a part of nature: "Only by hunting can man be in the country; I mean within a countryside which, moreover, is authentically countryside. And only the hunting ground is true countryside."[33] According to Ortega:

> Only the wild animal is properly in the countryside, not just on top of it, simply having it in view. If we want to enjoy that intense and pure happiness which is a 'return to nature,' we have to seek the company of the surly beast, descend to his level, feel emulation toward him, pursue him. This subtle rite is the hunt.[34]

Hunting is a way of acting which mediates between human beings and the natural world. It is a practice which is constitutive of a certain relationship with the environment. The hunter exists in a world which is different from the world of the farmer, the soldier, or the tourist. Not only is the individual constituted by the activity, but the countryside is as well.

This position presupposes a critique of other modes of relationship between human beings and nature. According to Shepard, the human species truly separated itself from a natural place in the world at the moment when the relationship became constituted by agriculture rather than hunting. Nineteenth-century landscape painting contributed to a further alienation from nature by constituting nature as a spectacle to be viewed. Thus, the tourist relationship to nature, which emphasizes the picturesque and the visually impressive, turns nature into the analog of a painting, and the human must necessarily live "outside" of it.[35] The farmer, on the other hand, perceives nature from the standpoint of the fertility of his or her crops. Plants and animals are divided into those which benefit and those which hinder the livelihood of the former. Such a view, according to Shepard, is selective and impoverished. Both Shepard and Ortega agree that hunting is the only practice which will reconstitute a natural and fully "alert" relationship with the countryside. In this sense, only hunting can bridge the gap between human beings and nature.

Ortega and Shepard's argument is distinguished from that of the land ethic by its aggressive attempt to define a place for human beings within the biotic communities of nature. While the land ethic has broader implications for human interactions with nature, it provides little support for hunting as a mode of moral intervention in the lives and habitats of wild animals. The argument we are presently considering, however, aligns the human with the natural by discovering the prehistoric hunter to be a continuous presence in human nature. If the predator is naturally and unavoidably within us, then refusing to hunt is an antinatural choice, one which separates us from our natural selves and produces alienation within the cultures that have dispensed with the hunting way of life. As a result, the happy life for human beings necessarily includes the activity of the hunt.

The claim that hunting will return us to nature by uncovering the natural within the human is an appealing one in the context of environmental ethics. Nonetheless, further analysis of the claim and its defense shows that Ortega and Shepard do not offer acceptable support for this view of hunting. I focus on Ortega in the remainder of this section.

As evidence of a submerged predatory presence in contemporary human nature, Ortega cites a description of an encounter between a group of hunters driving to their hunting grounds and two wolves who cross the road near their car. The hunters have packed their rifles away and are thus unprepared for this opportunity. "Braking, skidding, roars of 'Where's my rifle?' 'Give me my bullets.' Some jumped through a little door, others through a window. . . . There was one enthusiast who, in the face of his inability to get his rifle out of the sheath, thought seriously of pulling his knife and tearing the leather of the case."[36]

This uproar signals "the automatic discharge of the predatory instinct" still present even in the modern individual. The use of this example is puzzling. There is nothing of the hunt in what Ortega describes here, only the frenzied craving to destroy. There is no question of stalking, of engaging with nature, of giving the wolves a chance to exercise their natural talents for evasion. What is significant, however, is that Ortega places the stimulus for this "instinctual" response in the wolves themselves. The hunters are not the instigators of this encounter; rather, the wolves are. "It is not man who gives to those wolves the role of possible prey. It is the animal—in this case the wolves themselves—which demands that he be considered in this way, so that to not react with a predatory intention would be anti-natural."[37] Generalized, this way of thinking of the hunter exculpates him or her from any moral wrongdoing.

> Before any particular hunter pursues them they feel themselves to be possible prey. Thus they model their whole existence in terms of this condition. Thus they automatically convert any normal man who comes upon them into a hunter. The only adequate response to a being that lives obsessed with avoiding capture is to try and capture it.[38]

In other words, the animals that humans hunt "ask for it," hence they are "fair game." This rationalization of the hunter's relation to the prey is an instance of a more general style of argument. It places the responsibility for the harm that is done on the victim of that harm and thereby deflects the impetus to blame the perpetrator of violence. That a parallel can be seen to exist here between the hunter's discourse about the prey and stereotypical male discourse about women is not accidental, as I discuss in the next section.

Ortega's argument here is thin. He has blurred the distinction he drew earlier between hunting and mere extermination, seeing both now as natural expressions of our inherently predatory nature. Indeed, *Meditations on Hunting* contains many examples of hunting in which the essence of the hunt is paradoxically absent. Despite the denial that killing is the main end of hunting and the purpose of the hunter, Ortega's examples in fact demonstrate the exact opposite. The "hunter" in his discussions never does do anything more than wait for the opportunity to kill.

Ortega's defense of hunting also rests on historical claims about the importance of the hunt throughout Western history, especially for the nobility and the aristocracy. Since these classes are those most capable of choosing the life most appealing to human beings, their support for the hunt is evidence for the essential role of hunting in the good life.

It is beyond the scope of this essay to mount a complete historical critique of this aspect of the defense of hunting. However, a couple of observations may serve to justify our skepticism toward Ortega's romantic view of the nobility of the hunt. According to Ortega, the "fundamental task of all hunting" is not the killing of the prey, but the "bringing about of the presence of the prey." Indeed, "primitive" human tribes in which the paleolithic life is still present commonly distribute the largest share of the kill to the person who first sees and raises the game.[39] This practice is consistent with Ortega's notion that the pursuit of the game animal, the testing of skill in finding it and running it down, is what distinguishes hunting from mere extinction and destruction.

There is reason, however, to suppose that the nobility and aristocracy to whom Ortega appeals did not share his conception of what is important to the hunt. The aristocratic hunt has much more to do with war and the display of power than the return to nature. From ancient times onward, it has been typical of royalty to maintain game parks for their hunting pleasure. Our word *paradise* is derived from the word *paradeisos*, the name for such parks. The Latin historian Curtius, for example, describes the scene of a hunt by Alexander the Great in Central Asia:

> Of the wealth of the barbarians in those parts, there are no greater proofs than the herds of noble beasts of the chase they keep shut up in great woods and parks. . . . They surround the woods with walls and have towers as shelters for hunter. . . . Alexander now entered it with his whole army and gave orders for the game to be driven from all directions.[40]

During the course of the hunt, Alexander is attacked by a large lion and displays his bravery by "kill[ing] the beast with a single blow."[41] Alexander, in this description, does not engage in hunting as Ortega has defined it, since he does not pursue the game, nor allow the outcome of the hunt to be problematic.

The Assyrian king, Ashurbanipal, had animals trapped for him so that he might kill them at his convenience. J. K. Anderson quotes Xenophon: "I, Ashurbanipal . . . in my princely sport . . . with arrows I pierced at my feet a raging lion of the plain, which they had released from a cage, but he did not die. So . . . I then stabbed him with the iron dagger from my belt, and he died."[42] Hunting, under such conditions, bears little resemblance to Ortega and Shepard's descriptions of the hunt.

In *Man and the Natural World*, Keith Thomas writes: "Henry the VIII's manner of hunting did not differ very much from that of the eighteenth century King of Naples: he had two or three hundred deer rounded up and then loosed his greyhounds upon them."[43] Queen Elizabeth, he says, enjoyed hunting deer with her hounds, who killed the deer when they fled into the water.[44] In 1591, when the queen was older, she also shot captive deer with a crossbow in her park. In 1786, when Richard Colt Hoare "went after wild boar with the King of Naples . . . he was appalled to discover that the boar, so far from being wild, came when whistled for, and that the hunters stuck it with spears when it was held fast by dogs." Hoare writes, "I was . . . thoroughly disgusted with this scene of slaughter and butchery . . . yet the king and his court seem[ed] to receive great pleasure from the acts of cruelty and to vie with each other in the expertness of doing them."[45]

If these examples are representative, royal hunting bears no resemblance to the hunt which Ortega imagines. The pleasures that accrue to the nobility are precisely the pleasures of killing, not hunting, since the actual location and pursuit of the game is done either by servants or by hounds.[46] The justification of such hunting lies in its value for preparing men for war, and in its value as a symbol of the king's power and courage. But these justifications lie completely outside the domain of Ortega's and Shepard's argument.

In the next section I argue that Ortega and Shepard are mistaken to suppose that hunting is morally justified because it returns us to nature by restoring us to "that pristine form of being a man" that is not the corrupted product of civilization. This Rousseauistic picture of hunting in the innocence of a human prehistory ignores and obscures, as Rousseau himself did, the different roles played by men and women in hunter-gatherer societies.[47]

I have argued in this section that Ortega's appeal to historical fact is insufficient to justify his claim that hunting forms an essential part of the good life, and that his distinction between hunting and mere extermination and killing does not in fact explain the attraction which hunting exercises. In the next section, I argue that hunting, as a human practice, is embedded in a larger political context than that in which Ortega and Shepard have located it. Hunting is not fundamentally a human practice, but a male practice. Although Ortega and Shepard both acknowledge this fact, they fail to understand its significance.

The Patriarchal Hunt

The most important point in Ortega's defense of hunting is the claim that only hunting can constitute an authentic relationship with the countryside. Only the hunter is truly "alert." "Only the hunter, imitating the perpetual alertness of the wild animals, for whom everything is danger, sees everything and sees each thing functioning as facility or difficulty, as risk or protection."[48] To deny this and to argue against human participation in the slaughter of the hunt is, for Shepard, "a denial rather than an affirmation of the world, and affects an ethical bravado that gives the illusion of elevating man above the rest of nature."[49] This defense of hunting cannot, however, be separated from the patriarchal context in which it is embedded. Hunting has always been an essentially male activity despite the participation of some women in the hunt. This fact is often submerged in Shepard's book, but occasionally emerges in explicit statements. He writes, for example, that "man is in part a carnivore: the male of the species is genetically programmed to pursue, attack, and kill for food. To the extent that men do not do so they are not fully human."[50] Although Shepard's book is presented as a defense of the hunter-gatherer way of life, women, who are the gatherers, are quietly marginalized. Indeed, Shepard goes so far as to say that "the rhythm and physiology of [women's lives] are so different from those of men that it is almost as though they were another species."[51] These differences emerge on account of the exclusively male participation in the hunt.

Echoing Ortega, Shepard identifies hunting as one of the activities preferred by people of leisure: "hunting, dancing, racing, and conversing. It is no accident that these are precisely the 'true vocations' of primitive hunters, who, like the aristocrats, make no distinction between leisure and life."[52] But Shepard does not note how the aristocrats' power to live life leisurely presupposes a corresponding exploitation of people who are denied the power of leisure. Nor does he comment on his exclusion of women's "gathering" function from the list of "true vocations" of the primitive aristocrat. Women, it would seem, do not share in the fully human lifestyle even in the context of hunter-gatherer societies.

It is true that Shepard attributes the origins of the oppression of women to the farmer's preoccupation with fecundity, and thus sees the social and political subordination of women as yet another of the signs of the inferiority of an agricultural way of life.[53] Nonetheless, we cannot escape Shepard's silence on the role and significance of "gathering" to the hunter-gatherer way of life. Nor can we escape the fact that Shepard holds the skills involved in "gathering" in contempt. The manual skills which we value today, he says, such as surgical technique and piano playing, are the product of the dexterity derived from shaping tools for killing and dissecting animals. "The 'good hands' of the hunter is not a familiar image, yet he is the surgeon. Tuber grubbers, and soil tillers have hands calloused, arthritic, swollen, and otherwise deformed by their work."[54] The contempt expressed for these "agriculturalists" inevitably traces back to their precursors, female gatherers.

If we accept, therefore, that hunting is the only fully human activity and the only way to be in the countryside as opposed to simply on top of it looking on, then we are forced to conclude that women are not fully human, have never related authentically to the natural world, and have had little share in what Ortega calls the happy life. Interestingly, this conclusion is the immediate negation of the ecofeminist analysis.[55]

Ecofeminists argue that there is a connection between the patriarchal oppression of women and the economic and social habits and beliefs which underlie the exploitation of nature. Conceptually, the basis for the drive to dominate nature derives from the "normative dualisms" which lie at the heart of a patriarchal culture and provide the conceptual framework for a logic of domination. Karen Warren has suggested that dualistic thinking "conceptually separates as opposites aspects of reality that in fact are inseparable or complementary; e.g., it opposes human to nonhuman, mind to body, self to other, reason to emotion."[56]

Evidence for the centrality of dualistic thinking can be found throughout the history of Western philosophy.[57] The familiar distinctions between culture and nature, mind and body, reason and the emotions, ruler and ruled parallel the distinction between men and women. Men have been associated traditionally with culture, mind, and reason, while women have been linked to nature, the body, and the emotions. Men are said to be fundamentally active and aggressive, while women and nature are passive and receptive.

Dualistic oppositions such as these do not just make distinctions between different aspects of social and physical reality; they rank order the value of the differences. It is this disposition to distinguish and to rank the opposing terms hierarchically which provides a conceptual service to those in whose interests it is to subordinate women to men's control, and to reduce nature to merely instrumental values. Thus, from the standpoint of ecofeminism, dualistic thinking must be overcome if we are to move beyond patriarchal culture and end the degradation of nature which flows from it.

Ortega and Shepard have described hunting as the way to bridge the gap which we have opened between humans and nature. Yet hunting reasserts the superiority of humans over nature by defining the human as predator and the nonhuman as prey. Ortega and Shepard focus on killing and death as the modes of participation in nature which will bridge the gap. But in asserting the right and obligation to hunt, they assert the right to decide the fate of wild biotic communities despotically, thereby reaffirming the separation between the human who is faced with this power of choice and the nonhuman which is the object on which human choice operates.

The fascination with lethal rather than life-giving power is a feature of the defense of hunting which requires further scrutiny. In *Gyn/Ecology*, Mary Daly characterizes patriarchal culture as "necrophiliac."[58] Patriarchal culture is necrophiliac, in her view, because it displays an obsession with death and killing, giving value and status to those who are skilled in these rather than in life-affirming practices. Daly extends the sense of the word *necrophiliac* to include instances in which men create or value dead simulacra of living processes, while devaluing and destroying authentic creative and life-affirming processes. From the ecofeminist standpoint, modern sport hunting is a symptom of this necrophiliac orientation.

The charge that hunting is symptomatic of a necrophiliac culture, indeed of patriarchal culture, is confirmed by the work of Ortega and Shepard themselves. Participation in the life cycle of nature, for Ortega, is reduced to participation in the death of living beings. Hunting is the true relation of participation in the countryside, and one must kill in order to have hunted. Shepard is equally explicit. The two mysteries, he says, are the nature of animals and death. Both of these figure centrally in hunting.

The erotic connection with death, implicit in the concept of necrophilia, is visible in their defense of hunting as well. In describing the transformation which takes place in the hunter when the beaters have succeeded in raising some animal with their dogs, Ortega's language is suggestive: "[S]uddenly the orgiastic element shoots forth, the dionysiac, which flows and boils in the depths of all hunting. . . . There is a universal vibration. Things that before were inert and flaccid have suddenly grown nerves, and they gesticulate, announce, foretell."[59]

Paul Shepard pursues the connection between hunting and sexuality in

Environmental Ethics and the Case against Hunting

greater detail. The spear's interpenetration of the animal's body is seen as "the source of all new life," and the symbolic relationship he finds between marriage and eating customs in primitive peoples lies in the fact that "[b]oth have to do with the most profound of life's passions, the demonic moment of the kill and of orgasm. These two powerful expressions are related."[60] According to Shepard:

> [T]here is a danger in all [sic] carnivores of confusing two kinds of veneral aggression, loving and hunting. Among men particularly, it is important to protect other humans from malfunction of the hunting instinct. . . . Perhaps the razor's edge separates assault from love, requiring female postures that can instantly deflect attack. The female, whether the target of redirected fury or the sexual prey of the wild huntsman, has not only to deflect his violence but to entertain it.[61]

If the male hunter is now no longer merely a predator, it is because centuries of experience in "treating the woman-prey with love" have transformed the hunt itself.[62] Ortega and Shepard clearly find a parallel, then, between predation on animals and sexual predation against women.

Paradoxically, even the more positive contributions which hunters have made to conservation have the ring of the necrophiliac about them. Robert Loftin, for example, argues that hunting is morally defensible because hunters are the single most powerful constituency supporting the conservation of wilderness and natural habitats for wild animals.[63] The hunter's "conservation" defense is necrophilous, according to Daly's use of the term, because it favors the substitution of a managed and dependent system for a wild and free ecosystem. Autonomous living processes are replaced by the control and manipulation of human beings. Loftin's argument reflects the paradoxical idea that it is hunters who most fully love and respect animals, because it is they who pay most to preserve game species.

This argument will not work, however. Preservation of habitat and the lives of wild animals by hunters does not reveal a love for the wild, or for the independent existence of wild animals. It is, rather, a logical extension of the instrumental and anthropocentric relation to nature which made a managed environment necessary in the first place. Because the intention behind game management is to maximize the opportunity to kill the animals that are so artfully maintained, the hunter's interest in conservation paradoxically promotes human exploitation of the Earth as a set of natural resources to be harvested for the satisfaction of human whims and pleasures. Thoreau's critique of the logger's "love" for the trees applies equally well to this paradoxical, predatory love which the hunter is said to feel for the prey: the love which Shepard so narrowly distinguishes from assault:

> The character of the logger's admiration is betrayed by his very mode of expressing it. If he told all that was in his mind, he would say, it was so big that I cut it down and then a yoke of oxen could stand on its stump. He admires the log, the carcass or corpse, more than the tree. Why, my dear sir, the tree might have stood on its own stump, and a great deal more comfortably and firmly than

a yoke of oxen can, if you had not cut it down. What right have you to celebrate the virtues of the man you have murdered?[64]

The charge that hunting is a necrophilous and death-loving practice is supported by the defense of hunting which we have been discussing. Neither Ortega nor Shepard have been able to minimize the importance of killing the animal when they seek to explain why hunting is attractive and "normal." Both find links between hunting and sexual aggression that tend to sustain and justify predatory attitudes about the proper relationship between men and women. Both find little meaning in being in the world that is not derived from the passion to destroy wild animals. Rather than questioning inner violence and holding it up for critique, Ortega and Shepard romanticize and glorify it as a trait to be nurtured.

The consequence of linking the practice and discourse of hunting to gender relations is a recognition that the defenses of hunting which appeal to the significance of this way of relating to the natural world are illegitimate. By highlighting the vocabulary of sexuality and violence, ecofeminism links hunting to a broader cultural pattern characterized by male violence, both physical and social, against women. Who, then, is the audience for Ortega's and Shepard's defense of hunting? Who is it that can take it for granted that hunting is the only "authentic" way of relating to animals? From the ecofeminist standpoint it appears that this audience is paradigmatically male, not the audience of human beings generally. Should "we" then adopt the arguments in favor of hunting based on this "male" outlook? Or should "we" oppose hunting as a symptom of and a contribution to a general patriarchal cultural pattern?

Andree Collard suggests that there is an alternative way to conceptualize our relations with nature. A precondition of the hunting relation, she argues, is a process of objectification. In order to kill animals, hunters must first deaden their emotional sympathies toward the animal's pain. They must consider the animal as an object, as prey, rather than as a living being with a life of its own. The hunter necessarily abstracts the animal from the web of its living relationships, from the familial ties which link parent and offspring, from the pack or herd relations that constitute the animal's sociality, and from the predator-prey relations that sustain a healthy wild community. The hunter must learn not to care what happens to the young animals deprived of the experience, guidance, or protection of older animals; he must ignore the effects on the survivor of a mating pair.[65] In addition, he must ignore the effects of his own hunting on the wild who are either deprived of prey or themselves hunted because they compete for the object of the hunter's sport. Seen in this way, the hunter, far from being "alert," must in fact be blind to much that is present in the lives of animals, regarding the animal as a social atom, an object, living in no essential relations to other individuals. Predation, for Ortega and Shepard, is not just the primary animal relation; it is the only one which they notice.

This objectification of the prey, which is foundational to sport hunting in our culture, is clearly expressed in the language used to refer to game animals.

Game managers talk of replenishing the "stock" of some endangered game species, of "taking" or "culling" "surplus" animals, as if animals were cans on a grocery shelf or weeds in a garden that need to be pulled. Such language both hides and is symptomatic of the essential violence intrinsic to hunting. It thus both reveals the problem to be addressed, and hinders the task of moral appraisal with a bureaucratic and euphemistic veneer.

We need another way of conceptualizing what it means to participate in nature. Collard suggests that this new mode of participation should not violate and consume the nature to which we wish to return.[66] We need a way to be alert to the presence of nature without seeking to possess or dominate it. Alert participation in nature may take many forms, but the common factor in each form will be a renunciation of the need to control and manipulate. Such participation is incompatible with that represented by the hunt.

Hunting is not a simple, biologically necessary relationship between human beings and wild animals. It is a practice that gives priority to particular human potentialities while neglecting and marginalizing others. The alertness toward nature developed by hunting is a selective attention focused on killing and the prerequisites of killing. According to Daly, this deadly alertness, this fixation on the mystery of death, characterizes necrophilous culture. Men pursue game as they pursue women who evade them, as vigilantes pursue the escaping slave. The hunter must not only defend the hunt, therefore, but also the very desire to hunt. The will to violence demands moral justification no matter what its object, and if Shepard is right that the male of the species is "genetically programmed" to such violence, then ecofeminists are correct to see environmental morality as an integral part of an antipatriarchal, life-affirming morality whose inspiration will derive not from the hunter but from the gatherer.

Conclusion

In this essay, I have argued against the notion that sport hunting is morally justifiable. Although I think that Leopold's land ethic and the moral theories of the animal liberation movement support this conclusion, I have disagreed with the scope and structure of their arguments. While Callicott is right to be suspicious of moral individualism or atomism, the holistic turn does not serve the interests of a critical environmental ethics if it does not include a reflexive inquiry into the sources of our interpretation and understanding of the natural world and the place of human beings in it. One consequence of ignoring such an inquiry is the perpetuation of a stereotypically male understanding of what is important in reconstructing our relations with the environment, an understanding clearly articulated in the "primitivist" argument.

In order to accept the ecofeminist claim that hunting has a gendered dimension that reflects its relations to broader patterns of patriarchal culture, it is not necessary to accept the more radical claim which some feminists have made that women are closer to nature than men, or privileged by virtue of woman's nature to care for the environment in a way men cannot. Ecofeminism need not be implicated in essentialism. Its strength is that it refuses to abstract sport hunting from the cultural context within which such a sport is valued, promoted, and thought to be fundamentally human. Hunting thereby becomes a sign of cultural patterns other than itself, not a brute given in isolation from the context in which it is defended. Understood in this way, the ecofeminist critique of hunting is immune to the challenge that some women hunt too. In any cultural context in which inequalities exist between one group and another, we can expect that some, if not most, members of the oppressed group will agree with many of the interpretations of the world offered by the dominant group. Only if such ideological transference takes place can the dominant group hope to continue its privileged role. It is not surprising, therefore, that some women have adopted the construction of the natural world as something to be used and exploited for sport or comfort, since this forms an important part of the Western, patriarchal world view.

My argument in this paper is also immune to the charge that it speaks only of the hunter as a drunken and irresponsible lout. I have not presented a psychological profile of the hunter, but a moral critique of hunting. The only motivation in the hunter which I have presupposed is the desire to kill a wild animal for sport under conditions in which such killing is not necessary for survival. This desire need not be boorish and brutal. But if the argument of this paper is sound, it is a desire which is unjustified nonetheless.

My general conclusion, therefore, is that we cannot satisfactorily evaluate the individual acts of our moral lives without first understanding the contexts within which these acts are embedded. I have argued that hunting is not just pain and death undergone by an animal. Nor is it just the hunter's acts of stalking and killing. Hunting constructs a view of the natural world which informs the perception of the hunter and of those affected by the hunter's outlook. This view is itself in harmony with the view of nature which predominates in Western culture. Nature becomes the object of predation, a world to be pursued and violently subordinated to the hunter's will and desire. As such, nature retains its otherness only as the object of pursuit which may elude capture. The hunter is alert, but only to what may escape, lest it escape. His alertness carries deadly force, therefore, and plays no role in reconstituting the strained relations of the human species to the rest of the environment.

NOTES

1. A being is morally considerable if it deserves a standing in moral deliberation. The moral significance of a being depends on its relative standing in relation to other "morally considerable" beings (Kenneth E. Goodpaster, "On Being Morally Considerable," *Journal of Philosophy* 75 [1978]: 308–25).

2. Both Peter Singer and Tom Regan point out that some animals are more fully rational and self-conscious than a comatose person; yet we do not treat these human beings as if they possessed no moral standing (Peter Singer, ed., *In Defense of Animals* [Oxford: Basil Blackwell, 1985]).

3. J. Baird Callicott, "Animal Liberation: A Triangular Affair," in *In Defense of the Land Ethic: Essays in Environmental Philosophy* (Albany: SUNY Press, 1989).

4. Mary Daly, *Gyn/Ecology: The MetaEthics of Radical Feminism* (Boston: Beacon Press, 1978); Andree Collard with Joyce Contrucci, *The Rape of the Wild: Man's Violence Against Animals and the Earth* (Bloomington: Indiana University Press, 1989).

5. José Ortega y Gasset, *Meditations on Hunting* (New York: Charles Scribner's Sons, 1973); Paul Shepard, *The Tender Carnivore and the Sacred Game* (New York: Charles Scribner's Sons, 1973).

6. On the notion of a mixed human-animal community, see Mary Midgely, *Why Animals Matter* (Athens: University of Georgia Press, 1983). For Callicott's sympathetic revision of his thesis, see "Animal Liberation and Environmental Ethics: Back Together Again," in *In Defense of the Land Ethic*, pp. 15–38.

7. Major statements of Peter Singer's views can be found in Peter Singer, *Animal Liberation: A New Ethics for the Treatment of Animals* (New York: Random House, 1975); and Tom Regan and Peter Singer, eds., *Animal Rights and Human Obligations* (Englewood Cliffs, N.J.: Prentice-Hall, 1976).

8. On the distinction between act- and rule-utilitarianism, see Richard B. Brandt, "Toward a Credible Form of Utilitarianism," in *Morality and the Language of Conduct*, ed. Hector-Neri Castaneda and George Nakhnikian (Detroit: Wayne State University Press, 1963); see also John Rawls, "Two Concepts of Rules," *Philosophical Review* 64 (1955): 3–32.

9. James Whisker, *The Right to Hunt* (Croton-on-Hudson, N.Y.: North River Press, 1981), p. 90.

10. Tom Regan, *All That Dwell Therein: Animal Rights and the Case for Environmental Ethics* (Berkeley: University of California Press, 1982); Tom Regan, *The Case for Animal Rights* (Berkeley: University of California Press, 1983).

11. The notion that animals' rights depend on their possessing interests which are values for them is defended by Joel Feinberg, "The Rights of Animals and Unborn Generations," in *Philosophy and the Environmental Crisis*, ed. William T. Blackstone (Athens: University of Georgia Press, 1974), p. 51; and Donald Scherer, "Anthropocentrism, Atomism, and Environmental Ethics," *Enviromental Ethics* 4 (1982): 115–23.

12. Callicott, "Animal Liberation: A Trianglar Affair," pp. 32–34.

13. Aldo Leopold, *A Sand County Almanac: With Essays on Conservation from Round River* (New York: Oxford University Press, 1966), pp. 237–64.

14. Callicott, "Animal Liberation: A Triangular Affair," p. 17.

15. Leopold, *A Sand County Almanac*, p. 262.

16. Callicott, "Animal Liberation: A Triangular Affair," p. 22.

17. Ibid., p. 21.

18. Ibid., pp. 29–36.

19. Ibid., pp. 31–32.

20. Echoes of Ortega's argument can be found in Callicott, Shepard, and Whisker. A line from Ortega is also quoted in the hunting story by Anthony Acerrano: "The cadaver is flesh which has lost its intimacy, flesh whose interior has escaped . . . a piece of pure matter in which there is no longer anyone hidden" ("Pictures of Ourselves and Other Strangers," in *Seasons of the Hunter*, ed. Robert Elman and David Seybold [New

York: Alfred A. Knopf, 1985], p. 92). It seems to follow that hunting, which creates cadavers of living beings, is the destruction of intimacy and the rooting out and exposing of hidden interiority.

21. Ortega y Gasset, *Meditations on Hunting*, p. 31.
22. Ibid.
23. Ibid., pp. 110–11.
24. Ibid., p. 53.
25. Ibid., p. 58.
26. Ibid., p. 56.
27. Ibid., pp. 75–77.
28. A compelling depiction of this form of sport hunting can be found in Vance Bourjaily, *The Unnatural Enemy* (New York: Dial Press, 1963), chap. 3, "The Goose Pits."
29. Shepard, *The Tender Carnivore and the Sacred Game*, p. 146.
30. Ibid., p. 129.
31. Ortega y Gasset, *Meditations on Hunting*, p. 134.
32. Ibid., p. 136.
33. Ibid., p. 140.
34. Ibid., p. 141.
35. Shepard, *The Tender Carnivore and the Sacred Game*, pp. 140–41. See also Paul Shepard, *Man in the Landscape: A Historic View of the Aesthetics of Nature* (New York: Alfred A. Knopf, 1967); J. Baird Callicott, "Leopold's Land Ethic," in *In Defense of the Land Ethic*.
36. Ortega y Gasset, *Meditations on Hunting*, pp. 136–37.
37. Ibid., p. 137.
38. Ibid., p. 138.
39. Ibid., p. 76.
40. J. K. Anderson, *Hunting in the Ancient World* (Berkeley: University of California Press, 1985), p. 79.
41. Ibid.
42. Ibid., p. 63.
43. Keith Thomas, *Man and the Natural World: Changing Attitudes in England 1500–1800* (New York: Penguin Books, 1983), p. 145.
44. Ibid., p. 147.
45. Ibid., p. 143.
46. Ortega y Gasset, *Meditations on Hunting*, p. 90.
47. The primeval quality which Rousseau posits in his *Discourse on the Origin of Inequality Among Men* is notoriously contradicted by his misogynist arguments in *Emile*.
48. Ortega y Gasset, *Meditations on Hunting*, p. 151.
49. Shepard, *The Tender Carnivore and the Sacred Game*, p. 152.
50. Ibid., pp. 122–23.
51. Ibid., pp. 117–18.
52. Ortega y Gasset, *Meditations on Hunting*, p. 31; Shepard, *The Tender Carnivore and the Sacred Game*, p. 150.
53. Shepard, *The Tender Carnivore and the Sacred Game*, pp. 120–21, 244.
54. Ibid., p. 155.
55. For some discussions of the ecofeminist position, see Leonie Caldecott and Stephanie Leland, eds., *Reclaim the Earth: Women Speak Out for Life on Earth* (London: Women's Press, 1983); Susan Griffin, *Woman and Nature: The Roaring Inside Her* (New

York: Harper and Row, 1978); Marti Kheel, "The Liberation of Nature: A Circular Affair," *Environmental Ethics* 7 (1985): 138–63; Ynestra King, "The Ecology of Feminism and the Feminism of Ecology," *Harbinger: Journal of Social Ecology* 1 (1983); Carolyn Merchant, *The Death of Nature: Women, Ecology, and the Scientific Revolution* (San Francisco: Harper and Row, 1980); and Ariel Kay Salleh, "Deeper Than Deep Ecology: The Ecofeminist Connection," *Environmental Ethics* 6 (1984): 339–46.

56. Karen Warren, "Feminism and Ecology: Making Connections," *Environmental Ethics* 9 (1987).

57. Plato's *Phaedo* and Aristotle's *Politics* provide obvious examples of the importance of these dualisms in ancient Greek Philosophy.

58. Daly, *Gyn/Ecology*, p. 59.

59. Ortega y Gasset, *Meditations on Hunting*, p. 89.

60. Shepard, *The Tender Carnivore and the Sacred Game*, p. 170.

61. Ibid., p. 172.

62. Ibid., p. 173.

63. Robert Loftin, "The Morality of Hunting," *Environmental Ethics* 6 (1984): 246–47.

64. Henry David Thoreau, *The Maine Woods* (New York: Harper and Row, 1987), p. 314.

65. Collard, *The Rape of the Wild*, p. 39.

66. Ibid., p. 53.

REFERENCES

Acerrano, Anthony. "Pictures of Ourselves and Other Strangers." In *Seasons of the Hunter*, edited by Robert Elman and David Seybold. New York: Alfred A. Knopf, 1985.

Anderson, J. K. *Hunting in the Ancient World*. Berkeley: University of California Press, 1985.

Bourjaily, Vance. *The Unnatural Enemy*. New York: Dial Press, 1963.

Brandt, Richard B. "Toward a Credible Form of Utilitarianism." In *Morality and the Language of Conduct*, edited by Hector-Neri Castaneda and George Nakhnikian. Detroit: Wayne State University Press, 1963.

Caldecott, Leonie, and Stephanie Leland. *Reclaim the Earth: Women Speak Out for Life on Earth*. London: Women's Press, 1983.

Callicott, J. Baird. "Animal Liberation: A Triangular Affair." In *In Defense of the Land Ethic: Essays in Environmental Philosophy*. Albany: SUNY Press, 1989.

———. "Animal Liberation and Environmental Ethics: Back Together Again." In *In Defense of the Land Ethic: Essays in Environmental Philosophy*. Albany: SUNY Press, 1989.

———. "Leopold's Land Ethic." In *In Defense of the Land Ethic: Essays in Environmental Philosophy*. Albany: SUNY Press, 1989.

Collard, Andree, with Joyce Contrucci. *The Rape of the Wild: Man's Violence Against Animals and the Earth*. Bloomington: Indiana University Press, 1989.

Daly, Mary. *Gyn/Ecology: The MetaEthics of Radical Feminism*. Boston: Beacon Press, 1978.

Feinberg, Joel. "The Rights of Animals and Unborn Generations." In *Philosophy and the*

Environmental Crisis, edited by William T. Blackstone. Athens: University of Georgia Press, 1974.
Goodpaster, Kenneth E. "On Being Morally Considerable." *Journal of Philosophy* 75 (1987): 308–25.
Griffin, Susan. *Woman and Nature: The Roaring Inside Her*. New York: Harper and Row, 1978.
Kheel, Marti. "The Liberation of Nature: A Circular Affair." *Environmental Ethics* 7 (1985): 138–63.
King, Ynestra. "The Ecology of Feminism and the Feminism of Ecology." *Environmental Ethics* 7 (1985): 138–63.
Leopold, Aldo. *A Sand County Almanac: With Essays on Conservation from Round River*. New York: Oxford University Press, 1966.
Loftin, Robert. "The Morality of Hunting." *Environmental Ethics* 6 (1984).
Merchant, Carolyn. *The Death of Nature: Women, Ecology, and the Scientific Revolution*. San Francisco: Harper and Row, 1980.
Midgely, Mary. *Why Animals Matter*. Athens: University of Georgia Press, 1983.
Ortega y Gasset, José. *Meditations on Hunting*. New York: Charles Scribner's Sons, 1973.
Rawls, John. "Two Concepts of Rules." *Philosophical Review* 64 (1955): 3–32.
Regan, Tom. *All That Dwell Therein: Animal Rights and the Case for Environmental Ethics*. Berkeley: University of California Press, 1982.
―――. *The Case for Animal Rights*. Berkeley: University of California Press, 1983.
Regan, Tom, and Peter Singer, eds. *Animal Rights and Human Obligations*. Englewood Cliffs, N.J.: Prentice-Hall, 1976.
Salleh, Ariel Kay. "Deeper Than Deep Ecology: The Eco-feminist Connection." *Environmental Ethics* 6 (1984): 339–46.
Scherer, Donald. "Anthropocentrism, Atomism, and Environmental Ethics." *Environmental Ethics* 4 (1982): 115–23.
Shepard, Paul. *Man in the Landscape: A Historic View of the Aesthetics of Nature*. New York: Alfred A. Knopf, 1967.
―――. *The Tender Carnivore and the Sacred Game*. New York: Charles Scribner's Sons, 1973.
Singer, Peter. *Animal Liberation: A New Ethics for the Treatment of Animals*. New York: Random House, 1973.
Singer, Peter, ed. *In Defense of Animals*. Oxford: Basil Blackwell, 1985.
Thomas, Keith. *Man and the Natural World: Changing Attitudes in England 1500–1800*. New York: Penguin Books, 1983.
Thoreau, Henry David. *The Maine Woods*. New York: Harper and Row, 1987.
Warren, Karen. "Feminism and Ecology: Making Connections." *Environmental Ethics* 9 (1987).
Whisker, James. *The Right to Hunt*. Croton-on-Hudson, N.Y.: North River Press, 1981.

2

THE MEDICAL TREATMENT OF WILD ANIMALS

Robert W. Loftin

Taking care of sick and injured wild animals is commonplace in our society. Since I am an official of the Jacksonville chapter of the Florida Audubon Society and known as a birdwatcher, I often receive requests for advice on how to take care of helpless birds. Requests for information on the feeding and care of wild "patients" outnumber all others, both on the local Audubon Society telephone and at the national headquarters in New York. Several persons in my city are licensed by the federal government to hold wild birds in captivity for medical treatment. There are numerous "how-to-do-it" books on the medical treatment of wild animals and the rearing of wild orphans.[1]

Some wild animal hospitals are well financed and organized. Several have received financial support from foundations and corporations. The better ones have skilled professional veterinarians who sometimes undertake heroic measures to benefit injured animals. A case which attracted national media attention involved attaching artificial rubber flippers surgically to a sea turtle that had been injured in a shark attack. The attempt failed because there was too little bone for attachment.[2] In other cases, a team of surgeons tried to transplant a cornea from the eye of a badly injured eagle to the eye of another eagle, and a Laysan albatross found in San Francisco was fitted out with new feathers and flown back some five thousand miles to its home on Midway Island to be released.[3]

Those who undertake the medical treatment of wild animals are well intentioned, motivated by an understandable sympathy for their fellow creatures. I respect this, and these endeavors do have a certain value, but, as I argue in this paper, the value is quite limited, and, for the most part, not what people think it

---

This article originally appeared in *Environmental Ethics* 7 (fall 1985): 231–39. Reprinted by permission. Copyright © 1985 Pennye W. Loftin. All rights reserved.

is. Although taking care of helpless wild animals is not wrong, neither is it right. As a result, humans who refuse to extend medical treatment to wild animals have not failed in any moral duty, nor are they necessarily morally callous, for we have no moral obligations to suffering wild animals except to end their suffering.

Environmental ethics is variously divided, but one important watershed is between those who hold that individual nonhumans are the locus of value and those who hold that more corporate, systemic, or holistic entities such as species or entire ecosystems are the locus of value. Among the "individualists" are Tom Regan, Peter Singer, all of the "animal liberation" philosophers and fellow travelers, as well as those in the reverence-for-life tradition of Albert Schweitzer.[4] On the other slope of the watershed are thinkers such as Aldo Leopold, J. Baird Callicott, and many biologists who argue for systems or species as the locus of value.[5]

While proponents of these two approaches agree on many things, the treatment of injured wild animals clearly divides them. If an individual is what is valuable, if this individual has interests (which it surely does in a nontrivial sense) and even more strongly if it has rights, then humans have some obligation to provide assistance if they are able. I certainly have that kind of obligation to another human. If I am driving down a remote country road and I happen upon an injured person, I have failed in my moral duty if I merely drive on (assuming this person intends me no harm). This is so whether or not the person requests my assistance, or even, perhaps especially, if he is unconscious. If I happen upon a sick cormorant, have I failed in my moral duty if I merely drive on? If so, I contend, it is only because I have not stopped to put the suffering animal out of its misery.

Let us distinguish between negative and positive rights. Negative rights can be fulfilled by doing nothing. The cormorant has a negative right to life, liberty, and the pursuit of cormorant happiness, whatever that may be, which I can respect simply by leaving the bird alone (unless I have a good reason to interfere). The cormorant has this right not as an individual, but derivatively, as a part of a functioning system of interrelated organisms. If a third party is interfering with the cormorant, for example, by shooting at it illegally, I have a duty to intervene and stop the miscreant, if I am able—but this is a duty to the *system*, not to the individual bird.

As you may have suspected by now, I regard the ecosystem as more valuable than the individual animal. I don't approve of the medical treatment of wild animals because I locate myself more within the holistic than individualistic camp. As a result, I am extremely reluctant to get bogged down in the interminable morass of doctoring sick animals. It is better, as I see it, to spend what time and energy I have to save more habitat for the benefit of healthy animals. To treat individual animals is merely a one-shot, short-term action. Even if I can save the life of an individual, that animal, like all of us, is doomed to die. Unless I can somehow return it to the breeding population, I have done nothing that will survive the death of that particular individual.

The chance of getting a sick or injured animal back into the breeding population is a slim one. Even if I keep it alive, I may never be able to release it. If

I release it, it may not be able to care for itself in the wild. Only the fittest survive. Even if it can survive in fierce competition with healthier, more experienced animals, can it hold its own with them in the even fiercer competition to propagate its genes? If it does, is it perpetuating less fit genes, say by replacing the genes of a sea turtle that somehow knows how to *avoid* shark attacks? Other turtles know how to do that. Sharks and sea turtles have lived side by side in the oceans for eons. Most birds' nests fail. Most of the fledglings do not make it through the first year. The same is true of most other groups of animals, including human beings under natural conditions. That is how the system works; that's what *makes* it work.

Even an individualist could accept what I have said thus far. An individualist could readily agree that it was better to abandon one sick cormorant and work to establish a cormorant refuge, because more individual cormorants would then enjoy life, liberty, and the pursuit of cormorant happiness. There is nothing to prevent an individualist from calculating the greatest benefit for the greatest number of individual animals in budgeting his practical action. Careful economy of time and effort for maximum effect can characterize both individualists and holists. They might well agree about the best course of action in particular cases, while disagreeing about their respective value presuppositions.

Our human ethics, however, morally obligate us to try to keep all the bad human genes in the gene pool by keeping everyone alive that we possibly can. While the obligation to allow everyone to replicate his or her genes is less clear, we are reluctant to tell anyone that he or she cannot. I am willing to accept this burden in the case of humans. Sometimes I worry about the long-term effects of our well-organized and determined effort to undermine the human gene pool, but I see no alternative. I cannot simply stand aside and let human babies die. In this context, the idea of a "eugenics" program is not only unworkable, but unthinkable. There is no chance that any such program could adhere to objective biological criteria. Such considerations would supplant genuine biology from the outset. Those sterilized would not be the genetically defective, just those at the bottom of the scale of social status. Should we now develop a wildlife ethic which passes these kinds of problems on to nonhuman animals as well? Of course not, since such an approach ultimately harms both the system and the individual.

Wild animal hospitals do have value, but the value is indirect. They foster an attitude of sympathetic concern for wildlife. Unfortunately, however, they channel this concern in the wrong direction—toward individuals rather than systems. Some of this does spill over into a concern for healthy animals and the habitat necessary to sustain them. Part of this spillover is educationally valuable. A hawk or an owl that has been crippled by a gunshot wound can be used to teach youngsters not to shoot healthy hawks and owls that are part of a healthy, functioning natural system. Pictures of brown pelicans with their bills sawn off, when shown on national television, create a general sense of outrage toward such wanton acts of cruelty. Some of this indignation may even spill over into opposition to wearing furs and eating meat. I hope so. This spillover value, nevertheless, focuses on the protection of the system, not the individuals involved, and could more effectively be generated by efforts to protect the system directly.

In the case of endangered species, my position about individuals requires some qualification. If a species is so rare that every individual counts, then more is at stake than a single individual, and we act accordingly. Every species contributes to the diversity and stability of the whole system. Each species is an energy manager within the system. Each individual is, too, but the system is structured so that most are doomed to die early. Their lives however, are not wasted—sharks preying on sea turtles are part of the system, too. That *is* the system.

The system has to be managed. Man has long managed the system. Aboriginally he did it with fire. By setting fires at the right times and in the right places, he improved the forage for those animals he wished to encourage. With the coming of agriculture, man began to manage the ecosystem on a different scale. Every time one plows a field, one turns back ecological succession. This is management for the benefit of man himself, in his interests. There is another kind of management with a different goal, trying to undo some of the damage that man has done. We simply cannot afford the luxury of a nonmanagerial environmental ethics.

To try to save endangered species is to manage the system. This may entail medical treatment for those individuals whose survival potentially affects the whole system in the long term. But we should do it for the system in the long term, not for the individual in the short term.

Indeed, it seems to me that one of the major problems with the individualist approach to environmental ethics is that it has great difficulty accounting for the fact that we accord greater value to individuals which belong to species that have few members than to individuals which belong to species that have many members. An individual sandhill crane, which belongs to a species that has many members, has as many interests and presumably as many rights as a whooping crane. What reasons can we give for preferring the whooping crane? Any position that says there are none is in serious difficulty.[6] At the very least, that kind of individualism must be supplemented with another ethic that gives us reasons to be more concerned about endangered plants and animals.[7] (One can always drag in homocentric reasons—i.e., that people value rare species more than common ones—but that line of reasoning tends back toward the kind of human-centered ethical thinking which got us into this mess and should be avoided if possible.)

My major criticism of wild animal hospitals is that they compete for scarce resources (grants, corporate support, volunteer labor) which could be better spent.[8] Nothing I have said is intended to assert that doctoring animals or fostering wild orphans is in itself *wrong*. If one wishes to amuse oneself in this way, there is nothing wrong with it. In that respect, it is something like drinking a piña colada—you don't really need it, the money you spend doing it could do far more good elsewhere, but you owe yourself something in life. Yet, while it is not *wrong* to do it, it is also not right—not efficient, not even wise. It is far better to spend time, energy, and money working to set aside some undisturbed beaches for the benefit of healthy sea turtles, so that they will have a place to lay their eggs away from the distracting lights of beachfront condominiums, than to put rubber flippers on one turtle. (Perhaps it could be argued that money, time, and resources spent on doctoring animals or drinking piña coladas would just be wasted on something else if people stopped engaging in these activities. This kind of rationalization, however, is not a justification—just an excuse.)

Although it is not wrong to treat wild animals, it is a mistake to feel pious about it. Don't expect the animals to appreciate your efforts—they do not understand that you are trying to help them. All the sea turtle knows is that it hurts and these strange two-legged creatures, which it fears, keep messing around with its flippers. I wonder if a red-tailed hawk with only one wing enjoys life in a cage? I have no way of knowing, but I see no reason to believe that it does. It might as well be dead. If you want to go to the trouble of keeping it alive, that is permissible, but do not pretend that you are doing the *bird* a favor. You're keeping a wild pet.

Most bird hospitals seem to be based on biological illiteracy. Where there is life, there is death. Those who minister to wild animals, however, seem to fail to understand that. Birds are not little feathered people. Yet bird physicians act on that premise and treat the symptoms rather than the root causes of the disease.

A look at the activities of one of the more successful hospitals may help illustrate my point. In an article in *Camp-orama* (a magazine distributed in Florida to the owners of large motor homes), Ralph Heath, director of the Suncoast Seabird Sanctuary in Indian Shores, Florida, tells how God called him to the work of saving injured wildlife. He first picked up an injured cormorant, then people began to bring him injured gulls and the like. Word got around and soon he had four hundred sick birds on his hands: "These included a blue heron sickened by polluted water; ducks caught in an oil spill; an egret that had flown into a picture window; a blue jay ripped by a cat; an American bald eagle, victim of a power line; and a baby horned owl that had fallen from its nest, breaking a wing and leg."[9] Although Heath is aware that his work makes little difference in the long run, since he is only able to save a very small number of the birds actually injured, he defends his work on moral grounds: "We felt that our very being here in some mystical way helped offset the evil in the world."[10]

While it is a morally defensible position to argue that one's actions derive their moral worth from the mere performance of those actions and not from the good consequences derived from them, it is hard to accept that the treatment of sick birds should fall in this category. Heath, for example, claims to have been heartened one day to meet three teenage boys who had driven five hundred miles roundtrip from Miami Beach to bring him an injured pigeon! Those concerned about systems should blanch at the thought of the fossil fuel consumed to rescue a pigeon—a semidomesticated *nuisance* species. But why not? Once one accepts the premises—that wild (or semiwild) individual birds have a right to life—such actions become admirable rather than absurd. Perhaps the next step will be to require such actions of all of us—in short, to provide retirement homes for animals no longer able to live in the world on their own.[11]

If the moral worth of animal doctoring is grounded in the *principle* exemplified in the act and not in the consequences of the act, then special moral problems can arise with regard to conflicts of principle, motive, and interest. One needs to be morally consistent. In the case of animal hospitals, however, this is apparently not so easy to do. The fund-raising literature for Suncoast Seabird Sanctuary, for example, is sponsored by a restaurant which advertises steaks and chops on the back of the brochure. Apparently the deep concern for animal welfare expressed on the front of the brochure does not extend to the individual animals reared on factory farms.[12]

Animal hospitals also need to be careful about conflicts of interest. In 1982, for example, Exxon Oil Company gave the Suncoast Seabird Sanctuary a grant and then featured it in its advertising. In the advertisement, Heath is quoted as saying:

> About 90 percent of the birds we treat have injuries or ailments directly or indirectly related to man. They fly into, or they're hit by, or they get tangled up in something made by man. They get sick from contaminated water, or they may be deliberately attacked and cruelly hurt by people. But our records show that less than one percent suffer from oil contamination.[13]

While Heath may sincerely feel that all of these statements are true, the final remark, nevertheless, casts some doubt on the moral worth of the sanctuary's activities, and the moral worth of Exxon's donation, especially since in other literature the director of this sanctuary has pointed out that oil spills *are* a problem, for example, in the article in *Camp-orama* as quoted above. On the one hand, if the hospital changed its mind about the effects of oil spills because of the Exxon grant, then the moral worth of it activities is diminished. On the other hand, the moral worth of Exxon's donation would actually have been enhanced if the motivation had been based on the fact that oil spills are a problem. It would, of course, have no moral worth at all if the purpose of the grant had been nothing more than an attempt at improving public relations.

Is it really true that oil has little to do with animal welfare problems? Anyone trained to look at systems rather than individuals ought to be able to see instantly that oil is the fundamental driving force that fuels the industrial system that causes nearly all the injuries to the birds which Heath lists. I, for one, am not mollified by the fact that an oil company contributes money to patch up a few birds that petroleum-fueled automobiles run over. This superficial mitigation of very deep environmental sickness is somewhat like offering free hospitalization to a child who is dying of cancer contracted from exposure to industrial chemicals. It might make that individual child feel better, temporarily, but it doesn't get at the root of the problem. It reminds me of the "Christmas basket" approach to poverty relief. The best approach is to make every effort to eliminate the social problems that produce poverty, the pollution problems that expose people to industrial chemicals, and in the case of these kinds of animal injuries, to encourage oil conservation and the development of better ways to prevent oil spills and to clean them up when they do happen.

The basic philosophical or ethical defense of the medical treatment of animals usually focuses on animal rights arguments. But just what are the consequences of the position that animals have rights—implicit in animal doctoring—especially a right to life? It is tempting to say that the right to life is the most basic of all rights—without which all others would lose most of their meaning. It is possible, nevertheless, to hold that for wild animals the right to *liberty* is more fundamental than the right to life, and that therefore animal hospitals violate the right to liberty by confining animals in order to treat them. This position is a very strong one in that it suggests that animal hospitals are morally wrong because they place the right to life before the right of liberty. It is actually stronger than my own position, that such activity is neither right nor wrong. In this context, my position may appropriately be regarded as a kind of compromise.

Some defense of animal doctoring may be available in terms of Tom Regan's "preservation principle," which he defines as "a principle of nondestruction, noninterference, and generally, nonmeddling." He writes:

> By characterizing this in terms of a principle . . . I am emphasizing that preservation (letting be) be regarded as a moral imperative. Thus, if I regard wild stretches of the Colorado River as inherently valuable and regard these sections with admiring respect, I also think one ought not to meddle in the river's affairs, as it were.[14]

Regan distinguishes two versions of this preservation principle, prima facie and absolute versions. The absolute version precludes any management of the parts of the environment which one regards with admiring respect because they are inherently valuable. He wisely rejects this sweeping version of the preservation principle in favor of the prima facie version because "letting be what is at present inherently good in nature may lead to value diminution or loss in the future. For example, because of various sedimentary changes, a river which is now wild and free might in time be transformed into a small muddy creek; thus, it might be necessary to override the preservation principle in order to preserve or increase what is inherently valuable in nature."[15]

This version of the principle seems to leave room for all the management one would care to undertake, provided that management is directed toward increasing and preserving the inherent goodness of the system, rather than, say, providing enhanced opportunities for humans to enjoy the system. This version of the principle could countenance regulating the water level in a wildlife refuge for the benefit of the ducks, but not building a road so that people could get in and look at the ducks.

Whether the prima facie version of the "preservation principle" permits or forbids animal hospitals is not clear. Obviously it permits the treatment of endangered species, since greater inherent value may be gained in the long term, but whether the inherent value of an individual animal of a common species is sufficient to override the principle of nonmeddling is less clear. If it does not, then Regan too has to conclude that treating wild animals is a kind of meddling, and hence morally wrong.

How then should we treat sick or injured wild animals? Since pain is bad, it ought to be eliminated if that is feasible. Therefore, it is my moral duty to end their pain as quickly as I can. The best way to do this is simply to kill them, as quickly and humanely as possible.

In closing, I want to return to the "spillover" effect which, as I have argued, is virtually the only genuine benefit of wild animal hospitals Those of us who are dubious about the value of these efforts cannot afford to overlook this point. The people who are engaged in treating wild animals are well-intentioned, and deeply concerned about the welfare of the natural world, far more than most. Fundamentally, we are on the same side. Therefore, it is counter productive for the environmental holist to assume an adversarial stance against the animal physician. The best approach is to try to channel this genuine concern for the impact of man on the natural world in more productive directions which will result in more good in the long term.

Notes

1. For example, *Care and Feeding of Orphan Song and Garden Birds* and *Help For Hooked Birds*, both available from Suncoast Seabird Sanctuary, 18328 Gulf Boulevard, Indian Shores, FL 33535.

2. *New York Times*, January 24, 1984, p. A10.

3. Michael Baughm, "Perspectives," *Sports Illustrated* (November 21, 1983): 104.

4. Tom Regan, *The Case for Animal Rights* (Berkeley: University of California Press, 1983); Peter Singer, *Animal Liberation* (New York: Avon Books, 1976); Albert Schweitzer, *The Philosophy of Civilization* (New York: Macmillan, 1960).

5. Aldo Leopold, *A Sand County Almanac: With Essays on Conservation from Round River* (New York: Oxford University Press, 1949); J. Baird Callicott, "Animal Liberation: A Triangular Affair," *Environmental Ethics* 2 (1980): 99–120. See also C. H. D. Clarke, "Autumn Thoughts of a Hunter," *Journal of Wildlife Management* 22 (1958): 420–26.

6. See Peter Singer, "Not for Humans Only: The Place of Nonhumans in Environmental Issues," in *Ethics and Problems of the 21st Century* (West Bend, Ind.: University of Notre Dame Press, 1979); and Tom Regan, *The Case for Animal Rights* (Berkeley: University of California Press, 1983). Both Singer and Regan say there are no nonanthropocentric reasons to prefer very rare animals to common ones. As a result, their position is in serious difficulty. this is the Achilles heel of the animal liberation position. In "Why Species Matter," *Environmental Ethics* 3 (1981): 101, Lilly-Marlene Russow also falls back on anthropocentric reasoning in her argument that rare species should be preserved on aesthetic grounds, since these are presumably human aesthetic sensibilities which are to be considered. This forces her to the conclusion that there is little reason to save those animals which we find lacking in beauty, such as the snail darter.

7. See Brian Norton, *Why Preserve Natural Variety?* (Princeton: Princeton University Press, 1987).

8. In television commercials in the Jacksonville area Cynthia Mosling, head of a

local wildlife hospital, appeals for funds by pointing out that it costs her $1,000 per month just to feed the many animals she cares for.

9. Ralph Heath, "Children of the Air," *Camp-orama: Florida's Monthly Camping and RV Guide* 9, no. 10 (1983): 33.

10. Ibid., p. 47.

11. There are some already. The Wildlife Retirement Village in Waldo, Florida, is one, but I understand that they primarily take animals that have been reared in captivity and are no longer wanted by their former owners. This is often the case with animals such as the great cats, which are cute when young but grow up to be come dangerous. This is quite a different matter from taking it upon oneself to treat truly wild animals. Once humans have assumed the obligation of taking an animal or its ancestors out of the wild they have taken on a responsibility that cannot be lightly disregarded. See Eugene Hargrove's "Growing Old Chimpanzee," *Environmental Ethics* 3 (1981): 195–96.

12. Cynthia Mosling, a local wildlife doctor, stated in a presentation to the northeast Florida group of the Sierra Club, that she buys ten pounds of chicken every day to feed her crippled hawks. These chickens are reared in factory farms under appalling conditions. The continued existence (I hesitate to call it life) of a hawk that can never fly again is not worth the lives of the many chickens it will consume before its death, even on animal liberation grounds. Nor does it help to point out that if the hawk were healthy and free it would be killing an equal number of wild animals for food. I have already pointed out that this is the role of the predator system. Raising chickens on factory farms to feed captive hawks contributes nothing to the ecosystem, but tends to undermine it.

13. *Florida Naturalist* 55 (January–March 1982): 18.

14. Tom Regan, "The Nature and Possibility of an Environmental Ethic," *Environmental Ethics* 3 (1981): 31–32.

15. Ibid., p. 32.

References

Baughman, M. "Perspectives." *Sports Illustrated* (November 21, 1983): 104.
Callicott, J. Baird. "Animal Liberation: A Triangular Affair." *Environmental Ethics* 2, 1980: 99–120.
Clarke, C. H. D. "Autumn Thoughts of a Hunter." *Journal of Wildlife Management* 22 (1958): 420–26.
Florida Naturalist 55 (January–March 1982): 18.
Hargrove, Eugene. "Growing Old Chimpanzee." *Environmental Ethics* 3 (1981): 195–96.
Heath, Ralph. "Children of the Air." *Camp-orama: Florida's Monthly Camping and RV Guide* 9, no. 10 (1983).
Leopold, Aldo. *A Sand County Almanac: With Essays on Conversation from Round River*. New York: Oxford University Press, 1949.
New York Times. January 24, 1984, p. A10.
Norton, Bryan. *Why Preserve Natural Variety?* Princeton: Princeton University Press, 1987.
Regan, Tom. "The Nature and Possibility of an Environmental Ethic." *Environmental Ethics* 3 (1981): 19–34.
———. *The Case for Animal Rights*. Berkeley: University of California Press, 1983.

Russow, Lilly-Marlene. "Why Species Matter." *Environmental Ethics* 3 (1981): 101–12.
Schweitzer, Albert. *The Philosophy of Civilization.* Vols. 1 and 2. New York: Macmillan Press, 1960.
Singer, Peter. *Animal Liberation.* New York: Avon Books, 1976.
———. "Not for Humans Only: The Place of Nonhumans in Environmental Issues." In *Ethics and Problems of the 21st Century,* edited by K. E. Goodpaster and K. M Sayre. West Bend, Ind.: University of Notre Dame Press, 1979.

3

Ethical Considerations and Animal Welfare in Ecological Field Studies

R. J. Putman

Introduction

Research in population or community ecology is concerned primarily with seeking to explain patterns in the abundance or distribution of living organisms, looking for factors which affect the relative abundance of different species and promote or limit their coexistence.

In their search for the "rules" governing the dynamics of individual animal and plant populations, population ecologists in the early days for the most part withdrew to the laboratory where they could escape the confusing complexity of the wealth of interactions in any natural community and study the effects of individual factors in turn under strictly controlled conditions. Increasingly, however, it became apparent that the results from such contrived experiments might not truly reflect dynamics in the real world of more complex interaction in truly multispecies systems, where the effects of competition or predation might be altered through the presence of third parties involved in higher order interaction,[1] and population ecologists interested in dynamics of population interaction in the real world were forced to examine their dynamics in field situations.

However, quantifying the effects of such processes in the field—even proving that such interactions did have any effect upon the dynamics of the populations concerned—proved fraught with difficulty. The very complexity of the system made distinction of the effects of individual interactions extremely difficult. In addition, in studies of natural systems one is of necessity studying the

---

This article originally appeared in *Biodiversity and Conservation* 4 (1995): 903–15. Reprinted by permission of Kluwer Academic Publishers. Copyright © 1995 Chapman & Hall. All rights reserved.

end product of many years of interaction; presented with the end result it is commonly extremely difficult to attribute definitive cause. Community ecologists were faced with the same problem: While convinced that biotic interactions of various sorts were extremely important in generating the patterns they observed of species co-occurrence and relative abundance, they were by and large presented with a fait accompli and commonly could not distinguish which of a variety of alternative factors might have resulted in such arrays.

Recognizing the limitations of both laboratory-based population studies and community analyses based on simple observation of current pattern, ecologists were encouraged to embrace a more experimental approach to their field studies and embark on manipulative experiments to test explicit hypotheses rather than rely on post hoc rationalization of general theory, or "soft corroboration" of established theories.[2] Through the late 1960s up to the present this recognition that our ecological understanding could only advance through more formal experimentation has led to an increasing number of manipulative field experiments—in exploring, e.g., the importance and role of competition as a process at population or community level, or the idea of keystone species within community webs. Formal rigorous protocols have become established and the literature witnesses a proliferation of such manipulative experiments.[3]

But despite their scientific "power," such field experimentation in removal of key species from some natural system or artificial manipulation of relative abundances clearly have some ethical implications: What, we might ask actually happened to the individual *Ptethodon or Pisaster* removed in the classic experiments of Hairston or Paine?[4] And what are the moral issues raised by removal experiments involving keystone species which, if "successful," often result in dramatic repercussions throughout the rest of the community of dependent organisms? On the other hand, given that as rigorous scientists we may accept the need for experimentation if we are ever to gain an understanding of the rules and mechanisms governing the ecological patterns we observe, do we have any realistic alternatives? Opportunities to "cash in" upon past "experiments" (historical interventions contrived "before we knew any better") are perhaps few and far between (although Diamond, Pimm, and coworkers are to be commended for their continued exploitation of such past misdemeanors[5] and other workers might be exhorted to do the same[6]). At least such prior "experiments" have the merit that one does not have to wait a further fifty years for the result!

Such deliberate manipulative experiments in ecology are, however, something of a special case, are very explicit and overt instances of clear intervention and as such are immediately obvious candidates for ethical debate. A very useful review of the issues raised is presented by Farnsworth and Rosovsky.[7] In this paper I intend to focus upon the more widespread but less explicit intervention implicit in more general ecological fieldwork, in the application of general *techniques* often in common, everyday use during the course of collecting basic data in a variety of population, community, or behavioral ecology studies.

Even if they are not explicitly constructed as manipulative *experiments*, many ecological field studies involve some degree of intervention during routine monitoring programs: through disturbance caused merely by the presence of an observer or where specific sampling techniques themselves involve capture, handling, and marking. Such interventive techniques may cause discomfort, distress, or loss of fitness, even in the extreme may result in incidental mortality—and the ethical scientist should critically evaluate the implications of each methodology before adopting any procedure—even in the limit abandoning the research proposal if he/she cannot justify its value in relation to the welfare costs attached.

Objective information is readily available for a lot of these techniques for many species: thus decisions can often be taken against a background of hard data, and there is no easy "excuse" for not addressing these issues formally in relation to one's own work. In the rest of this paper, therefore, I will attempt to illustrate some of the types and sources of information now available for various vertebrates in relation to (1) distress and mortality during capture operations; (2) mortality or distress caused at the time by marking; (3) longer-term consequences of handling and marking in terms of subsequent mortality or loss of fitness (e.g., implications among social species of wearing a collar/eartag), before suggesting a "decision-making" framework of questions which ethical scientists might wish to ask themselves before embarking on any program of interventive fieldwork.

LEGISLATIVE CONSTRAINTS AND COST-BENEFIT ANALYSES

Many of the techniques employed in such routine field study are of course covered by restrictive legislation and need special license (e.g., invasive procedures, such as blood sampling, are explicitly covered in the United Kingdom by Home Office regulations; many forms of live trapping/live capture need special license of exemption under the Wildlife and Countryside Acts or other relevant legislation). Within the United Kingdom, the main legislative orders affecting ecological and behavioral fieldwork are the provisions of the Wildlife and Countryside Act of 1981 (and as amended in 1985 and 1991); where work involves administration of anesthetic or procedures which may cause pain or suffering or result in long-term damage, such procedures also require licensing under the Protection of Animals (Anesthetics) Acts of 1954 and 1964, and the subsequent Animals (Scientific Procedures) Act of 1986. I should stress that these are by no means the only acts containing clauses relevant to the field ecologist, and it is beyond the scope—and the intention—of this paper to provide a comprehensive or exhaustive review of all relevant legislation. In my own research on deer, for example, I came under specific provisions of the Deer Acts of 1967 and 1991, and individual researchers should carefully check through all relevant legislation affecting their study systems or intended procedures. Useful summaries of current legislation affecting British wildlife are provided by, e.g., Cooper.[8]

Licensing authorities have considerable discretion in whether or not to issue a license in the first place, and considerable room for maneuver to enforce welfare issues through the imposition of special conditions. The Wildlife and Countryside Act and its amendments protect all wild plants which occur naturally in the wild in Great Britain (with some exceptions under specific, defined circumstances); in addition the act protects all birds (with similar exception), the majority of land vertebrates (including amphibians) and some fish and sea mammals. Taking or killing of protected creatures for scientific or educational purposes may be permitted under license. Such restriction under the Wildlife and Countryside Acts applies inter alia to capture of deer (protected also under the Deer Acts), bats, and shrews, and to mist-netting of birds or ringing birds at the nest.

In granting license, the appropriate authority (English Nature [EN], Countryside Commission for Wales [CCW], Scottish Natural Heritage [SNH]) may make the license general or specific to any degree and may choose to impose special conditions. Licenses are generally issued only where the applicant has proven experience in capture and handling of the relevant species and where the scientific case is strong, although as noted, licenses may also be granted on purely educational grounds.

Costs—in terms of animal life and suffering—and benefits are also explicitly considered before a project license may be granted by the Home Office under the Animals (Scientific Procedures) Act: where the calculated or incidental loss of life and level of short- and longer-term suffering are weighed against the perceived benefits of the research for animal or human life. Here too the experience of the individual researcher as well as the implications of the procedure proposed are considered carefully before license may be granted. In all cases, in reaching their decision about what license to grant—and under what conditions—the Home Office, or other statutory body (EN, CCW, SNH, and so on), considers the experience and "attitude" of the researcher and explicitly assess the likely costs and benefits of the proposed research.

By no means all research procedures need Home Office license, or licensing under the various wildlife acts; in any case, simply relying on someone else's decisions on whether or not to grant you a license to determine your ethical position is simply passing the buck. Some professional organizations and societies now issue guidelines to their members;[9] a number of journals now refuse to publish papers where data appear to have been collected by methods/protocols considered unethical, but again we should not allow restrictions in law or guidelines proffered by professional institutes to relieve us of the need to consider the ethics of our own activities; it is up to each individual to determine whether or not procedures he/she proposes to adopt are justifiable. However, the factors now considered by the statutory authorities in determining and controlling the granting of such licenses establish at least evaluation systems by which we might assess other, unrestricted activities.

Nor should it be thought that such self-scrutiny need be restricted to research activity alone. While I focus attention here primarily on issues raised at research level, many of the same problems also apply to educational exercises such as school or college field studies. Particularly in the case of invertebrate sampling, destructive sampling methods are often employed; further, some incidental mortality to target or nontarget species is inevitable even in the use of nondestructive techniques of sampling, such as live trapping of small mammals. Professional researchers may be becoming increasingly aware of the ethical implications of their own research procedures, but how many small mammals and invertebrates die each year—and unconsidered—in the cause of "education" in student field courses?

Assessing "Costs" and "Benefits"; Value Judgements and Objective Data

Although some possible approaches to weighing costs in animal suffering against research benefit are explored by, for example Smith and Boyd,[10] in any such cost/benefit analysis both "costs" and "values" to an extent must be subjectively determined. Perceived benefits and costs also depend on whether one adheres to a *scientistic, utilitarian,* or *moralistic* code of ethics,"[11] or, to whether one (1) believes in the pursuit of science for its own sake, (2) feels one may justify costs to individual animal life or welfare in the expectation of future benefits to the conservation, management, or welfare of that species or of some perceived benefit to humans, or (3) feels the rights of the individual animal are paramount.

While evaluation of the *significance* of impact, or the relative balance of cost/benefit, may be affected in this way by a personal ethical perspective, and costs in terms of psychological/emotional stress and suffering are rather harder to quantify, at least some of the explicit costs (as measured mortality, measured loss of individual fitness) may be objectively determined for many species for a number of these routine procedures of ecological fieldwork. In this paper I review some of that objective information to show the kinds of information that are available to the inquiring scientist, against which he/she may make a careful assessment of the precise welfare implications of each step of any such procedure—and then make a reasoned decision on whether or not it may be justified. While the paper will address general issues, inevitably (and in some case to save embarrassment to others!) it will be biased towards my own experience.

Impact of Simple Behavioral Observation

Disturbance to wildlife caused by simple intrusion into their environment (in walked census counts, for example, or continuous behavioral observation), while apparently intervention at the lowest possible order, is not without impact. The effects of human disturbance (largely recreational disturbance, but the same principles apply) on wildlife have been explicitly examined by Freddy et al. Jeppesen, Tyler, and van der Zande et al.[12] Both short-term responses to disturbance with no lasting effect and long-term changes in breeding performance, for example, have been reported. Ornithologists have long acknowledged that human intrusion can influence social behavior, reproductive performance of adults, and survival of chicks.[13]

Distress and Mortality During Capture and Tagging Operations

As noted earlier, it is generally extremely hard to determine what may constitute "distress" in animals—and certainly difficult to quantify this rather nebulous concept of "suffering"—a problem frequently recurring in debates over the welfare implications in housing, handling, or management of farm animals and animals in zoos or circuses.[14] A number of behavioral protocols have been suggested however[15] and some progress has been made in domestic mammals in the measurement of levels of cortisol production[16] or blood levels of breakdown products such as LDH-5 (lactic dehydrogenase).

Initial attempts to apply these same physiological indicators in measurements of the perceived stress of capture and handling procedures for fallow deer (*Dama dama*) have been made by Jones and Price.[17] Results of these studies show that levels of LDH-5 are indeed elevated immediately after capture and decline as the animals become more and more quiescent after handling and prior to release. However, the LDH levels recorded are tremendously variable and it proves impossible at present to develop any predictive model of levels of stress implied by a given recorded blood concentration of LDH-5.[18] Further difficulties in interpretation arise in that levels of this enzyme increase as the result of prolonged physical exertion, as the result of physical bruising, as well as in response to "psychological" stress, and thus it seems unlikely to develop as a reliable indicator against which to calibrate the stress levels imposed by different procedures.

Levels of actual physical injury or mortality contingent upon different procedures are more readily quantified. Continuing the cervid theme, I choose as my illustrations here data on injuries and mortality rates associated with different methods of live capture of British deer. The statutory organization responsible for issuing licenses for live capture of deer, the Nature Conservancy Council, and its daughter organizations (EN, CCW, SNH) all require a written return from licensees after each exercise, but few comments in these reports refer explicitly to injury and mortality rates. Data presented here derive from analysis of my own various catches of fallow deer in recent years. Only physical methods of capture were employed rather than chemical methods of immobilization.[19] Fallow deer were caught either in long-nets[20] or within purpose-built wooden handling units. Between 1985 and 1992, we caught and handled nearly 5500 animals during winter catching exercises; 670 of these were handled in fixed catchups, 4750 were caught by netting. Both methods result in some bruising and minor lacerations to a proportion of animals.

Major injuries resulted in humane destruction of a total of fifty-six animals during the entire seven-year period (1.03 percent); serious injury or death was higher (1.33 percent) in purpose-built handling units than where animals were caught in long-nets (0.99 percent).

Equivalent data are available from the United States: In review of levels of mortality or serious injury associated with different methods of physical capture of white-tailed deer (*Odocoileus virginianus*) Sullivan et al. recorded mortality rates of between 2.1 percent and 16.2 percent among deer caught by box trap or corral trap (all statistics here are restricted to studies with large samples > 100), mortalities of around 6–7 percent with the use of drop nets or cannon nets, and mortalities of around 0.9 percent associated with drive-netting techniques where deer are driven by beaters or helicopter into long-nets. Mortalities experienced in all these methods were considerably higher when deer were tranquilized postcapture and certainly mixing chemical methods of restraint with physical capture is almost always counterindicated.[21]

One of the main problems here is accurate assessment of mortality arising from such capture operations *because not all consequential mortality is immediately apparent at the time of the catch*. Deer, like a number of other ungulates, hares, and many birds, are subject to trauma-related stress-shock, most commonly manifested in the form of a progressive posttraumatic myopathy in which rapid and irreversible changes in muscle tissues lead to prostration, progressive depression and death after a period of up to four days after capture.[22] Clearly, there are marked differences in mortality rate associated with different capture methods available, but my main point in this context is rather that *the data on which to draw such conclusions are readily available if we choose to seek them out*. Even the most favored of the techniques (drive-netting) was accompanied by mortality rates of about 1 percent in both the studies summarized by Sullivan et al.[22]

In the United Kingdom, licenses are also required under the Wildlife and Countryside Act for use of live-capture traps to take small mammals if the traps are set in such a way that possibly they could take shrews. Such provisions were enacted because whatever efforts are made to provide adequate or appropriate food for shrews taken within such traps, mortality rates of animals trapped are very high. Even when used just for sampling more abundant species such as voles and mice, some incidental mortality is accepted as inevitable. Sadly little formal data are available on such mortality among small mammals during population studies by, for example, Longworth trapping; but my own experience suggests once again that while with care it may be kept to low levels,[23] it is nonetheless significant and certainly not zero.

Small mammals frequently also perish by drowning in pitfall traps set out to sample ground-active insects. Such observation, together with the incidental mortality of shrews in Longworth traps set for other species, already noted, make the additional point that evaluation of the impact of any procedure should extend to consideration of injury and mortality within nontarget species. (These latter two procedures, live-trapping of small mammals and pitfall trapping for invertebrates, are methods commonly employed during educational field courses as well as in primary research, and emphasize again the point that even where no specific license may be required permitting exercise of some restricted procedure for educational purposes, everyday sampling regimes used in teaching exercises may well raise the same ethical issues of justification as academic research. Here is an opportunity to explore these same issues while introducing the techniques themselves to students.)

Ethical issues associated with catching and ringing birds are considered by Greenwood.[24] Although the British Trust for Ornithology as the main body responsible for regulating such activities does not routinely gather mortality data, incidental mortality is considered infrequent because training standards required for ringers in the United Kingdom are extremely high and individual licenses are not issued to ringers until they have served an extensive apprenticeship under the supervision of more experienced workers. Further, new methods are introduced cautiously through small numbers of very experienced ringers, exploring carefully the difficulties and dangers before they are adopted more widely and thus ensuring that potential problems are generally identified without mortality incidents.

Capture and Handling of Juveniles

One special case of capture for marking perhaps warrants separate consideration here: the case where mammalian neonates are marked soon after parturition or birds are ringed at the nest either as pulli or as adults. The relative immobility and defenselessness of neonates, coupled with close seasonal synchronization of breeding, facilitates the capture and marking of large numbers of individuals in a short space of time. For a number of bird species capture even of adults at the nest may be appropriate because the birds cannot conveniently be approached or captured at any other time. Clearly a potential problem here is maternal rejection of juveniles handled and thus scented by humans or nest desertion due to disturbance.

Drawing once again from my own experience, where we have routinely marked under license large numbers of red deer calves and fallow fawns within hours of birth (although always restricting such marking until after the neonate has been licked clean of birth membranes and is fully dry), subsequent monitoring of mortality rates among marked and unmarked fawns showed no significant difference in mortality; similar conclusions are reported by Ozoga and Clute for white-tailed deer.[25] The implications of catching and marking adult birds at the nest are reviewed by Kania.[26]

Loss of Fitness Caused by Marking, Tagging, or Sampling

Handling of any animals for sampling, whether for simple measurements of body weight, tibia or wing length, examination of reproductive status, or more intrusive sampling such as taking blood or fecal samples, involves a certain measure of stress. Many of the more intrusive procedures properly require special licensing, but all demand restraint and manipulation of wild animals to whom close proximity to humans is necessarily traumatic. In the extreme, prolonging handling for such procedures may cause sufficient trauma as to result in death directly, or as a result of later myopathic degeneration.

Marking of captured animals prior to release may also involve extended handling. Such marking may serve a variety of purposes: Animals may be marked to distinguish animals previously captured from new capture in capture-mark-recapture techniques, they may be marked to ensure future recognition of known individuals over the short or long term, or they may be fitted with radiotelemetric devices for future location/triangulation. Requirements of the marking program may determine the length of handling required to affix the tag; equally the mark itself may cause immediate or subsequent distress.

Most marking of vertebrates for simple distinction of previously captured individuals from new captures or to provide identification at the individual level in the short term can be achieved by nonintrusive means: fur clipping or dyeing for small[27] or larger mammals, scale clipping or paintmarking in reptiles or amphibians.[28] Such marks are lost, however, during any subsequent molt or skin change and will not persist from year to year.

Permanent marking for identification, for birds and larger terrestrial mammals, is most commonly by attachment of some form of visible tag (leg ring, wing tag, ear tag or colored collar); marking of smaller mammals was traditionally achieved through toe clipping—in amputation of particular combinations of digits.[29] Toe-clipping historically has also been the method most extensively used in permanent marking of reptiles and amphibians.[30] But use of such procedures is now generally regarded with disfavor and actively discouraged by professional societies. (Toe clipping is, in any case, not an infallible technique from the researcher's viewpoint; toes may be lost naturally, confusing subsequent identification; in amphibians, amputated digits regenerate over a period of as little as seven months.[31])

Alternatives to toe clipping which have been exploited include ear tagging, leg ringing, or, for medium-sized mammals such as rabbits or squirrels as well as for larger reptiles and amphibians, hot—or cold—branding.[32] More recently, some experimental work has been done to assess the potential for field use of subcutaneous electronic tags based on transponders. Such tags have been used successfully for individual mice and rats in the laboratory, although not yet proven for field use. Certainly subcutaneous tags seem the only practicable way of marking animals with no obvious external appendages or with highly streamlined bodies (such as snakes or aquatic mammals) or others living in dense habitats who might otherwise regularly snag neck collars or ear tags.

While some instantaneous injury is inflicted in hot or cold branding, insertion of external ear tags, or the surgery involved in implantation of internal tags, the trauma seems short-lived. More problematic seems subsequent injury or loss of efficiency through wearing such markers. Frequent injuries used to be reported from ill-fitting leg rings among birds, and more particularly damage caused in wading or water birds during winter from ice accumulation. Williams et al. provide a recent analysis of the effects of mass ringing on body condition and gosling survival of Lesser Snow geese, while Calvo and Furness consider the effects of different marking devices.[33]

Southern considers in detail the difficulties associated with using leg rings in marking studies of small wild rodents, and it is clear that here too ill-fitting rings can lead to injury, with swelling of the limb around the ring and even subsequent loss of the foot or leg.[34] Fullagar and Jewell also offer a comparison of the effectiveness of a range of different tagging methods for small mammals and highlight problems associated with the use of such leg rings (though neither study gives precise figures).[35] In a study on the survival of individual woodmice and bank voles marked by toe clipping or small metal ear tags—and the survival of the mark itself—Hill noted that there was no significant difference in survival (survival rate or actual longevity) of animals marked by ear tag against those marked traditionally by toe clipping, that ear tags were retained by 91 percent of all animals marked, and that in those cases where tags were subsequently lost, the average time to loss was 1.6 months in *Apodemus sylvaticus* and 3.0 months in *Clethrionomys glareolus*.[36]

Although the tag survival rate was high in this particular case, Hill's study draws explicit attention to the compromise that must be faced by any fieldworker between using the marking system that will minimize any risk to the animal and yet fitting a tag or marker that will persist for a sufficient period to permit ready identification.

One additional problem associated specifically with the fixing of radiotelemetric transmitters to study an animal is caused by the necessary weight of such devices and the increased load carried by marked animals which can further seriously restrict movement or impede locomotion, reducing fitness and possibly increasing probabilities of mortality by reducing efficiency of foraging or hindering escape from predators. In radiotracking studies it is necessary to give thought to the size of the transmitter package and the period over which the animal will have to carry it (which may well be its natural life span if recapture is unlikely). As a general rule of thumb it is argued that a transmitter package should weigh no more than 3 percent of the animal's own unladen body weight. A number of authors have attempted to investigate the effect of affixing radio collars to their study animals in this way by investigating body weight changes or changes in behavior. Excellent reviews have been offered by Kenward and White and Garrott.[37]

One final issue, rarely considered explicitly by those involved in marking studies of any kind, is the influence of capture and handling, or any mark applied, on other conspecifics and thus the released individual's social relationships. Human scent associated with handling and any transient or longer-term changes in behavior due to stress of handling, the after affects of drugs, or simply associated with the wearing of a collar or tag may well result in a wary reception among conspecifics, or possible rejection. The mark itself may directly affect social acceptance by conspecifics. In animals with more formally structured social organization, such marks may affect dominance rank; and for both solitary and social species may possibly influence future reproductive fitness—in both cases conferring advantage through the appeal of novelty or disadvantage through suspicion of the unfamiliar.

I have no evidence for social advantage or disadvantage resulting from marking. As noted earlier, we have had no cases of rejection of deer fawns handled by us and marked as neonates, though it should be stressed that handling was always kept to a minimum and no animals were ever marked before being licked dry by the mother. Among the social species with which I have worked we have never seen any suspicion by other conspecifics of adult fallow marked by us with ear tags or collars. However, the major problem experienced by one researcher working with Reeves's muntjac is that the life span of the ear tags is limited due to them being chewed by other muntjac until they are lost or unreadable. I have found similar problems in populations of park fallow deer. Such observations certainly suggest the tag is noticed and is an object of special attention, but are hardly consistent with social rejection.

Conclusion

This paper has inevitably covered a wide range of issues. My purpose throughout has been to enhance awareness of the issues involved in handling and marking wild animals during routine ecological field studies: to make explicit many of the implications which many may not appreciate, or at best overlook. While we may not yet formally assess the extent of distress experienced by animals during such procedures, I hope to draw attention to the fact that objective data are frequently available on the risk of actual physical injury/mortality associated with any particular operation, and the relative merits and demerits of alternative methodologies, so that the responsible field-worker may formally assess the risks involved in adopting a given procedure. Equally, it is clear that for many of the procedures, while choice of the best available alternative may *minimize* risk of mortality/injury, *all* available methods are associated with at least some level of risk greater than zero. Therefore there remains always the subsidiary question of what is an "acceptable" level of "suffering," of mortality or fitness loss beyond which one should abandon the research?

Any framework for individual decision making must mimic the cost-benefit analysis which is carried out in assessment of any procedures formally requiring statutory license, but I believe the same assessment of risks and benefits should be applied to any piece of work whether requiring license or not. In practice, relatively few researchers explicitly confess to doing so.[38] As noted earlier, more formal approaches to cost-benefit analyses have been presented,[39] but evaluation of even the statistics which may be collated about the risks associated with any given procedure and the benefits which may be gained depends on one's individual moral viewpoint (as, for example, "scientistic," "utilitarian," or "moralistic").

To focus the issues, therefore, I present a review of the decision stages I pass through in reaching ethical decisions in my own work, where I regularly capture, measure and mark deer. While I must satisfy myself on each point before I even embark on the whole program of research, I rehearse them over and over again—even in the car *every* time I drive to each new catch; I believe one should never stop asking them.

(1) What are the objectives of the research program on which I am embarking/in which I am engaged?
(2) Can those objectives be realized in any way other than by (in turn) (a) catching and measuring, (b) marking, and/or (c) blood sampling or other interventive procedure?
(3) What alternative methods are available for capture, marking, restraint during sampling, and sampling?
(4) What are the actual risks of injury/death/discomfort associated with the different methods available for catching, marking, restraint and sampling for my particular species? Am I sure I have selected those methods which minimize such risks/costs?
(5) Given that some residual risk attaches to each procedure, does the value of the data I will collect during (a) catching and measuring, (b) marking, (c) sampling justify the possible cost in terms of injury or death?
(6) If I am honest with myself, is my research simply self-indulgent, or do I honestly believe that the work that I am doing and the data that I will collect will genuinely enhance our understanding in an important area of biology (or, in my case, do I believe that the work will improve management for the future and thus are the risks I am taking with current animals' lives justified against the returns to be gained in the welfare of others in the future)?

If I cannot satisfy myself on these points, I turn the car around.

NOTES

1. E.g., S. Y. Strauss, "Indirect Effects in Community Ecology: Their Definition, Study and Importance," *Trends in Ecology and Evolution* 6 (1991): 206–10.

2. See for example, D. R. Strong et al., *Ecological Communities: Conceptual Issues and the Evidence* (Princeton: Princeton University Press, 1984).

3. E.g., E. A. Bender, T. J. Case, and M. E. Gilpin, "Perturbation Experiments in Community Ecology: Theory and Practice," *Ecology* 65 (1984): 1–13; J. A. Wiens, *The Ecology of Bird Communities*, vol. 2: Processes and Variations (Cambridge: Cambridge University Press, 1989).

4. N. G. Hairston, "The Experimental Test of an Analysis of Field Distributions: Competition in Terrestrial Salamanders," *Ecology* 61 (1980): 817–26; R. T. Paine, "Food Web Complexity and Species Diversity," *American Naturalist* 100 (1966): 65–75.

5. K. L. Crowell and S. L. Pimm, "Competitions and Niche Shifts of Mice Introduced onto Small Islands," *Oikos* 27 (1976): 251–58; M. P. Moulton and S. L. Pimm, "The Introduced Hawaiian Avifauna: Biogeographic Evidence for Competition," *American Naturalist* 121 (1983): 669–90; M. P. Moulton and S. L. Pimm, "The Extent of Competition in Shaping an Introduced Avifauna," in *Community Ecology*, ed. J. M. Diamond and T. J. Case (New York: Harper and Row, 1986), pp. 80–97; J. M. Diamond et al., "Rapid Evolution of Character Displacement in Myzomelid Honeyeaters," *American Naturalist* 134 (1989): 675–708.

6. J. M. Diamond, "Laboratory, Field and Natural Experiments," *Nature* 304 (1983): 586–87.

7. E. J. Farnsworth and J. Rosovsky, "The Ethics of Ecological Field Experimentation," *Conservation Biology* 7 (1993): 463–72.

8. M. E. Cooper, *An Introduction to Animal Law* (London: Academic Press, 1987); M. E. Cooper, "British Mammals and the Law," in *The Handbook of British Mammals*, 3d ed., ed. G. B. Corbet and S. Harris (Oxford: Blackwell, 1991).

9. E.g., Association for the Study of Animal Behaviour, "Guidelines for the Use of Animals in Research," *Animal Behaviour* 29 (1981): 1–2.

10. J. Smith and K. Boyd, *Lives in the Balance: The Ethics of Using Animals in Biomedical Research* (Oxford: Oxford University Press, 1991).

11. S. R. Kellert, "Japanese Perceptions of Wildlife," *Conservation Biology* 5 (1991): 297–308; Farnsworth and Rosovsky, "The Ethics of Ecological Field Experimentation."

12. D. J. Freddy, W. M. Bronaugh, and M. C. Fowler, "Responses of Mule Deer to Persons Afoot and Snowmobiles," *Wildlife Society Bulletin* 14 (1986): 63–68; J. L. Jeppesen, "The Disturbing Effects of Orienteering and Hunting on Roe Deer (*Capreolus cepreolus*)," *Danish Review of Game Biology* 13 (1987): 1–24; N. J. C. Tyler, "Short-Term Responses of Svalbard Reindeer to Direct Provocation by a Snowmobile," *Biological Conservation* 56 (1991): 179–94; A. N. van der Zande et al., "Impact of Outdoor Recreation on the Density of a Number of Breeding Bird Species in Woods Adjacent to Urban Residential Areas," *Biological Conservation* 30 (1984): 1–39; and see review by R. J. Putman and J. Langbein, "Behavioural Responses of Park Red and Fallow Deer to Disturbance and Effects on Population Performance," *Animal Welfare* 1 (1992): 19–38.

13. E.g., D. C. Duffy, "Human Disturbance and Breeding Birds," *Auk* 96 (1979): 815–16; D. W. Anderson and J. O. Keith, "The Human Influence on Seabird Nesting Success: Conservation Implications," *Biological Conservation* 18 (1980): 65–80; additional references reviewed in Farnsworth and Rosovsky, "The Ethics of Ecological Field Experimentation."

14. D. M. Broom and K. G. Johnson, *Stress and Animal Welfare* (London: Chapman & Hall, 1994).

15. E.g., M. S. Dawkins, *Animal Suffering: The Science of Animal Welfare* (London: Chapman & Hall, 1980); M. S. Dawkins, "From an Animal's Point of View: Motivation, Fitness and Animal Welfare," *Behavioral and Brain Science* 13 (1990): 1–61; P. Bateson, "Assessmant of Pain in Animals," *Animal Behaviour* 42 (1991): 827–39; P. R. Wiepkema and J. M. Koolhaas, "Stress and Animal Welfare," *Animal Welfare* 2 (1992): 195–218.

16. R. F. Smith and H. Dobson, "Effects of Preslaughter Experience on Behaviour, Plasma Cortisol, and Muscle pH in Farmed Red Deer," *Veterinary Records* 126 (1990): 155–58.

17. A. R. Jones and S. E. Price, "Measuring the Responses of Fallow Deer to Disturbance," in *Biology of Deer*, ed. R. D. Brown (New York: Springer-Verlag, 1992), pp. 211–16.

18. Ibid.

19. See for example reviews by D. M. Jones, "The Capture and Handling of Deer," in *The Capture and Handling of Deer*, ed. A. J. B. Rudge (Peterborough, England: Nature Conservancy Council, 1984), pp. 34–85; and R. Harrington, "Guidelines for the Capture and Handling of Deer," in *Methods for the Study of Large Mammals in Forest Ecosystems*, ed. G. W. T. A. Groot Bruinderink and S. E. van Wieren (Arnhem, The Netherlands: Rijksinstituut voor Natuurbeheer, 1991).

20. R. H. A. Cockburn, "Catching Roe Deer Alive in Long-Nets," *Deer* 3 (1976): 434–40; R. H. Smith, "The Capture of Deer for Radio-Tagging," in *A Handbook on Biotelemetry and Radiotracking*, ed. C. J. Amianer and D. W. Macdonald (Oxford: Pergamon Press, 1980).

21. But see N. G. Chapman et al., "Techniques for the Safe and Humane Capture of Free-Living Muntjac Deer (*Muntiacus reevesi*)," *British Veterinary Journal* 143 (1987): 35–43, in relation to capture of muntjac deer.

22. A. M. Haarthorn, K. van der Walt, and E. Young, "Possible Therapy for Capture Myopathy in Captured Wild Animals," *Nature* 247 (1976): 577; G. A. Chalmers and M. W. Barrett, "Capture Myopathy," in *Non-infectious Diseases in Wildlife*, ed. G. L. Hoff and J. W. Davies (Ames: Iowa State University Press, 1982), pp. 84–94; R. J. Putman, "The Care and Rehabilitation of Injured Wild Deer, I *Deer* 8 (1990): 31–35.

23. J. Gurnell and J. R. Flowerdew, *Live Trapping Small Mammals—A Practical Guide* (London: Occasional Publications of the Mammal Society, 1982).

24. J. J. D. Greenwood, "Research on Wild Birds: Ethical Issues of Ringing," *Proceedings of the Internatioal Ornithological Congress*, Vienna, 1994.

25. J. J. Ozoga and R. K. Clute, "Mortality Rates of Marked and Unmarked Fawns," *Journal of Wildlife Management* 52 (1988): 549–51.

26. W. Kania, "Safety of Catching Adult European Birds at the Nest," *The Ring* 14 (1992): 5–50.

27. Gurnell and Flowerdew, *Live Trapping Small Mammals*.

28. C. P. Blanc and C. C. Carpenter, "Studies on the Iguanidae of Madagascar: Social and Reproductive Behaviour of *Chalarodon madagascariensis*," *Journal of Herpetology* 3 (1969): 125–34; I. F. Spellerberg, "Marking Live Snakes for Identification of Individuals in Population Studies," *Journal of Applied Ecology* 14 (1977): 137–38; C. A. Simon and B. E. Bessinger, "Paint Marking Lizards: Does It Affect Survival?" *Journal of Herpetology* 17 (1983): 184–86; reviewed by I. R. Swingland, "Marking Reptiles," in *Animal Marking: Recognition Marking of Animals in Research*, ed. B. Stonehouse (London: Academic Press, 1978); J. W. Ferner, "A Review of Marking Techniques for Amphibians and Reptiles," Society for the Study of Amphibians and Reptiles, *Herpetological Circular* 9 (1979): 1–41.

29. G. I. Twigg, "Marking Mammals," *Mammal Review* 5 (1975): 101–16, gives a full review; see also J. S. Fairley, "Short Term Effects of Ringing and Toe Clipping on Recaptures of Woodmice," *Journal of Zoology* 197 (1982): 295–97.

30. A. d'A. Bellairs and S. V. Bryant, "Effects of Amputation of Limbs and Digits of Lacertid Lizards," *Anatomical Record* 161 (1968): 489–96.

31. Ferner, "A Review of Marking Techniques for Amphibians and Reptiles."

32. E.g., C. H. Daugherty, "Freeze-Branding as a Technique for Marking Anurans," *Copeia* 4 (1976): 836–38.

33. B. Calvo and R. W. Furness, "A Review of the Use and Effects of Marks and Devices on Birds," *Ringing and Migration* 13 (1992): 129–51.

34. H. N. Southern, ed., *The Handbook of British Mammals*, 1st ed. (Oxford: Blackwell, 1964).

35. P. J. Fullagar and P. A. Jewell, "Marking Small Rodents and the Difficulties of Using Leg Rings," *Journal of Zoology* 147 (1965): 224–28.

36. S. D. Hill, *Influences of Large Herbivores on Small Rodents in the New Forest, Hampshire*, Ph.D. thesis, University of Southampton, 1985.

37. R. E. Kenward, *Wildlife Radio-Tagging: Equipment, Field Techniques and Data Analysis* (London and New York: Academic Press, 1987); G. C. White and R. A. Garrott, *Analysis of Wildlife Tracking Data* (London and New York: Academic Press, 1990).

38. But see for example, F. A. Huntingford, "Some Ethical Issues Raised by Studies of Predation and Aggression," *Animal Behaviour* 32 (1984): 210–15; I. C. Cuthill, "Field Experiments in Animal Behaviour: Methods and Ethics," *Animal Behaviour* 42 (1990): 1007–14; and discussions of M.S. Dawkins and M. Gosling, *Ethics in Research on Animal Behaviour* (London: Association for the Study of Animal Behaviour/Academic Press, 1992).

39. For example, Smith and Boyd, *Lives in the Balance*.

REFERENCES

Anderson, D. W., and J. O. Keith. "The Human Influence on Seabird Nesting Success: Conservation Implications." *Biological Conservation* 18 (1980): 65–80.

Association for the Study of Animal Behaviour. "Guidelines for the Use of Animals in Research." *Animal Behaviour* 29 (1981): 1–2.

Bateson, P. "Assessment of Pain in Animals." *Animal Behaviour* 42 (1991): 827–39.

Bellairs, A. d'A., and S. V. Bryant. "Effects of Amputation of Limbs and Digits of Lacertid Lizards." *Anatomical Record* 161 (1968): 489–96.

Bender, E. A., T. J. Case, and M. E. Gilpin. "Perturbation Experiments in Community Ecology: Theory and Practice." *Ecology* 65 (1984): 1–13.

Blanc, C. P., and C. C. Carpenter. "Studies on the Iguanidae of Madagascar: Social and Reproductive Behaviour of *Chalarodon madagascariensis*." *Journal of Herpetology* 3 (1969): 125–34.

Broom, D. M., and K. G. Johnson. *Stress and Animal Welfare*. London: Chapman & Hall, 1994.

Calvo, B., and R. W. Furness. "A Review of the Use and the Effects of Marks and Devices on Birds." *Ringing and Migration* 13 (1992): 129–51.

Chalmers, G. A., and M. W. Barrett. "Capture Myopathy." In *Non-infectious Diseases in Wildlife*, edited by G. L. Hoff and J. W. Davies. Ames: Iowa State University Press, 1982.

Chapman, N. G., K. Claydon, M. Claydon, and S. Harris. "Techniques for the Safe and Humane Capture of Free-Living Muntjac Deer (*Muntiacus reevesi*)." *British Veterinary Journal* 143 (1987): 35–43.

Cockburn, R. H. A. "Catching Roe Deer Alive in Long-Nets." *Deer* 3 (1976): 434–40.

Cooper, M. E. *An Introduction to Animal Law*. London: Academic Press, 1987.

———. "British Mammals and the Law." In *The Handbook of British Mammals*, 3d ed., edited by G. B. Corbert and S. Harris. Oxford: Blackwell, 1991.

Crowell, K. L., and S. L. Pimm. "Competition and Niche Shifts of Mice Introduced onto Small Islands." *Oikos* 27 (1976): 251–58.

Cuthill, I. C. "Field Experiments in Animal Behaviour: Methods and Ethics." *Animal Behaviour* 42 (1990): 1007–14.
Daugherty, C. H. "Freeze-Branding as a Technique for Marking Anurans." *Copeia* 4 (1976): 836–38.
Dawkins, M. S. *Animal Suffering: The Science of Animal Welfare*. London: Chapman & Hall, 1980.
———. "From an Animal's Point of View: Motivation, Fitness and Animal Welfare." *Behavioral and Brain Science* 13 (1990): 1–61.
Dawkins, M. S., and M. Gosling. *Ethics in Research on Animal Behaviour*. London: Association for the Study of Animal Behaviour/Academic Press, 1992.
Diamond, J. M. "Laboratory, Field and Natural Experiments." *Nature* 304 (1983): 586–87.
Diamond, J. M., S. L. Pimm, M. E. Gilpin, and M. LeCroy. "Rapid Evolution of Character Displacement in Myzomelid Honeyeaters." *American Naturalist* 134 (1989): 675–708.
Duffy, D. C. "Human Disturbance and Breeding Birds." *Auk* 96 (1979): 815–16.
Fairley, J. S. "Short Term Effects of Ringing and Toe Clipping on Recaptures of Woodmice." *Journal of Zoology* 197 (1982): 295–97.
Farnsworth, E. J., and J. Rosovsky. "The Ethics of Ecological Field Experimentation." *Conservation Biology* 7 (1993): 463–72.
Ferner, J. W. "A Review of Marking Techniques for Amphibians and Reptiles." *Society for the Study of Amphibians and Reptiles, Herpetological Circular* 9 (1979): 1–41.
Freddy, D. J., W. M. Bronaugh, and M. C. Fowler. "Responses of Mule Deer to Persons Afoot and Snowmobiles." *Wildlife Society Bulletin* 14 (1986): 63–68.
Fullagar, P. J., and P. A. Jewell. "Marking Small Rodents and the Difficulties of Using Leg Rings." *Journal of Zoology* 147 (1965): 224–28.
Greenwood, J. J. D. "Research on Wild Birds: Ethical Issues of Ringing." *Proceedings of the International Ornithological Congress*, Vienna, 1994.
Gurnell, J., and J. R. Flowerdew. *Live Trapping Small Mammals—A Practical Guide*. London: Occasional Publications of the Mammal Society, 1982.
Haarthorn, A. M., K. van der Walt, and E. Young. "Possible Therapy for Capture Myopathy in Captured Wild Animals." *Nature* 247 (1976): 577.
Hairston, N. G. "The Experimental Test of an Analysis of Field Distributions: Competition in Terrestrial Salamanders. *Ecology* 61 (1980): 817–26.
Harrington, R. "Guidelines for the Capture and Handling of Deer." In *Methods For the Study of Large Mammals in Forest Ecosystems*, edited by G. W. T. A. Groot Bruinderink and S. E. van Wieren. Arnhem, The Netherlands: Rijksinstituut voor Natuurbeheer, 1991.
Hill, S. D. *Influences of Large Herbivores on Small Rodents in the New Forest, Hampshire*. Ph.D. thesis, University of Southampton, 1985.
Huntingford, F. A. "Some Ethical Issues Raised by Studies of Predation and Aggression. *Animal Behaviour* 32 (1984): 210–15.
Jeppesen, J. L. "The Disturbing Effects of Orienteering and Hunting on Roe Deer (*Capreolus capreolus*)." *Danish Review of Game Biology* 13 (1987): 1–24.
Jones, D. M. "The Capture and Handling of Deer." In *The Capture and Handling of Deer*, edited by A. J. B. Rudge. Peterborough, England: Nature Conservancy Council, 1984.
Jones, A. R., and S. E. Price. "Measuring the Responses of Fallow Deer to Disturbance." In *Biology of Deer*, edited by R. D. Brown. New York: Springer-Verlag, 1992.

Kania, W. "Safety of Catching Adult European Birds at the Nest." *The Ring* 14 (1992): 5–50.

Kellert, S. R. "Japanese Perceptions of Wildlife." *Conservation Biology* 5 (1991): 297–308.

Kenward, R. E. *Wildlife Radio-Tagging: Equipment, Field Techniques and Data Analysis.* London and New York: Academic Press, 1987.

Moulton, M. P., and S. L. Pimm. "The Introduced Hawaiian Avifauna: Biogeographic Evidence for Competition." *American Naturalist* 121 (1983): 669–90.

———. "The Extent of Competition in Shaping an Introduced Avifauna." In *Community Ecology*, edited by J. M. Diamond and T. J. Case. New York: Harper and Row, 1986.

Ozoga, J. J. and R. K. Clute. "Mortality Rates of Marked and Unmarked Fawns." *Journal of Wildlife Management* 52 (1988): 549–51.

Paine, R. T. "Food Web Complexity and Species Diversity." *American Naturalist* 100 (1966): 65–75.

———. "A Note on Trophic Complexity and Community Stability." *American Naturalist* 103 (1969): 91–93.

Putman, R. J. "The Care and Rehabilitation of Injured Wild Deer." *Deer* 8 (1990): 31–35.

Putman, R. J., and J. Langbein. "Behavioural Responses of Park Red and Fallow Deer to Disturbance and Effects on Population Performance." *Animal Welfare* 1 (1992): 19–38.

Simon, C. A., and B. E. Bessinger. "Paint Marking Lizards: Does It Affect Survival?" *Journal of Herpetology* 17 (1983): 184–86.

Smith, J., and K. Boyd. *Lives in the Balance: The Ethics of Using Animals in Biomedical Research.* Oxford: Oxford University Press, 1991.

Smith, R. F., and H. Dobson. "Effects of Preslaughter Experience on Behaviour, Plasma Cortisol, and Muscle pH in Farmed Red Deer." *Veterinary Records* 126 (1990): 155–58.

Smith, R. H. "The Capture of Deer for Radio-Tagging." In *A Handbook on Biotelemetry and Radiotracking*, edited by C. J. Amianer and D. W. Macdonald. Oxford: Pergamon Press, 1980.

Southern, H. N., ed. *The Handbook of British Mammals*, 1st ed. Oxford: Blackwell, 1964.

Spellerberg, I. F. "Marking Live Snakes for Identification of Individuals in Population Studies." *Journal of Applied Ecology* 14 (1977): 137–38.

Strauss, S. Y. "Indirect Effects in Community Ecology: Their Definition, Study and Importance." *Trends in Ecology and Evolution* 6 (1991): 206–10.

Strong, D. R., D. Simberloff, L. G. Abele, and A. B. Thistle. *Ecological Communities: Conceptual Issues and the Evidence.* Princeton: Princeton University Press, 1984.

Sullivan, J. B., C. A. De Young, S. L. Beasom, J. R. Heffelfinger, S. P. Coughlin, and M. W. Hellickson. "Drive-Netting Deer: Incidence of Mortality." *Wildlife Society Bulletin* 19 (1991): 393–96.

Swingland, I. R. "Marking Reptiles." In *Animal Marking: Recognition Marking of Animals in Research*, edited by B. Stonehouse. London: Academic Press, 1978.

Twigg, G. I. "Marking Mammals." *Mammal Review* 5 (1975): 101–16.

Tyler, N. J. C. "Short-Term Responses of Svalbard Reindeer to Direct Provocation by a Snowmobile." *Biological Conservation* 56 (1991): 179–94.

van der Zande, A. N., J. C. Berkhulzen, H. C. van Latesteijn, W. J. ter Keurs, and A. J. Poppelaars. "Impact of Outdoor Recreation on the Density of a Number of Breeding Bird Species in Woods Adjacent to Urban Residential Areas." *Biological Conservation* 30 (1984): 1–39.

White, G. C., and R. A. Garrott. *Analysis of Wildlife Tracking Data*. London and New York: Academic Press, 1990.

Wiens, J. A. *The Ecology of Bird Communities*. Vol. 2: Processes and Variations. Cambridge: Cambridge University Press, 1989.

Wiepkema, P. R., and J. M. Koolhaas. "Stress and Animal Welfare." *Animal Welfare* 2 (1992): 195–218.

Williams, T. D., F. Cooke, E. G. Cooch, and R. F. Rockwell. "Body Condition and Gosling Survival in Mass-Banded Lesser Snow Geese." *Journal of Wildlife Management* 57 (1993): 555–62.

Part II

Strategies for Conservation and Management

4

LAUNCHING THE NATURAL ARK

James R. Udall

"Free the lobo."

A petition bearing this slogan recently circulated through the biology department at the University of New Mexico, gathering signatures with ease. Supported by the Mexican Wolf Coalition and the Wolf Action Group, its aim was to introduce Mexican gray wolves bred in captivity to parts of their historic range in New Mexico and Arizona.

After some reflection, James Brown, a biologist best known for his landmark study of mammal extinctions in the Great Basin, decided not to sign. "I'm all for the lobo, " Brown explains. "But there are only thirty-nine adults left. Their former habitat is so fragmented that reestablishing a viable population may not be possible. I'm just not sure there's a place for the wolf in the Southwest anymore."

This statement—which, in essence, consigns one of North America's rarest subspecies to imprisonment in zoos—is considered heresy by many local conservationists, who argue that parts of the Southwest are still big and wild enough to support wolves. But many of those who disagree with Brown about the lobo share his larger concerns about the progressive degradation of wildlife habitat in this country. In symposia and scientific journals, in Sierra Club and gun-club meetings, the traditional approach to wildlife conservation is being reevaluated—and found wanting.

"Our emphasis on saving critically endangered species," says conservation biologist Blair Csuti, "has too frequently resulted in crisis management for individual plants and animals."

This article originally appeared in *Sierra* (September/October 1991): 80–89. Reprinted by permission of the author. Copyright © 1991 James R. Udall. All rights reserved.

Although Csuti doesn't advocate making an omelet out of the next clutch of California condor eggs, he is one of many scientists who believe that the U.S. Endangered Species Act, which focuses on life forms on the brink of extinction, must be broadened to include endangered habitats and ecosystems. The logic is simple and pragmatic: Last-ditch efforts to resuscitate a vanishing plant or animal are heroic but expensive; conservation is easiest when a species is still common, before most of its habitat has been destroyed.

Conservationists are quick to defend the Endangered Species Act; in the nearly twenty years since its passage it has been one of their most powerful tools. The act has been instrumental in protecting the northern spotted owl and portions of the Northwest's ancient forests, for example, and has restored to healthy numbers once-faltering species, such as the bald eagle, peregrine falcon, and American alligator. Despite these successes, however, the loss of wildlife habitat continues.

"We've been winning a few battles, but losing the war," admits biologist and Sierra Club staffer Gene Coan. "It's time to broaden the focus."

For biologists and conservationists, broadening the focus is a recurring theme. There is, they say, an urgent need to change the mindset underlying our current approach to wildlife conservation. We need to shift the emphasis from next year to next century, continuing to build lifeboats for critically endangered species, but also beginning the larger task of relaunching the natural ark.

"Over the past century, we've managed to preserve many parks and wilderness areas, but they were selected largely on the basis of their scenic and recreational importance," says John Hopkins, chair of the Sierra Club's Public Lands Committee. "Now activists must turn to the enormous challenge of preserving lands of biological significance."

The push to revise current conservation strategies stems in part from the forecast for the next century: The world's population will continue to mount by 250,000 each day. Our forests will be cut down, and then global warming will kick in. Rainfall patterns will shift. Crops will fail. Governments will dither as billions starve, dragging millions of species down with them. Today the Sahel; tomorrow Brazil, Mexico, Kenya.

The scenario is too grim to sugarcoat. The Era of Life's Impoverishment, ecologist Norman Myers calls it. Michael Soulé of the University of California at Santa Cruz dedicated his 1986 book, *Conservation Biology*, "to the students who will come after, who will witness the worst, and accomplish the most."

If current trends continue, according to a recent report to the National Science Foundation, "the rate of extinction over the next few decades is likely to rise to at least one thousand times the normal background rate . . . and will ultimately result in the loss of a quarter or more of the species on earth."

Although the loss of biological diversity will be greatest in the tropics, where most of the planet's animal and plant species reside, our continent will not be spared. Indeed, ongoing habitat destruction makes the loss of thousands of North American species a certainty—unless we fundamentally change the way we use the land.

Biodiversity is a thorny term, most commonly associated with rainforest preservation. Asked to define it, an ecologist will usually say, "It's more than a species," before diving into a thicket of complexities. To explore the different layers and levels of biodiversity, I did a little bushwhacking with conservation biologist Larry Harris, author of *The Fragmented Forest*, a blueprint for preserving biodiversity in the heavily logged forests of Oregon.

There are three levels, he explained to me. The first, "genetic diversity," refers to genes within a species. A century ago, biologist C. Hart Merriam identified dozens of populations of grizzly bears: the well known Plains grizzly, for one, but also distinct grizzly populations in California, Arizona, Texas, Oregon, and elsewhere. Today all these animals are gone. "Some people will tell you that we really didn't lose anything, because we still have grizzlies in Yellowstone," said Harris, now at the University of Florida. "Nonsense. We lost a lot."

What we lost was genetic information, the DNA coding that enables organisms to cope with change. The bears that vanished were all the same species, but because they had adapted to different environments, they weren't all the same. Genetic diversity can also be lost through habitat fragmentation and subsequent inbreeding among a stranded population. This is why the long-term survival of Yellowstone's grizzlies is problematic, and why even captive breeding may be too little, too late to save the Florida panther.

"Species diversity" refers to the variety of species within a habitat or broader area. One place rich in this respect is the 502-square-mile Gray Ranch in Southwestern New Mexico. Recently purchased by The Nature Conservancy, the site contains more mammal species than any national park or wildlife refuge in the lower forty-eight.

The Gray Ranch also has an abundance of the third type of diversity biologists look for, "ecosystem diversity." This term refers to the variety of habitats within a region. "One reason the Gray Ranch is so rich in species is that it has so many different kinds of habitats," says Harris.

Maintaining a healthy ecosystem requires protecting all three levels—genetic, species, and ecosystem diversity. Otherwise, significant losses at one level are liable to produce cascading losses at others. For example, some scientists attribute Yellowstone's small beaver population to the fact that the park's willows—an important food source for beavers—have been overbrowsed by elk. The elk in turn have proliferated because their chief predator, the gray wolf, has been exterminated. (Large predators often function as "keystone" species whose impacts profoundly affect entire plant and animal communities—one argument for reintroducing them, whenever possible, to habitats from which they've been removed.)

To preserve biodiversity over long periods of time, natural processes must also be protected, including the nutrient, hydrologic, and fire cycles that shape ecosystems. "The Everglades are in trouble because of human-caused changes in the water cycle," says Harris. "Within a few decades, all of the romantic elements we associate with that national park—wading birds, red wolves, roseate spoonbills, Florida panthers—may be gone." The park will still exist, of course, but, in the words of conservation biologist Reed Noss, "Scenery is a hollow virtue when ecological integrity has been lost."

What is happening in the Everglades is not unusual—with some exceptions, this continent's ecosystems have been losing natural diversity since the first hunters crossed the Bering Strait. Although many of our landscapes are still scenic, and some are still reasonably healthy, on the whole they have been hollowed out like so many ripe pumpkins by trapping, plowing, logging, mining, damming, poisoning, and other forms of human intrusion.

What if, by some quirk of history, the first European settlers had wanted to conserve North America's natural wealth rather than plunder it? If they had known what conservation biologists know today, they would have set aside at least two big blocks of land, four times bigger (say) than Yellowstone Park, in each of the country's bioregions. (Big, to function properly over long periods of time; two blocks, because you don't want all your bears or wolverines or redwoods stranded in one spot.)

Historically, of course, the only place we had an opportunity to adopt such a farsighted plan in advance of large-scale human impact was Alaska. But in the other forty-nine states we still have a chance, and a need to work backward toward that ideal. Though we may never achieve it all, we can achieve a great deal. According to Forest Service ecologist Hal Salwasser, "If biodiversity has a chance anywhere in the world, it is in North America."

But the task is immense. We've been beating back the forest for centuries, and it will take an equal effort to recover what we've lost. "The object," says landscape architect Leslie Sauer, "is to gradually shift the impact of our actions from being largely negative to being largely positive."

One of the first things we must do is reevaluate our existing system of national parks, wilderness areas, and wildlife refuges. Because these areas were chosen based on criteria that have little biological relevance, Harris calls them "an agglomeration of artifacts." These wildlands fail to promote biodiversity in two ways. First, regions and habitat are unequally represented. The West is relatively rich in protected areas; the East, depauperate; the Midwest, bankrupt. We've saved plenty of snowcapped peaks and scree slopes, but lower-elevation forests, wetlands, grasslands, and coastal areas are underrepresented or, too often, absent altogether.

Second, many reserves are too small to maintain all the species once found within them. According to biologist James Brown, "Refuges of less than 125,000 acres in which animals are tightly confined may lose more than half their species in a few thousand years." Thirty percent of the United States national parks and 93 percent of its wildlife refuges fall into that vulnerable category.

The significant qualifier is "tightly confined." Nearly all reserves were originally embedded in a natural matrix; their boundaries were permeable. But now, as population growth and development continue, many parks and refuges are gradually being transformed into islands in a sea of humanity. (Roads are a chief culprit. "The highway system is just a killing machine," says Harris, noting that automobiles are now the leading cause of mortality among Florida's endangered mammals. The sole exception there is the seagoing manatee; its leading cause of death is collision with motorboats.)

Worldwide, habitat fragmentation weighs most heavily on wide-ranging carnivores—the wolves and big cats. But fragmentation also harms large herbivores, primates, and bears; habitat specialists like the giant panda and black-footed ferret; and species that require virgin forest, such as the lynx and red-cockaded woodpecker. Small creatures are not immune. As forests in the East shrink under development pressure, even songbirds and plants with modest spatial needs find themselves threatened.

Among wildlife managers there is a growing realization that fragmentation puts at risk everything conservationists thought they had saved. The fallacy of viewing the national park as a kind of ark has been exposed. Warns conservation biologist David Western, "If we can't save nature outside protected areas, not much will survive inside."

To prevent our parks and refuges from slowly losing their natural richness, it is essential to buffer them from development. According to William Penn Mott, former director of the National Park Service, "Stopping our concern at the boundary has got to be replaced with practical applications of buffer zones, regional planning, and consideration of the social and economic conditions adjacent to reserves." The benefits and difficulties of implementing these ideas are being demonstrated in the Rockies, where federal and state agencies, prodded by concerned citizens, are attempting to fashion a management strategy for the Greater Yellowstone ecosystem. The same grand experiment needs to be tried elsewhere—in the Everglades, the Cascades, the Sierra Nevada, and the Appalachians, for example—to create cohesive landscape units rather than mosaics of bureaucratic turf.

Wildlife corridors and protected lands called "stepping stones" can also help strengthen isolated refuges. By connecting existing parklands, stepping stones can aid migrating birds and wide-ranging predators and herbivores, while helping to link together patchwork landscapes.

"Creating corridors will be of particular importance in the East," says Harris, "where many existing reserves are simply too small to contain even a single panther or a viable population of black bears." An encouraging recent example was the Forest Service's purchase of twenty-nine thousand acres in a corridor that will eventually connect the Okefenokee National Wildlife Refuge and the Osceola National Forest along the Florida/Georgia border, a move that will dramatically enhance the ecological integrity of both areas.

Wildlife corridors are not a panacea, however. They help animals more than plants, and some animals more than others. They can also pass along pests and diseases, and their effectiveness decreases as they become narrower. Despite these limitations, corridors are still one of the most effective tools we have for buttressing existing nature preserves.

Another urgent task is to "map the gaps and buy the hot spots"—to locate and purchase rare natural communities that aren't currently protected. In many countries, including China and England, biological inventories are a top priority. Not here. If you want to know how the land is shaped, the U.S. Geological Survey will sell you a topographic map. If you want to know what minerals are underground, fine—they have that mapped too. But if you want to find out what's on top of the ground, to locate an undisturbed stretch of riparian habitat in Arizona or a rare plant community in California, the government can't help you.

Federal agencies have provided limited support to some private efforts, however. With federal help, The Nature Conservancy has made great strides toward completing an inventory of the nation's rare and endangered species. "Gap analysis," a satellite-mapping program developed by the U.S. Fish and Wildlife Service, has also proven its worth in Idaho and may soon be employed nationwide.

Although new buffer zones, wildlife corridors, and protected hot spots will help bolster existing parks, wildlife refuges, and national-forest wilderness, our biological heritage cannot be preserved solely by these means, for such lands cover only 7 percent of the United States. About 18 percent is currently used for agriculture. Cities and pavement claim another 3 percent. Biologist James Brown believes that the biodiversity battle will be won or lost on the remaining 72 percent of our land—the area he calls the seminatural matrix or the semiwild. In the West, much of the semiwild is public land, managed by the Forest Service and Bureau of Land Management (BLM); in the East, it is mostly private. The semiwild now hosts a variety of human activities; the challenge facing conservationists is to balance biodiversity protection with grazing, timber-cutting, recreation, and mining. According to the Forest Service's Salwasser, it should be possible to "protect genetic resources, sustain viable populations, perpetuate natural biological communities, and maintain a full range of ecological processes while also meeting human needs."

The key will be enlightened management, based on the principles of conservation biology. The Forest Service, BLM, and private landowners will have to start, in Aldo Leopold's words, "thinking like a mountain." Whether the "crop" these lands provide is sawlogs, grass, or recreational user-days, the objective should be to harvest a sustainable yield without damaging, or further fragmenting, the wildlife habitat.

What about the semiwild and "metroforest" along the densely populated eastern seaboard? Although these areas no longer meet the classic definition of wildness, they still have a role to play in the preservation of biodiversity. Indeed, as the human population of the East continues to grow, the remaining undeveloped lands become even more valuable as wildlife havens. Fortunately, nature is resilient—after an absence of many decades, cougars have returned to New England—and the eastern landscape has not reached the point of no return. Says Zev Naveh, an Israeli ecologist, "My part of the world has been heavily grazed for thousands of years. Compared to the goatscape we are working with, your country offers wonderful opportunities."

To seize those opportunities, says Leslie Sauer, "We in the East must set our sights higher. There should be no place where there is no wildness at all, no site where natural values are not deemed important. It's time to make a habit of restoration, to take streams out of pipes, rivers out of channels, pavement off of meadows."

Large tracts of wilderness in the East are rare, but Reed Noss says "wilderness recovery areas" could be created in the Appalachians by closing roads, restoring habitats, allowing natural fires to burn, and reintroducing extirpated animals such as the panther, bison, and elk. In the Northeast, the Sierra Club and other conservation groups are working to establish extensive reserves within the twenty-six million acres of forest that carpet Vermont, New Hampshire, Maine, and northern New York.

Such ideas may seem visionary. But so was the notion, once upon a time, of even a single national park.

The movement to preserve biodiversity represents a natural evolution of conservation strategy. It is both the next step on the road to a land ethic and the ultimate grassroots challenge. Meeting that challenge will require sweat, dedication, and a willingness to take a new look at the American landscape. Nature preserves, new forestry and grazing practices, wildlife corridors, more enlightened land management, public participation at all levels of government, restoration—each will have a role to play.

For those who decide to take on these tasks, the work could be satisfying—just as it was for a handful of Sierra Club volunteers who began trying to restore a degraded park on the outskirts of Chicago fourteen years ago. They prowled railroad grades gathering native seeds, cut thickets that were choking ancient oaks, lit fires to kill invading weeds, and perused nineteenth-century journals for botanical insights.

As news of their labors spread, more and more people turned out to help. This year more than three thousand people, a veritable Noah's army, will volunteer their time and talents. Their mission is to salvage something of tremendous significance, a landscape that was said to exist no longer in North America—a tallgrass savanna, a grassland with trees.

Nature Conservancy staffer Steve Packard has been one of the project's guiding lights. Asked what his experiences have taught him about preserving biodiversity, he replies, "Restoring the prairie has been like building a cathedral. Thousands of people are involved, and none of us understands more than our own little part of it. Sometimes I visit the site, and the birds are flying around or the flowers are blooming, and I think to myself 'My goodness, all this depended on us.' And it did. It feels wonderful."

5

WEEDING THE GARDEN

Andrew Neal Cohen

"Eider nest. Eider nest."
"Yup."
"Gull nest."
"Yup."
"Eider nest."
"Gull nest."
"Yup."

Four men advanced in a ragged line across Green Island, a low, treeless bit of windswept land emerging from the Gulf of Maine. Eyes to the ground, the men called out to a fifth who trailed behind them, taking notes. Gulls floated lazily just overhead, while waves of sibilant chatter rose from a raft of eider ducks sheltered in the lee of the island, waiting for the men to depart. The rocks and grassy hummocks were thick with nests, and the searchers stepped gingerly.

"Gull nest."
"Yup."
"Eider nest. Eider flush."

A female eider leaped into the air and shot low over a tangled ridge to join the waiting raft. Four olive-drab eggs snuggled in a bowl of down that she had plucked from her own breast. Passing these nests, the men plucked at the down themselves in a token gesture at covering the unprotected eggs.

"Two eider nests."
"Gull nest."

This is a revised version of an article that originally appeared in *Atlantic Monthly* 270, no. 5 (November 1992): 76–86. Reprinted by permission of the author. Copyright © 1992 Andrew Neal Cohen. All rights reserved.

"Gull nest."

The gull nests also held eggs, the larger eggs of great black-backed gulls and the slightly smaller eggs of herring gulls, both splotched with brown. Into each gull nest, surgical-gloved hands deposited two or three "baits"—little sandwiches made of margarine spread between cubes of bread.

"One eider nest."

"Gull nest. Wait—make that two gull nests."

Mixed into the margarine was a white, powdery substance that Tom Goettel, the U.S. Fish and Wildlife Service biologist in charge of the operation, had carefully measured out from a canister the day before. The label on the canister read:

> RESTRICTED USE PESTICIDE
> 1339 GULL TOXICANT 98% CONCENTRATE
> DANGER—POISON

The label went on to describe human health hazards, proper methods of application, and the toxicant's effect on gulls, causing death from kidney failure within twenty-four to forty-eight hours. It was Goettel's hope that the gulls flying overhead would return to their nests, eat the poisoned bait, and quietly keel over and die.

"You know, I didn't join Fish and Wildlife to kill gulls," says Goettel. He just wanted to save terns.

Striking Back

Herring gulls and great black-backed gulls are notorious predators of the eggs and young of terns and other seabirds, and easily outcompete the smaller and less aggressive terns for island nesting sites. Petit Manan Island, just half a mile south of Green Island and connected to it by a cobble beach at low tide, was historically one of the most important tern colonies in Maine. Petit Manan's flat terrain and low vegetation provided abundant nesting sites for terns, and the island's lighthouse keepers, perennially at war with the pestiferous gulls, killed any that tried to nest there. But when the Coast Guard automated Petit Manan Light in 1972, and the last lighthouse keeper departed, the gulls took over and chased out the terns. Within a few years, the Petit Manan tern colony had been wiped out.

In 1984, the Fish and Wildlife Service struck back. That spring, Goettel and his colleagues poisoned the Petit Manan gulls with Toxicant 1339, and in two weeks several hundred terns were claiming nesting sites on the island. More terns arrived in the following years, and Petit Manan is now once again one of Maine's largest tern colonies, with three species of nesting terns—common terns, arctic terns, and the endangered roseate terns—along with hundreds of nesting guillemots and thirteen nesting pairs of rare Atlantic puffins. It is in order to protect these birds that Tom Goettel routinely poisons the gulls on Green Island.

The use of Toxicant 1339 to kill gulls in New England was promoted by the National Audubon Society, which in 1971 asked Fish and Wildlife to clear the gulls off of Matinicus Rock, on the coast of Maine, where they had been preying on a colony of Atlantic puffins. Since then, gulls have been killed on several other islands in Maine and Massachusetts to allow for the recovery of puffin and tern colonies.

Ironically, in the past wildlife enthusiasts were kept busy protecting gulls. In the late nineteenth century, after egg gatherers and plume hunters had nearly wiped out New England's gulls, local Audubon societies contributed money to hire wardens to guard the few colonies that remained. Despite these efforts, by the turn of the century only about eight thousand nesting pairs of herring gulls were left, confined to the outer islands of Maine. But with the passage of laws that banned seabird hunting and egg collecting, and that established seabird refuges, and with abundant new sources of food from landfills, sewage outfalls, and the discarded wastes of an expanding fishing industry, gull populations exploded. Gulls extended their nesting range down the Maine coast and into New Hampshire and Massachusetts by 1920, to New Jersey and Maryland by 1950, and to North Carolina by 1960. Current estimates placed the nesting population at 150,000 pairs, and the total population at more than a million birds.

As their numbers soared, gulls increasingly became a nuisance. Flocks of gulls at airports posed a danger to planes on landing and takeoff, at reservoirs they were suspected of contaminating water supplies, and at garbage dumps and sewage ponds they were considered vermin—rats with wings. But most attempts to reduce their numbers have been dismal failures. From 1940 to 1953 the largest attempt ever made to control gulls began in Maine. Teams of workers sprayed nearly a million eggs with a mixture of oil and formaldehyde, which suffocated the developing embryos but preserved the eggs so that the parent gulls continued incubating rather than laying new ones. At best, however, this massive effort only delayed the gulls' increase, and some researchers, noting a contemporaneous surge in the Massachusetts gull population, believe that the main effect was to encourage the gulls to spread southward.

Then in the 1960s Dr. William Drury, the research director for the Massachusetts Audubon Society, began a series of studies on methods of lethal gull control. He and his coworkers tore apart nests, broke eggs, harassed and shot gulls, introduced predators into gull colonies, and administered chemical sterilizers and poisons. Drury showed that although these methods would not reduce the overall gull population, under favorable conditions they could be used to remove some or all of the gulls from specific sites. He concluded that the most efficient and humane approach would be poisoning with Toxicant 1339, which was originally developed to kill starlings. Over the past two decades gull colonies on several islands have been treated with 1339 in order to allow other, rarer seabirds to flourish—an action that Drury likened to "weeding a garden."

A Plague of Snakes

In recent years, wildlife "gardeners" have been hacking at an increasing number of weeds. In California, red foxes, hawks, and other predators have been killed to protect the eggs and chicks of endangered least terns and clapper rails. Coyotes that kill San Joaquin kit foxes have been gunned down from helicopters. In Alaska, the Fish and Wildlife Service has trapped and poisoned arctic foxes that prey on Aleutian Canada geese. Coyotes have been trapped and shot to protect whooping cranes in Idaho and greater sandhill cranes in Oregon. Ravens have been poisoned and shot to protect greater sandhill cranes and California desert tortoises. Cowbirds threaten many songbird populations through nest parasitism—removing songbird eggs from nests and laying their own eggs in place, which the unsuspecting songbirds then raise—and tens of thousands of cowbirds have been exterminated to protect endangered birds in California, Michigan, Oklahoma, Puerto Rico, and Texas. In Washington's Olympic National Park, rangers have proposed shooting hundreds of mountain goats whose disturbance of the soil threatens several rare plants that are unique to the Olympic Peninsula. In other places raccoons, skunks, opossums, ground squirrels, mountain lions, badgers, pigeons, meadowlarks, crows, shrikes, owls, northern harriers, and kestrels have been killed to prevent them from harming rare species.

When wildlife managers discuss these programs, certain themes recur. Oftentimes the population of a species was initially reduced by some direct human activity—overhunting or excessive collecting, destruction or degradation of critical habitat—and the current predator merely threatens to deliver the final blow. "This is the kind of mess we get into when we push animals to the brink of extinction," says Ronald Schlorff, an endangered-wildlife specialist with the California Department of Fish and Game. "Predator control is a necessary human intervention in a system that's out of balance. Predation is a normal part of the natural scene, but it's been concentrated, accentuated, and exacerbated by human activities."

Chief among the activities leading to a predatory imbalance are intentional or accidental introductions of predators to new regions where they devastate native species that have few defenses against them. In Alaska, for example, eighteenth- and nineteenth-century fur trappers stocked hundreds of islands with arctic foxes and red foxes, which ravaged the many species of ground-nesting and burrow-nesting seabirds that migrated to the isolated islands in order to breed in safety. Similarly, Midwestern red foxes that were introduced to the interior valleys of California by hunters and fur farmers have spread to the coast, where they threaten endangered clapper rails and least terns. In other cases, changes that humans have made in the landscape, such as creating landfills, chopping up extensive forests into smaller wooded areas, or grazing livestock, have enabled a harmful species to expand its population and range dramatically. "What we're seeing is a general phenomenon of what we call 'garbage' animals," says Dave Wilcove, an ecologist with the Environmental Defense Fund, referring to the spread of gulls, ravens, raccoons, foxes, and coyotes. "We've made the world very nice for scavenging omnivores."

Whatever the ultimate causes, wildlife managers charged with protecting certain rare species believe that the imminent risk of extinction from predation is real. Events on the island of Guam provide a chilling example. The brown tree snake, a native of New Guinea and Australia, was accidentally brought to the island by cargo ships in the 1950s. With no natural predators and few competitors, the mildly venomous snake flourished, eating its way through the island's unique avifauna. "The brown tree snake has virtually wiped out the native forest birds of Guam," according to a Fish and Wildlife Service report. "Nine species of birds, some found nowhere else, have disappeared from this island, and several others persist in precariously low numbers close to extinction." Biologists managed to save and successfully breed one of these birds—the Guam rail—but are reluctant to return it to the snake-infested island, where it would have little chance of survival.

As far as predatory wildlife goes, the Department of Agriculture's Animal Damage Control (ADC) unit holds the nation's principal license to kill. ADC's 1931 enabling act instructed it "to conduct campaigns for the destruction or control" of a long and nonexclusive list of predators and pests that farmers and ranchers found bothersome. Over the years, ADC pursued this mission with a single-minded zeal that, not surprisingly, provoked unrelenting hostility from wildlife organizations. Today, however, ADC's savvier "gopher chokers" sense that using their skills to protect endangered species may be a means of gaining credibility with their environmental adversaries.

Pete Butchko, an ADC district supervisor, estimates that his former southern California ADC division, which just a few years ago concentrated on eradicating sheep-eating coyotes and cleansing the suburbs of skunks and raccoons, now spends at least a third of its time eliminating the predators of endangered prey. In a speech at the Fourteenth Annual Vertebrate Pest Conference, Butchko argued that these efforts have "allowed ADC to expand its influence and demonstrate its professionalism in new areas and to people not traditionally receptive to predator control." Indeed, when endangered prey are involved, it is no longer uncommon for environmental organizations to support the killing of wild predators. In these cases the traditional environmental demand of "Keep your hands off Nature" has been replaced by an endorsement of the use of lethal force.

The Tortoise and the Raven

But this change has not come without conflict, ranging from disputes within organizations to the filing of lawsuits to stop control programs that some environmentalists believe are urgent. The reasons for conflict have included technical concerns over the methodology of predator control, distrust of ADC and fears of overkill, reluctance about getting involved in the killing of wildlife and the manipulation of nature, philosophical differences over animal rights and the expendability of non-native species, and political fears that predators may be blamed for problems that human society has created.

Science and politics were both at issue when the Humane Society of the United States sued the U.S. Bureau of Land Management (BLM) to stop a project initially supported by Defenders of Wildlife and the Kern County Audubon chapter. With the consent of these two organizations, the BLM was shooting and poisoning ravens to prevent them from preying on the young of California desert tortoises, whose population was rapidly declining.

"We're concerned about tortoises," says John Grandy, a vice president of the Humane Society and former vice president of Defenders of Wildlife, "but we're absolutely opposed to random, unnecessary destruction of ravens." Grandy doesn't agree that there is adequate scientific justification for a broad program of raven killing, and argues that efforts should target only those ravens specifically known to eat juvenile tortoises. "There's urgency, but it's not so severe that we need to rush off and slaughter ravens willy-nilly. This is a problem that is associated with individual ravens. The raven experts that we talked to suggest that raven predation on tortoises is a learned behavior, probably engaged in by a few resident birds."

The Humane Society also charged that by focusing on ravens, the BLM was ignoring more important threats to the tortoise's survival. Cattle grazing, off-road vehicle use, military maneuvers, highway construction, and encroaching urban development have all been cited as contributing to the degradation and destruction of tortoise habitat. "One of our concerns is that the raven is being made a scapegoat for all the problems with the tortoises. The least politically powerful thing out there is the raven, the easiest thing to divert attention to."

Grandy also got involved when biologists at San Francisco Bay National Wildlife Refuge proposed to trap and kill red foxes that feed on endangered California clapper rails. The proposal created a temporary rift in the local Sierra Club when the club's wildlife committee sided with the foxes and its wetlands committee sided with the rails. "The red fox is a victim just as much as the clapper rail," one Sierra Club member wrote. "Trying to further manipulate nature by killing one species because another is favored only adds to the mistakes of the past."

One hotly debated issue was whether red foxes, having been introduced to California from the Midwest, merit less consideration than native species. "They couldn't ever just say 'the red fox,'" complained a Wildlife Committee member. "It was always 'the *alien* red fox.' That kind of labeling really fanned up people's emotions." Wildlife jingoism aside, Grandy argues that it may make sense to remove nonnative species from otherwise pristine areas, but in California "where virtually everything is introduced, red foxes are nearly as natural as you're going to get." He adds, "There's something deeply troubling about us—we who've only been on this continent for 350 years—talking about eradicating something because it's nonnative. If we're so concerned about native species, then where are the Indians?"

Changing Attitudes

The resolutions of these two situations say different things about the necessity, urgency and appropriateness of predator control programs. A year after the Humane Society's lawsuit forced a temporary halt in the killing of desert ravens, the BLM transferred control of the project from a herpetologist to an ornithologist. The BLM "went after ravens first, in part because it was easy," according to Bill Boarman, the new project manager. "Off-road-vehicle use and cattle grazing are politically charged and controversial issues, and it will take many years to get anywhere with them. Ravens had no constituency, and were easy to jump on." Although Grandy claims this amounted to unfairly targeting the raven, to the BLM it was simply a wise use of resources. Without enough funding and people to do all that it would like for the tortoise, the BLM chose to work on an issue which promised quick results.

But, Boarman says, the BLM has now rethought its priorities. "It's important in the long run to control predation of juvenile tortoises. But in the short run the most important thing is saving the reproducing adults." The immediate threat to adults is a respiratory disease brought on by poor nutrition, apparently due to destruction of vegetation by six years of drought, by cattle grazing, and by off-road vehicles. While the agency focuses on fighting the disease, raven control has been put on indefinite hold.

In the other case, the San Francisco Bay National Wildlife Refuge is going ahead with a predator control plan, albeit a modified one that will try to use nonlethal methods wherever possible. The precarious situation of the rails, whose population is down to fewer than five hundred birds, ultimately persuaded the red fox's defenders to close ranks with other environmentalists and support the plan.

The difficulty of developing a consistent position on predator control is clearly illustrated by Massachusetts Audubon Society's thirty years of experience with gulls and terns. In the 1960s and 1970s, Massachusetts Audubon and the Fish and Wildlife Service experimented with eradicating gulls from islands that were potential tern colonies. Through these investigations, William Drury, Massachusetts Audubon's research director, became the country's leading expert on killing and harassing gulls.

In 1968 Drury and his colleagues began studying nesting terns at the southern tip of Monomoy Island, a thirteen-mile-long sandspit hanging off the elbow of Cape Cod which was home to one of the largest tern colonies in the Northeast. In the 1970s herring gulls began moving onto the island in ever-increasing numbers, reaching a population of eighteen thousand pairs within a few years. They eliminated terns from the island except for a remnant colony at the northern end.

To make more room for terns, Fish and Wildlife proposed clearing gulls off the ends of Monomoy and, with the knowledge and apparent consent of Massachusetts Audubon (whose symbol is the tern), began to poison gulls in 1979 and 1980. But when accounts of the killing appeared in the press, Audubon did an abrupt about-face. "We got stabbed in the back," one Fish and Wildlife biologist says. "Audubon made the original request for us to do the work, but then they came out in the newspapers and said we were terrible people. They said they would sue." Largely in response to Audubon's protests, the program was halted, and the tern population continued to decline. Now, however, Audubon seems a little uncertain about its role in this. When I asked Gerard Bertrand, the president of Massachusetts Audubon, about his organization's turnabout, he downplayed it. "We may have questioned the program, but I don't think we ever officially opposed it. I never threatened a lawsuit."

Today there are no plans to further disturb Monomoy's gulls, but in 1990 biologists began poisoning gulls on Ram Island, in Buzzards Bay, to create nesting space for endangered roseate terns. This time Massachusetts Audubon prepared a position statement supporting the project, but included a peculiar warning note that suggests the organization's continued wariness: "This position is for response to public inquiry only. We will make no announcement to the press or other media before or after the Ram Island action." Thus, over a period of thirty years Massachusetts Audubon has gone from being a leading researcher and promoter of lethal gull control to an aggressive opponent to a surreptitious supporter.

Drury, who left Audubon in 1976, believed that the organization's change of heart at Monomoy was due to a change in cultural attitudes. "In my early years with the Audubon societies, the Audubon leaders were mainly hunters—businessmen and moneymakers. These men looked on killing as much less of an anathema than environmentalists do today." Such aggressive intervention in nature, he argued, is necessary to preserve the species we care about. "Humans have killed other organisms in their self interests ever since we were hunter-gatherers, and I think that the philosophical question of killing one species to favor another was answered by the early agriculturalists who pulled up plants that inhibited the growth of their crops—they weeded their garden."

NOT JUST ANOTHER TOOL

In a world where population growth, economic pressures and technological advances are continually degrading ecosystems and accelerating extinctions, and are beginning to alter the genetic basis of life, there are no easy answers to questions about managing nature. Although distasteful, the killing of predators may be necessary in some cases to preserve species brought to the very brink of extinction. I do not believe, however, that it is simply one more wildlife management tool. Each time that we resort to it is a sign of our continuing failure to live in harmony with the needs of our planet.

Two days after the nests were baited on Green Island, I returned with the Fish and Wildlife crew to collect poisoned gulls. Some looked as though they had merely gone to sleep on the nest, eyes closed, bills tucked gently under their wings. Others appeared to have died in greater distress, falling forward onto their breasts with wings askew and necks outstretched. Some of the baits hadn't been eaten—perhaps the gulls had learned from three prior years of poisoning that margarine sandwiches can be dangerous.

Despite the caution of a few gulls, this island that had so teemed with life a few days before was now permeated by death. We all, I think, just wanted to finish our work and leave. Earlier, I had been exhilarated by the sight of the thousands of birds that fly here each spring to mate and lay their eggs, doing so not in response to the goals and objectives of a federally sanctioned management plan, however well intentioned and scientifically defensible, but in response to animal impulses beyond human control or understanding. Now, trudging across the island with an armload of dead gulls, I felt less like a gardener tending his beds than like a vandal trampling on them.

6

THE OLYMPIC GOAT CONTROVERSY: A PERSPECTIVE

Victor B. Scheffer

INTRODUCTION

The twenty-year experience of the National Park Service (NPS) in dealing with the mountain goats of Olympic National Park (ONP) is a useful case history in wildlife management. It is well documented.[1] The NPS has concluded that, if nonlethal means of preventing damage by goats should prove infeasible, goats must be shot. The Fund for Animals disagrees.[2] In this paper I examine the arguments offered by both sides in the debate.

BACKGROUND

Goats were translocated during the 1920s from Canada and Alaska to the Olympic Peninsula. In 1937 they numbered about 25[3] and by 1983 about 1200.[4] Although 155 goats are known to have been removed between those years, the population grew at an average rate of about 9 percent a year.

But the soils and biotas of the park had evolved on a goat free "land-bridge island."[5] By the late 1980s, the park's drier regions were beginning to show changes as a result of goat grazing, wallowing, and trampling. Goats were even "mining" bare soil where hikers had urinated! Most conspicuous were changes in floral composition, such as the disappearance of lichen and moss cover, which stabilizes bare soil surfaces in the absence of vascular plants.[6] And the NPS per-

---

This article and the subsequent responses originally appeared in *Conservation Biology* 7 (December 1993): 916–19 and 954–58. Reprinted by permission. Copyright © 1993 Blackwell Science, Inc.

ceived threats to certain unique endemic plants—nine species and varieties—growing in areas used by goats.

Between 1981 and 1989, humans removed 509 animals from the goat population.[7] Of these, 360 were captured alive, 28 accidentally killed during capture, 19 shot for research, 99 killed by sport hunters outside the park, and 3 killed by poachers. At the end of 1990 the estimated population on the peninsula was only 389 plus or minus 106 (172 goats seen). That total showed clearly that removals by humans outnumbered natural recruitment.

The NPS also tested population control by contraception.[8] Later, an independent five-member panel comprised of veterinarians, wildlife biologists, and a reproductive physiologist evaluated the potential of goat control by contraception. Panel members visited the goat range, studied past research by the NPS, and brought to bear their collective experience with the application of contraceptives to overabundant wild or feral animals. They concluded that "current contraceptive or sterilant technologies will not eliminate mountain goats from ONP."[9]

Although the reestablishment of wolves (Canis lupis) in the park would impose a degree of control on the goat population, the NPS has never included this possibility in its management plans. (The last Olympic wolf was killed in the 1920s.) Students at Evergreen State College have suggested that the Peninsula could support at least forty to sixty wolves.[10]

In 1987 the NPS released an environmental assessment that gave preference to settling the goat controversy by removing all goats from the core of the park and thereafter removing—by capturing or killing—any that appeared along its borders.[11] Later, the NPS announced that it would release a Final Environmental Impact Statement in 1993.[12]

Conflict: Factual Considerations

The Fund for Animals, a national society of 150,000 members, claims that goats occupied the Olympic Peninsula into the nineteenth century.[13] If so, the present population is a replacement or "restoration" (my term) entitled to protection. The Fund builds its case partly on a model drawn by anthropologist R. Lee Lyman[14] and partly on "documented and scientific historical evidence." Lyman examined the known distribution of goats in five northwestern states in relation to the postulated distribution of Pleistocene ice lobes. From a "dispersal model" of goat occurrences at various times and places, he concluded that by ten thousand years ago goats could have reached the Olympics. He suggested that, if goat remains dating from the recent thousand years ever *should* be found here, the NPS should rethink its policy, quit calling the planted animals exotics, and leave them undisturbed. The Fund for Animals also points to narratives published between 1844 and 1917 that mentioned the goat as a member of the Olympic fauna.

NPS rests its case on the present distribution of mammals in western Washington and on the unreliability of reports of Olympic goats before 1925.

First, the goat is one of eleven species of mammals native to the Cascade Range of Washington that are not native in the Olympic Range only 120 kilometers away.[15] Among the missing are six species characteristic of alpine or subalpine habitats. Conversely, one mammal species (*Marmota olympus*) native to the Olympics is not recorded from the Cascades. Geologic clues indicate that continental ice in the Puget Sound Basin would have isolated the high Olympics from the high Cascades long before the first goats reached North America, perhaps forty thousand years ago.[16]

Second, early reports of Olympic goats cannot be taken seriously. For example, John Dunn visited the Indians living near Cape Flattery and reported that they "manufacture some of their blankets from the wool of the wild goat."[17] But ethnologist Erna Gunther later learned from descendants of those Indians that "the mountain goat does not occur on the Olympic Peninsula. . . . Mountain-goat wool was bought in Victoria [British Columbia] through the Klallam."[18] Albert B. Reagan, Indian agent at Lapush in the early 1900s, excavated middens along the seacoast, where he found remains of bighorn sheep and mountain goat "usually only in the ladle form of the horns."[19] These, again, would surely have been trade goods. Eight years earlier, Reagan had published a list of the animals of the Olympic Peninsula; it did not include the goat.[20]

Two other narratives briefly mentioned Olympic goats. The first, composed after a five-month crossing of the Olympic Range in winter and spring (the first crossing ever) stated simply that "one goat was seen by the party."[21] The second included "mountain goat" and "pelican," among other species, as "game animals" of the Olympics.[22] These narratives can hardly be taken as zoological records.

The strongest evidence—albeit negative—that goats were not indigenous comes from the published accounts of the dozen or more zoologists who explored the Olympics between 1895 and 1921 on expeditions of the U.S. Biological Survey and the Field Museum of Natural History. These explorers reported no goats.[23]

ETHICAL CONSIDERATIONS

But the goat controversy is basically a clash of human values—the sort of controversy that is settled through agreement rather than discovery. Informed public opinion will ultimately determine whether Americans want a goat-free Olympic Park at the cost of routinely exiling or killing goats. The Fund for Animals has chosen unwisely to offer what it calls "historic and scientific evidence" in defending its case.[24] Would not the Fund gain wider public support by relying purely on moral persuasion? Philosopher Mary Midgley has asked, "What does it mean to say that scruples on behalf of animals are merely emotional or emotive or sentimental? What else ought they to be?"[25]

Two lessons can be read in the Olympic Park experience with its unwanted goats.

First, national park managers will increasingly deal with exotic species as they deal with wildfires, hurricanes, and floods: with patience yet with steady resolve to maintain indigenous biosystems as nearly natural as possible. While "natural" as a state unperturbed by humans has long been an unreality—an abstraction—it is still useful as a goal. And all land managers need goals, however visionary or remote.

Second, *animal welfare*, an umbrella term for kindness to animals, humaneness, animal protection, anticruelty, and (lately) animal rights, will continue to grow in American thought. As a societal endeavor to win greater consideration for the interests of all living things, animal welfare began in the 1960s to draw energy from the "liberation" and "ecology" movements of the era.[26] The significance of the animal welfare ethic for national park managers is that they will increasingly become more sensitive to public opinion—a set of preferences compounded of sentiment (or emotion) and logic (or reason). Park managers will increasingly turn for advice to social scientists, who will sample public attitudes and preferences with respect to park uses; will develop new technologies for interpreting park values; will assist in the drafting of regulations; and will join in mediating disputes over the status of exotic species, such as goats.

Conclusions

The planting of foreign goats in the Olympics seemed a good idea at the time and even ten years later (1935) when I first worked as a biologist in the Olympic National Forest. But today, public attitudes toward natural areas and their biota are changing. The more we humans shape and color the landforms around us according to the designs of each new generation, the more we treasure those fragments kept undesigned. Wild places. Places to which we respond with all our senses, places where we bond with the earthly systems that nourish our civilization and our species. If a personal thought may be injected here it is this: The humane removal of goats is a small price to pay for keeping the Olympics wild.

Response to Scheffer

Cathy Sue Anunsen and Roger Anunsen

The Fund for Animals has had a longstanding and unwavering commitment to the protection of endangered species. The Fund demonstrated that commitment in December 1992 in a legal settlement with the Interior Department that "will expedite federal protection for hundreds of imperiled species," both plants and animals.[27] The Fund supports humane management when a need for control is documented, but it opposes the extermination of animals simply on the basis of their having been arbitrarily labeled an exotic species.

It is the position of the Fund that officials of Olympic National Park have not substantiated their claims that (1) the goat is an exotic species or (2) the goats' impact on park flora warrants the radical solution they propose: total extermination. If a need for control is established and verified by disinterested scientists, the Fund recommends, as it has since its 1988 appointment to the park's Technical Advisory Committee, the use of nonlethal control methods such as contraception.

The park's 1978 Management Policies state that "control programs will most likely be taken against exotic species which have a high impact on protected park resources and where the program has a reasonable chance for successful control."[28] There are no park plants federally listed as threatened or endangered. The park has not been able to document that the goat is exotic or that it has a significant, let alone "high impact," on park plants. A panel of four scientists, including two who are colleagues of the author, conducted what the park purports to be a "peer review" of the vegetation impact data in April 1992. The author, Ed Schreiner, had not completed the manuscript at the time of the "peer review." This vegetation study was supposed to support the park's damage claims. It was reportedly in its "last phase" in July 1990, but three years later it still had not been released to the public.

In 1987, however, when the goat population was at its highest—twelve hundred—the park's Environmental Assessment stated that "there is no apparent danger that these (plant) species will be extirpated."[29] In an August 1992 interview with the National Geographic News Service, Schreiner conceded that "None of the estimated sixty alpine and subalpine plant species are in danger from extinction."

Finally, excerpts from the as yet unreleased vegetation report include the following admission: "Specific relationships between mountain goats and rare plants were difficult to assess because few rare taxa occurred in either recon plots or permanent plots at Klahhane Ridge, Mount Dana, and Avalanche Canyon."[30]

It is the Fund's belief that park officials have grossly exaggerated the impact on park flora from the 386 goats that remain in the 900,000-acre park.

Significant omissions occur in Scheffer's materials, particularly in the areas of (1) soil erosion; (2) the viability of contraception as a nonlethal means to control the goat population; and (3) the "land-bridge island" analysis.

Scheffer reports that goats were "mining" the soil in a particular area. But the role park biologists played in creating the most severely damaged area was not discussed. In the late 1960s, park and college researchers placed artificial salt licks at the park's Klahhane Ridge to lure the goats to be counted, studied, and later trapped. The fourteen-year-long program resulted in an artificial concentration of goats, with one third of the entire park's population living on 0.5 percent of the park. The population there soared from 29 goats in 1971 to 229 goats in 1981. As the licks, which had been placed directly on the ground, dissolved, the salt leached deep into the soil and the mineral-starved goats dug into the ground, displacing the soil and trampling the surrounding vegetation. Park officials continue to characterize this area as a typical product of "destructive" goat behavior.[31]

In his discussion of contraception, Scheffer quotes the scientific panel as determining that contraception "will not eliminate mountain goats from the

ONP." He omits the next sentence of the report: "Indeed, even with the use of lethal shooting, it will likely be very expensive and difficult to totally eliminate mountain goats from ONP."[32] The proposal for total extermination becomes even more far-fetched in light of the position of Washington State's Department of Wildlife that the goats are native to the state and that they should stay—at least on Forest Service land surrounding the park.

Scheffer also fails to acknowledge a highly publicized letter to the park from a member of that contraception panel, Jay Kirkpatrick, a senior research scientist at the Deaconess Research Institute and a nationally recognized authority on wildlife fertility. In a letter of February 8, 1993, Kirkpatrick "scolded" the park for misleading the public about the potential use of contraceptives for mountain goat control. In an interview with *Seattle Times* reporter Ron Judd, Kirkpatrick said he and others were taken aback by the parties' seeming predisposition to shoot the goats, when logic suggested that simply limiting herds might suffice. Kirkpatrick added:

> We were given a very narrow charge: Can contraception be used to eliminate goats from the park? We had the answer in about five minutes—no. The thing that bothered us was, we were given no latitude to make other recommendations. . . . Can contraceptives be used to control wildlife populations? Clearly, yes. The technology has advanced to the point it could be tried on Olympic goats. The odds of it working on a mountain goat are very, very high.[33]

Dr. Kirkpatrick said he wrote the park because Olympic officials repeatedly took his study panel's conclusion out of context, suggesting that contraception to control goats was not a viable alternative to shooting. That wasn't true then, and it's even less true now, he said.

Scheffer's conclusion that the peninsula evolved on a goat-free land-bridge island is convenient but suspect. As noted by R. Lee Lyman, much has been learned since Dalquist outlined his hypothesis in 1946.[34] The more recent information "suggests the Puget lowland served as a biogeographic filter rather than a barrier for mammalian taxa."[35]

Scheffer's admonition to the Fund is twofold: (1) He asserts that the Fund should stay out of the scientific kitchen, or in this case the library, and not use the information we have discovered in our defense of the goats, and (2) he advises the Fund to limit its involvement in ecological issues to emotional pleas based on moral persuasion. Surely he doesn't mean to say that only an elite few should have access to archival records?

Scheffer argues that an "informed public opinion will ultimately determine whether Americans want a goat-free park." The public can only become informed if it has all of the facts, not just those spoon-fed to them by an anointed few who feel they have an exclusive right to find and interpret information relating to the public's wildlife. It was Scheffer's scientists and historians who failed to find the early references to the existence of mountain goats in the

Olympics. If the public had to rely on those professionals, they would still believe there was no prior record of mountain goats in the Olympic range, not even a suggestion.

An examination of the record of those scientists, local authors, and park officials charged with the responsibility of informing the public about the history of the mountain goat on the Olympic peninsula provides the best argument for the inclusion of all ideas and information in controversies involving public resources:

> *Scheffer.* In *Mammals of the Olympic Peninsula,* Scheffer's discussion of the history of the mountain goat is predicated on the assumption that the first goats on the peninsula were those that were planted. He relied on "scattered notes" in manuscript files and on "a few published references" to the planting.[36] Scheffer dismisses the Fund's references to the accounts of early explorers because they can "hardly be taken as zoological records," but he accepted "scattered notes" for his own paper.
>
> *Robert L. Wood.* The author of *Across the Olympic Mountains,* a book about the *Seattle Press* expedition, failed to find the reference to a goat in the 1890 *Seattle Press* account by Captain Barnes. The article contains an account of the wildlife encountered during the famous expedition and includes a specific reference to an 1890 sighting of "one goat" by the entire party.[37] In late 1991, Mr. Woods reiterated to Fund researchers the often-repeated conclusions of his research: "There is no record of anyone in the *Press* party ever seeing mountain goats."[38] Fund researchers found the *Seattle Press* reference a few months later.
>
> *Park officials.* In an effort to support the park's contention that the goats were not native, park spokesperson Paul Crawford told the *Bremerton Sun* in January 1992 that "none of the early peninsula explorers reported seeing goats."[39] Mr. Crawford made this statement on the same day that the paper reported that the Fund had discovered an 1896 *National Geographic* magazine account of an 1889 exploration of the Olympics by the respected explorer Samuel C. Gilman. In the article Gilman lists the wildlife of the peninsula, including "mountain goat."[40]

The degree of credibility accorded the reports of the early explorers by the park and their allies has totally depended on the perceived content of those reports. When the park believed that none of the early explorers saw goats, they repeatedly cited this "fact" as a cornerstone of their proof that the goat was not native. After the Fund found that two of the first three major expeditions into the Olympics reported mountain goats, the park reversed its position and dismissed these early explorers' reports as unreliable.

Scheffer diminishes Reagan's credibility as a scientist when he identifies the Stanford Ph.D. as "Albert B. Reagan, Indian agent." A fellow of the American Association for the Advancement of Science, a member of many learned societies, including the American Ethnologists Society, the American Anthropological Association, and the New York and California Academy of Sciences, Reagan conducted archaeological research for the Laboratory of Anthropology Institution at

Santa Fe, New Mexico. He authored over five hundred papers, which he contributed to various publications both in the United States and abroad.

Scheffer did not cite Reagan's report in his 1946 work,[41] and he now dismisses the mountain goat bones identified by Dr. Reagan as "surely" being the result of trade because the reference says "usually only in the ladle form."[42] Scheffer overlooks the significance of the word "usually." Clearly, some were not in ladle form or Reagan would not have made this distinction. The important point for this discussion, however, is that had the public record been limited to the conclusions of the park biologist, there would be no debate.

The primary author of the park's publications regarding the mountain goat, including the 1987 Environmental Assessment, is research biologist Bruce Moorhead. Moorhead claimed that his research was "exhaustive," but he had to confess at a public meeting in January 1992 that he did not know of the Reagan paper, found by Fund researchers. Moorhead maintained for over a decade that "there is no evidence to suggest that they [the goats] ever inhabited the Olympic Peninsula."[43]

At the time the 1987 statement was made, Moorhead's files contained a computer list entitled, "A chronology of early mountain goat reports." This list included the Gilman reference as well as the quote "one goat was seen by the party." He did not cite these references and discuss their merit; he said they did not exist. Regarding the reliability of Gilman's work, the *Seattle Post-Intelligencer* wrote that "[n]o work of equal importance and value in a practical sense has been accomplished on the American continent in the past quarter of a century,"[44] and in a 1983 study the National Park Service proclaimed the Gilman accounts as "accurate and thorough."[45] Neither goat report reference was pursued by park biologists or researchers.

Contrary to Scheffer's contention that the public should stay out of the scientific arena of ecological debates, the goat controversy and the Fund's discoveries show the importance of the checks and balances provided by the involvement of citizens and wildlife advocacy organizations who have no scientific turf to protect nor jobs to perpetuate. The Fund's role in the policy debate over the goat issue is not and should not be limited to mere sentimentalism, but to full exposure of all issues—scientific, historical, archaeological, and ethical.

Archaeological documents and historical accounts cast strong doubt on the contention that goats are exotic to the Olympic peninsula. An absence of any scientific evidence documenting ecological damage by goats stretches the faith of interested parties who are told that the animals imperil rare plants. And statements from respected scientists about a high probability of success in applying immunocontraceptive techniques to goats put the lie to the suggestion that killing the goats is the only feasible management option. Despite Dr. Scheffer's valiant, if misguided, defense of the Park Service, the agency has not met its burden of proof.

Reply to the Anunsens

Victor B. Scheffer

I'm pleased that the Fund for Animals has commented on the Olympic goat controversy. The Fund is a New York-based anticruelty organization with 150,000 members. I respond to important points raised by the Anunsens.

"There are no park plants federally listed as threatened or endangered." True, though I do not use this argument. I emphasize that "changes in floral composition" have resulted from heavy grazing by goats.

"The park has not been able to document that the goat is exotic...." I reiterate that the expeditions of the Biological Survey between 1897 and 1921, and the expedition of the Field Museum in 1898, found no evidence of goats. Published allusions to goats in the park before 1925 are bare statements without elaboration.[46] There exist no specimen records of endemic goats, with the improbable exception of "big horn" and "mountain goat" remains that Regan found in seacoastal middens, "usually only in the ladle form of the horns."[47] That either species was endemic to the Olympic Mountains is equally unlikely. Ethnologist Erna Gunther's statement that "the mountain goat does not occur on the Olympic Peninsula"[48] is important. Working in the Washington State Museum for over 30 years, she spent hundreds of hours interviewing the Clallam and Makah about their lifeways, traditions, and knowledge of the Olympic environment.

With respect to Lyman's theory,[49] I agree that goats could have reached the Olympics in the late Quaternary. Lyman elaborates on the theory in an unpublished 1993 paper, "Indirect Evidence for the Pre-1925 Presence of Mountain Goats on the Olympic Mountains of Washington State."

"Scheffer also fails to acknowledge [the comments of Jay Kirkpatrick on the value of goat contraceptives]." I first learned of these comments in September 1993. I accept Kirkpatrick's opinion that a vaccine administered by dart might keep a female sterile for up to three years. But Kirkpatrick's voice was only one in a five-man panel which concluded that "treating mountain goats with these agents [sterilants] would represent a very expensive, never-ending program that, at best, would only partially control the population." If goats are to be eliminated, not simply controlled, "lethal shooting appears to be the only feasible option."[50]

The National Park Service's Final Environmental Impact Statement on goat management may appear while the present article is in press. I trust that it will recommend what is best for Olympic National Park and, in the long run, best for people.

Notes

1. See *Environmental Assessment: Mountain Goat Management in Olympic National Park* (Port Angeles, Wash.: National Park Service, 1987); *Decision Record: Mountain Goat*

Management in Olympic National Park (Port Angeles, Wash.: National Park Service, 1988); B. Carlquist, "An Effective Management Plan for the Exotic Mountain Goats in Olympic National Park," *Natural Areas Journal* 10, no. 1 (1990): 12–18; D. B. Houston, B. B. Moorhead, and R. W. Olson, "Mountain Goat Population Trends in the Olympic Mountain Range, Washington," *Northwest Science* 65, no. 5 (1991): 212–16; and D. B. Houston, B. B. Moorhead, and R. W. Olson, "Mountain Goat Management in Olympic National Park: A Progress Report," *Natural Areas Journal* 11, no. 2 (1991): 87–92.

2. "Historic and Scientific Evidence That Mountain Goats are Native to the Olympic Peninsula," Press Release, Fund for Animals Northwest Region, Salem, Ore.

3. V. B. Scheffer, *Mammals of the Olympic National Park and Vicinity* (Seattle, Wash.: U.S. Fish and Wildlife Service, 1949), p. 237.

4. D. B. Houston, B. B. Moorhead, and R. W. Olson, "An Aerial Census of Mountain Goats in the Olympic Mountain Range, Washington," *Northwest Science* 60, no. 2 (1986): 131–36.

5. W. D. Newmark, "A Land-Bridge Island Perspective on Extinctions in Western North American Parks," *Nature* 325, no. 6103 (1987): 430–32.

6. *Environmental Assessment: Mountain Goat Management in Olympic National Park*, pp. 7–8.

7. Houston, Moorhead, and Olson, "Mountain Goat Management in Olympic National Park," p. 89.

8. *Environmental Assessment: Mountain Goat Management in Olympic National Park*, pp. 46–47.

9. *The Applicability of Contraceptives in the Elimination or Control of Exotic Mountain Goats from Olympic National Park: Final Report to the National Park Service*, January 1992, unpublished.

10. Students of Evergreen State College, *A Case Study For Species Reintroduction: The Wolf in Olympic National Park*, Olympia, Wash., 1975.

11. *Environmental Assessment: Mountain Goat Management in Olympic National Park*, pp. 52–54, 65–67.

12. Interagency Goat Management Team, Newsletter no. 1 (January 1992): 6.

13. "Historic and Scientific Evidence that Mountain Goats Are Native to the Olympic Peninsula."

14. R. L. Lyman, "Significance for Wildlife Management of the Late Quaternary Biogeography of Mountain Goats (*Oreamnos americanus*) in the Pacific Northwest, U.S.A.," *Arctic and Alpine Research* 20, no. 1 (1988): 13–23.

15. W. W. Dalquest, "Mammals of Washington," *Museum of Natural History* 2 (1948): 1–144; V. B. Scheffer, *Mammals of the Olympic National Park and Vicinity*.

16. *Environmental Assessment: Mountain Goat Management in Olympic National Park* pp. 7, 17; A. R. Kruckeberg, *The Natural History of Puget Sound Country* (Seattle: University of Washington Press, 1991), pp. 2-33.

17. John Dunn, *History of the Oregon Territory and British North American Fur Trade* (London: Edwards and Hughes, 1844), p. 231.

18. Erna Gunther, "A Preliminary Report on the Zoological Knowledge of the Makah," *Essays in Honor of Alfred Louis Kroeber* (Berkeley: University of California Press, 1936), p. 117.

19. Albert B. Reagan, "Archeological Notes on Western Washington and Adjacent British Columbia," *Proceedings of the California Academy of the Sciences*, 4th Series, 7, no. 1 (1917): 16.

20. Albert B. Reagan, "Animals of the Olympic Peninsula, Washington," *Proceedings of the Indiana Academy of Science* (1908): 193–99.

21. "The Olympics: An Account of the Explorations made by the 'Press' Explorers," *Seattle Press*, July 16, 1890, p. 20.

22. S. C. Gilman, "The Olympic Country," *National Geographic Magazine* 7, no. 4 (1896): 138.

23. F. S. Hall, " A Historical Resume of Exploration and Survey: Mammal Types and Their Collectors in the State of Washington," *Murrelet* 13 (1932): 74.

24. "Historic and Scientific Evidence That Mountain Goats are Native to the Olympic Peninsula."

25. Mary Midgeley, *Animals and Why They Matter* (Athens: University of Georgia Press, 1983), p. 33.

26. Victor Scheffer, *The Shaping of Environmentalism in America* (Seattle: University of Washington Press, 1991), pp. 29–30.

27. *The Washington Post*, December 16, 1992, p. 1.

28. National Park Service, *Environmental Impact Statement, Olympic National Park* (1987), p. 4.

29. Ibid., p. 22.

30. E. Schreiner, A. Woodward, and M. Grace, *Vegetation in Relation to Introduced Mountain Goats in Olympic National Park: A Technical Report*, National Park Service, Port Angeles, Wash., 1993, p. 84.

31. C. S. Anunsen, "Saving the Goats: A Showdown in Washington State," *Animals Agenda* (January/February 1993): 24.

32. *The Applicability of Contraceptives in the Elimination or Control of Exotic Mountain Goats from Olympic National Park*, p. 25.

33. *Seattle Times*, April 8, 1993, p. 1.

34. Lyman, "Significance for Wildlife Management of the Late Quaternary Biogeography of Mountain Goats (*Oreamnos americanus*) in the Pacific Northwest, U.S.A.," p. 16.

35. Ibid., p. 13.

36. Victor Scheffer, *Mammals of the Olympic Peninsula, Washington* (Fish and Wildlife Service, 1946), p. 124.

37. "Resume of the Natural Resources of the Explored Region" *Seattle Press*, July 16, 1890, p. 20.

38. Robert L. Wood, personal communication, 1991.

39. *Bremerton Sun*, January 9, 1992, p. 1.

40. S. C. Gilman, "The Olympic Country," *National Geographic Magazine* 7, no. 4 (1896).

41. Scheffer, *Mammals of the Olympic Peninsula*, Washington.

42. Reagan, "Archeological Notes on Western Washington and Adjacent British Columbia," p. 16.

43. *Environmental Assessment: Mountain Goat Management in Olympic National Park*, p. 17. For other statements that there was no evidence that goats were native, see ibid., pp. 5, 6; and B. Moorhead, "Non-native Mountain Goat Management Undertaken at Olympic National Park," *Park Science* 9, no. 3 (1989): 10–11.

44. *Seattle Post-Intelligencer*, June 5, 1990, p. 9.

45. G. Evans, *Historic Resources Study* (Seattle: National Park Service, Pacific Northwest Region, 1983), p. 25.

46. See Dunn, *History of the Oregon Territory and British North American Fur Trade*, p. 231; "Resume of the Natural Resources of the Explored Region," p. 20; and Gilman, "The Olympic Country," p. 133–40.

47. Reagan, "Archeological Notes on Western Washington and Adjacent British Columbia," p. 16.

48. Gunther, "A Preliminary Report on the Zoological Knowledge of the Makah," p. 117.

49. Lyman, "Significance for Wildlife Management of the Late Quaternary Biogeography of Mountain Goats (*Oreamnos americanus*) in the Pacific Northwest, U.S.A.," pp. 13–23.

50. *The Applicability of Contraceptives in the Elimination or Control of Exotic Mountain Goats from Olympic National Park*, p. 2.

REFERENCES

Anunsen, C. S. "Saving the Goats: A Showdown in Washington State." *Animals Agenda* (January/February 1993): 24.

Bremerton Sun, January 9, 1992: 1.

Carlquist, B. "An Effective Management Plan for the Exotic Mountain Goats in Olympic National Park." *Natural Areas Journal* 10, no. 1 (1990): 12–18.

Dalquest, W. W. "Mammals of Washington." *Museum of Natural History* 2 (1948): 1–144.

Dunn, J. *History of the Oregon Territory and British North American Fur Trade*. London: Edwards and Hughes, 1844.

Evans, G. *Historic Resources Study*. National Park Service, Pacific Northwest Region, Seattle, 1983.

Finnerty, Maureen. Letter to author, July 20, 1990.

Gilman, S. C. "The Olympic Country." *National Geographic* 7, no. 4 (1896): 133–40.

Gunther, E. "A Preliminary Report on the Zoological Knowledge of the Makah". In *Essays in Honor of Alfred Louis Kroeber*. Berkeley: University of California Press, 1936.

Hall, F. S. "A Historical Resume of Exploration and Survey: Mammal Types and Their Collectors in the State of Washington." *Murrelet* 13 (1932): 63–91.

"Historic and Scientific Evidence That Mountain Goats are Native to the Olympic Peninsula." Press Release, Fund for Animals, Northwest region, Salem, Ore, 1992.

Houston, D. B., B. B. Moorhead, and R. W. Olson. "An Aerial Census of Mountain Goats in the Olympic Mountain Range, Washington." *Northwest Science* 60, no. 2 (1986): 131–36.

———. "Mountain Goat Population Trends in the Olympic Mountain Range, Washington." *Northwest Science* 65, no. 5 (1991): 212–16.

———. "Mountain Goat Management in Olympic National Park: A Progress Report." *Natural Areas Journal* 11, no. 2 (1991): 87–92.

Interagency Goat Management Team (IGMT). Newsletter no. 1. IGMT, Olympic National Park, Port Angeles, Wash. (January 1992).

Kruckeberg, A. R. *The Natural History of Puget Sound Country*. Seattle: University of Washington Press, 1991.

Lyman, R. L. "Significance for Wildlife Management of the Late Quaternary Biogeography of Mountain Goats (*Oreamnos americanus*) in the Pacific Northwest, U.S.A." *Arctic and Alpine Research* 20, no. 1 (1988): 13–23.

Midgley, Mary. *Animals and Why They Matter.* Athens: University of Georgia Press, 1983.

Moorhead, B. "Non-native Mountain Goat Management Undertaken at Olympic National Park." *Park Science* 9, no. 3 (1989): 10–11.

National Park Service. *An Environmental Assessment on the Management of Introduced Mountain Goats in Olympic National Park.* Port Angeles, Wash., 1981.

———. *Environmental Assessment: Mountain Goat Management in Olympic National Park.* Port Angeles, Wash., 1987.

———. *Environmental Impact Statement,* Port Angeles, Wash., 1987.

———. *Decision Record: Mountain Goat Management in Olympic National Park.* Port Angeles, Wash., 1988.

Newmark, W. D. "A Land-Bridge Island Perspective on Extinctions in Western North American Parks." *Nature* 325, no. 6103 (1987): 430–32.

"The Olympics: An Account of the Explorations Made by the Press Explorers." *Seattle Press,* July 16, 1890, pp. 1–11.

Reagan, A. B. "Animals of the Olympic Peninsula, Washington." *Proceedings of the Indiana Academy of Science* 1908 (1909): 193–99.

———. "Archeological Notes on Western Washington and Adjacent British Columbia." *Proceedings of the California Academy of Sciences.* 4th Series, 7, no. 1 (1917): 16.

"Resume of the Natural Resources of the Explored Region." *Seattle Press,* July 16, 1890, p. 20.

Scheffer, V. B. *Mammals of the Olympic Peninsula, Washington.* U.S. Fish and Wildlife Service, Seattle, 1946.

———. *Mammals of the Olympic National Park and Vicinity.* U.S. Fish and Wildlife Service, Seattle, 1949. Unpublished.

———. *The Shaping of Environmentalism in America.* Seattle: University of Washington Press, 1991.

Schreiner, E., A. Woodward, and M. Grace. *Vegetation in Relation to Introduced Mountain Goats in Olympic National Park: A Technical Report.* National Park Service, Port Angeles, Wash., 1993.

Seattle Post-Intelligencer. June 5, 1890, p. 9.

Seattle Times. April 8, 1993, p. 1.

Students of Evergreen State College. *A Case Study for Species Reintroduction: The Wolf in Olympic National Park.* Olympia, Wash., 1975.

The Applicability of Contraceptives in the Elimination or Control of Exotic Mountain Goats from Olympic National Park: Final Report to the National Park Service. January, 1992. Unpublished.

Washington Post. December 16, 1992, p. 1.

Wood, R. L. *Across the Olympic Mountains.* Seattle: University of Washington Press and the Mountaineers, 1967.

7

Captive Breeding of Endangered Species

Robert W. Loftin

It is an increasingly common practice throughout the world to take rare animals into captivity for the purpose of breeding them to augment the world population. This has been done for many purposes, including producing rare and economically valuable animals for commercial sale, producing game animals to be released and hunted, and saving animals that are threatened with extinction in the wild. This chapter will concentrate on the latter practice—captive breeding of species that are threatened with extinction in the wild. The goal of captive breeding in this sense is, or should be, the production of animals that can be reintroduced into the wild and survive there ultimately without human assistance. This practice will be referred to simply as "captive breeding" for the sake of brevity.

Many captive-breeding programs have been highly publicized as well as sharply criticized in recent years. I will begin, in the first two sections, by surveying the arguments for and against captive-breeding programs. Then several visible breeding and reintroduction programs will be examined in light of the arguments for and against captive breeding. Finally, I will consider the extent to which captive-breeding and reintroduction programs can be justified on the philosophical foundations of Aldo Leopold's land ethic and discuss general rules to guide future reintroduction programs.

From *Ethics on the Ark*, edited by Brian Norton, Michael Hutchins, Elizabeth F. Stevens, and Terry Maples. Reprinted by permission. Copyright © 1995 Smithsonian Institution Press. All rights reserved.

Arguments Supporting Captive Breeding

The first, and perhaps the strongest, argument for captive breeding is the argument that there is simply no alternative for some species. It seems clear that without human intervention, some animals are certainly doomed in the wild either through habitat loss, genetic inbreeding, overharvesting, environmental contamination, epidemic diseases, some combination of these, or perhaps other causes. This is indeed a powerful argument. If there is no alternative to extinction other than captive breeding, then few would oppose the practice, because few are willing to accept the extinction of a species without expending every effort possible to avoid this outcome; but there are always other alternatives. One could attempt to address the causes of endangerment rather than initiate desperate, stopgap captive breeding. Obviously, this becomes an exercise in fine drawing: When and how does one decide that there is no alternative? There can be no formula for that. Informed humans must make a judgment call. Despite the frightful pragmatic difficulty of determining exactly when there is no feasible alternative to captive breeding, there is little difficulty conceptually. If it is the case that there is no alternative to a captive-breeding program, if that is the only way a species can be preserved, then we ought to at least attempt it. Sincere and informed humans will differ on when that point is reached, but in those cases (if any) in which breeding in captivity is the sole hope of survival, that is a sufficient condition to justify such a program.

As I will argue below, sometimes captive breeders have apparently rushed to the judgment that there are no alternatives, but this is not to say that there are alternatives in every case. Every animal and every situation is different and it is hard to generalize. Cases in which there was probably no alternative to extinction in the wild include the Arabian oryx, the California condor, the red wolf, the black-footed ferret, and probably others. In cases with truly no alternative to extinction in the wild, taking the remnant into captivity for the purpose of augmenting the population through captive breeding is justified. The difficulty is to discern when this is and is not the case.

The second argument supporting captive breeding is that even if reintroduction proves to be unfeasible for whatever reason, it is better to have the species in captivity than not to have it at all. Some authors dispute this claim,[1] but advocates of captive breeding can make the telling point that extinction is irreversible. Once the species is gone there is nothing humans can do save lament its passing. So long as the species lives even in captivity there are at least some alternatives, few though they might be. Conditions could conceivably change, more habitat might become available, public attitudes might shift, or environmental contamination might decrease. Unlikely as these scenarios are for some animals, at the very least keeping the biological species in existence in some form, even in a cage, keeps some future alternatives open to some extent.

A third argument supporting captive breeding is somewhat more philosophical. According to this argument many species are near extinction because of human activities; therefore, humans have a moral duty to restore what we have damaged and degraded. According to this philosophical argument, principles of justice demand that we humans redress the imbalance by rebuilding what we have destroyed. While this is a compelling argument, it is not an argument for captive breeding specifically, unless it is used in conjunction with the first argument. Provided there are alternatives to captive breeding, the justice argument may support the much more general practice of environmental restoration. If captive breeding is the only (or the best available) means to accomplish this end, then it is justified. But it may or may not be the best means to this end. Some other program such as habitat preservation might be more effective. Hence, this argument supports captive breeding only in some cases.

Perhaps the major factor undermining this argument is the fact that the wild does not and cannot exist anymore. The idea of putting things back like they were, of restoration of the wild, is not feasible. The best that we might hope for is to manufacture some facsimile of the wild as it once was. As Jan DeBlieu has shrewdly observed, "The predominant goal of [captive-breeding] programs is not to restore animal populations to their original conditions but to reshape them so they can exist in a thickly populated, heavily developed, economically expanding nation."[2] There are two ways in which animals have been reshaped by captive-breeding programs: genetically and behaviorally.

In some cases, captive-breeding programs have reshaped animals genetically by deliberately introducing genes from populations that are alien to the region in question. It has been pointed out that the reintroduction of the European bison to Poland was accomplished with animals that were "hybrids of native and nonnative subspecies mixed with a little blood from American plains bison . . . and domestic cattle." The same thing is true of the much publicized reintroduction of the peregrine falcon to eastern North America, which has been widely hailed as a good example of a highly successful captive-breeding project.

In 1946, when DDT came into wide use, there were about 350 pairs of peregrines of the subspecies *anatum* nesting in eastern North America. The aeries were located almost exclusively in mountainous areas well inland. By 1965 the peregrine was gone from eastern North America—the subspecies *anatum* was extinct. After the extinction of the eastern subspecies, the captive-breeding program began under the leadership of Tom Cade and the Peregrine Fund at Cornell. Some falcon chicks were taken from aeries in Alaska, Canada, and the western United States. Some adult falcons were contributed to the program by falconers, who took a strong interest in the program from the outset. Birds from Chile, Scotland, and Spain were used. According to Cade, "We purposely wanted to create a mishmash population. We felt that was the best way to guarantee enough genetic diversity for the birds to survive and reproduce in the eastern environment."[3]

Obviously Cade was right because the hybrid falcons bred very well in captivity, so well that in 1974 two chicks were hacked back into the wild from the top of a tall building (note the location) in New Paltz, New York. Plans were also made to release falcons from hacking towers placed in coastal marshes, a habitat where falcons had not bred historically. By 1991 more than sixty pairs were breeding along the Atlantic Coast on skyscrapers in cities and on towers in salt marshes far from the traditional mountain cliffs where the extinct subspecies *anatum* had had its aeries. Everyone seems pleased, but it is worth noting that the falcon flying free in the East today is not an example of restoration of what once was but rather of introduction of another bird, of the same species to be sure, but with a somewhat different genetic makeup and different behavior patterns. What we have are facsimile falcons. So long as most people, especially reporters, cannot tell the difference, it does not seem to matter.

A fourth argument is that captive-breeding programs raise consciousness regarding the plight of whole natural systems. Captive-breeding programs are highly dramatic and make good press. Proponents of this argument point to the educational value of captive-breeding programs and conclude that these programs are justified by the attention they bring to the plight of the natural world.[4] Although it is true that humans learn from captive-breeding programs, exactly what they learn is more problematical. Captive breeding is an extreme example of single-species management. It is clear that single-species management directs attention away from entire systems and toward single species, usually the more conspicuous species preferred by humans. The publicity given to captive breeding also fosters the high-tech managerial approach to environmental problems. It creates the impression that omnipotent humans can cope with the global decrease in biological diversity through in vitro fertilization. In the midst of a recent controversy in Florida over speed limits for powerboats that were hitting manatees, one observer suggested that the Florida Game and Freshwater Fish Commission simply capture the manatees, place them in a large freshwater lake where they would be safe, and permit the powerboaters unrestricted use of the waterways. Although this suggestion is clearly absurd, the managerial attitude that lies behind it is typical of an unfortunate number of citizens—if some species of animals are making things inconvenient for humans, the problem can be solved by developing new techniques and by manipulation of populations. Do publicity campaigns for captive-breeding programs encourage or discourage this unfortunate attitude?

A fifth argument is that captive-breeding programs provide additional incentive for habitat preservation. Once captive populations reach a certain level there will be no place to release the animals unless additional habitat is set aside. In a sense, we might end up with a species all dressed up with no place to go. Since we clearly need some place to release captive-bred animals, more habitat will be preserved. The counterargument is that this has not happened to any great extent so far. Supporters of this argument can point to few projects where additional habitat has been obtained specifically to release animals bred in captivity. The approach has been to look around for some place to release animals that is already protected by public ownership or to strike an agreement with private landowners to permit release on their lands. Contrary to this argument, it could also be argued that captive-breeding programs are frightfully expensive and these programs divert money from field studies and from habitat acquisition programs.

A sixth argument for captive breeding is that in the future, human attitudes will change; rather than less habitat for wild animals in the future, there may well be more. Captive breeding is a stopgap to buy time until the human social and political climate changes to the point that conditions will improve the prospects for reintroduction. Several discussions I have seen on this subject speak in terms of a fifty-year period. The general argument seems to be, if we can preserve a species for fifty years, conditions will be more favorable for the species at that point than they are now. This argument seems to fly directly in the face of current trends, especially in human population growth. While some habitat is being set aside, the net effect of human activities globally is a sharp decrease in available habitat for wildlife. In short, this argument seems to be whistling in the dark.

A final argument supporting captive breeding is that there simply is no wild anymore, no natural systems that are as they once were. Systems have been altered to the point that animals must change in order to survive. Change and adaptation are natural processes because nature is a dynamic system rather than the static system some seem to imagine it to be. Those animals that are adaptable are usually not the ones that are endangered. Since change is a natural process and the environment has changed to the point of irreversibility, management is fully justified in helping animals to adjust and adapt to changed ecosystems. By training animals to adapt to new conditions and extinguishing their traditional behaviors, humans are merely abetting natural processes and benefiting the animals as well as the systems.

The main objection to this argument is that it is based on human arrogance. A species is what it is only because of its traditions, its culture. Once these traditions are extinguished, the animal is no longer wild even if the morphological aspects of the species are unaffected. In essence, the wild animal is extinct, having been replaced by a quasiwild animal.

Arguments Against Captive Breeding

Having looked at some of the reasons that have been given to support captive-breeding programs, let us briefly survey some of the arguments that can be made against the practice.

Captive breeding is an extreme example of single-species management, in which an attractive species is singled out for an enormous amount of attention and effort. This is biologically misleading in that it creates the impression that systems consist of mere collections of entities, each of which can stand alone. An equal amount of effort expended on the preservation of complete ecosystems through habitat preservation would be far more valuable in the long term.

Captive breeding represents an anthropocentric, high-tech approach of manipulative intervention in nature. This attitude of anthropocentric manipulation is what caused the decrease in biological diversity that the planet is experiencing at the present time; thus captive breeding reinforces and perpetuates the attitudes that make such desperate action necessary in the first place. As Michael Fox has put it, "With the high-tech innovations of operant training devices, behavioral monitoring, ova transplantation, and genetic engineering, the contemporary zoo is fast becoming an endorsement of capitalist industrial technology."[5] Rather than fostering this managerial approach to nature, we should be following a philosophy of "Nature knows best." It is the attitude of manipulation that got us into the difficulty; we will escape the current situation only when we renounce that attitude.

Captive breeding and subsequent reintroduction are excessively cruel to individual animals that are born and reared in captivity and thus lack the skills needed to survive in the wild. Breeding animals in captivity is, in some sense, breeding the wild out of the animal. Those traits that make it likely that the animal will thrive in captivity are usually precisely the opposite of the traits needed to make it in the wild. Moreover, cultural traditions and social structures are disrupted in captivity, altering those behaviors that make a wild animal what it is. Mortality among captive-reared animals is high when they are released into the wild because they have lost the information and behavioral patterns necessary to get by out there. Hence, captive breeding is merely producing disposable animals, doomed to spend their lives in a cage somewhere in a breeding facility or to be exposed to a quick, cruel death in the wild.

In some cases, the genetics of the species or subspecies have been altered in captive-breeding programs. Animals with different genetic backgrounds are bred into the population, sometimes deliberately, in order to increase the variation in the breeding population. The end result is an animal that may look like the ancestral form but is subtly different biologically. In these specific cases, critics of zoos and other captive-breeding programs can argue that the end—of preserving a naturally occurring species—is not achieved by the program.

It has also been argued that by encouraging the man-manipulates-nature attitude, and by suggesting that alternatives to habitat protection exist, captive breeding gives corporate culture an excuse to continue its rapacious ways. Through the concept of mitigation in one form or another, critics argue, society feels that the present course of action, while not ideal, is at least acceptable. After all, we are making heroic efforts to save the animals, so it is permissible to proceed along the present path of plundering the planet.

Captive breeding is an impractical waste of effort, it is sometimes argued. Why produce animals in captivity to be put back out into the same situation that brought them to the brink of extinction in the first place? If conditions were such that the species could survive in the natural situation, it would not be endangered. On the other hand, unless something changes in the habitat, the same situation will merely repeat itself. Someone has defined stupidity as doing the same thing over again while expecting different results.

Captive breeding can actually be counterproductive, if taking animals out of the wild disturbs the remnant wild population and actually hastens the demise of the wild population. When the species is dangerously reduced to a remnant population, it becomes a self-fulfilling prophecy to take a significant portion of the remnant into captivity. This not only reduces the actual number of wild animals but the harassment and disruption involved also alter social structure and hasten the downward spiral toward extinction.

Captive breeding shifts resources away from field studies to labwork. While captive breeding may produce greater numbers of individuals, the knowledge of how the species fits into its environmental niche goes begging. Since knowledge of the causes of extinction and understanding of the factors that enhance the chances that the animal will make it in the wild are more important than actual numerical count, this argument suggests that captive breeding ought to be deemphasized in favor of field studies of the organisms in their natural environment.

Captive breeding sacrifices the interests of individual animals to the interests of the species. Breeding facilities cannot keep all animals, so animals with common genes are often surplused for the sake of those with rare genes. Those whose bloodlines are well represented become dispensable.[6] One recent writer on this subject favors killing the animals once they have replaced themselves in the captive population.[7]

Captive rearing lowers fitness by pampering animals with easily available food and protection from predators, parasites, and disease, so that the process of natural selection is disrupted. Animals that are produced have a lower level of fitness compared to animals reared in the wild, thus the fitness of the species is lowered overall. Hence, the species has a better chance if it is left alone in the first place. Captive breeding may produce a higher number of animals in terms of simple numerical count, but since those individuals are less fit to survive in the wild, the net effect may be harmful on the whole.

Evaluation of Some Visible Reintroduction Programs

Having surveyed the arguments for and against captive breeding, I believe we can conclude that there are situations in which captive-breeding and reintroduction programs are indeed justified. But the numerous arguments and concerns stated also suggest that each particular case must be examined carefully before a judgment can be made. In this section, I will present in detail several highly visible, and purportedly successful, reintroduction programs. We will see that these programs may not always fulfill the conditions implied by the arguments surveyed above.

It is important to point out that in many of the most publicized cases of captive breeding, the criterion of no alternative has not been met. Let us first consider the case of the golden lion tamarin, which is especially interesting since it has been so often put forward as a paradigm of a successful captive breeding program. In 1972, when the Wild Animal Propagation Trust became interested in captive propagation of this tamarin, deforestation and the capture of wild animals for medical research, the pet trade, and exhibition in zoos had reduced the species from its original range in eastern Brazil to a few small areas northeast of Rio Janeiro. Only about 2 percent of the original rainforest habitat of the species remained intact.[8] No one knew exactly how many tamarins were left in the wild. The trust located 70 tamarins in zoos throughout the world, and over the next decade scientists, led by Devra Kleinman, developed much improved methods of propagating tamarins in captivity. But in 1974 a much more significant event took place when the Brazilian government, at the urging of Brazilian primatologist Adelmar Coimbra-Filho, set aside 4,856.4 hectares (12,000 acres) of habitat at the Poco das Antas Preserve.

By 1984 tamarins were propagating so well in captivity that 20 were released into the wild at Poco das Antas. At that point there were about 100 wild tamarins on the reserve. The way for the release had been paved by an education program carried out by James and Lou Ann Dietz. This program educated the local people on the rarity and importance of tamarins. Despite the careful training given to released animals by Benjamin Beck to prepare them to survive in the wild, only 3 out of the 20 were alive (plus a pair of twins born in the wild to one pair of released tamarins) in June 1985. Releases have continued at Poco das Antas and on private ranches nearby up to the present time. Of the 91 tamarins freed between 1984 and 1990, only 32 were still alive in the summer of 1991, but the good news is that 38 young animals born to former captives in the wild were also a part of the total population of about 450 to 500 animals. My point is that about four-fifths of the total population are wild animals that continued to make it on their own in the wild without human assistance.

It therefore seems that the most important things done for the golden lion tamarin have been setting aside the Poco das Antas preserve and educating the local people, rather than breeding tamarins in captivity, although the wild population may have been enhanced genetically by the introduction of additional bloodlines from the captive population. Proponents of captive breeding will argue that the preserve and the educational program would not have been established without the attention focused on the captive breeding program, but my point is that captive breeding was not the only alternative available in the case of the golden lion tamarin. If the money spent on the captive breeding program had been expended on the acquisition of additional habitat and the educational program alone, the tamarin would no doubt be approximately as well-off as it is today with the captive-breeding program.

Another case that has been widely acclaimed as a victory for captive breeding is the case of the whooping crane. Here again, close scrutiny of the history of the program reveals that the no-alternative argument does not apply to this case. In the 1930s this crane had been reduced to two flocks, one that wintered at Aransas on the coast of Texas and one nonmigratory flock of about 13 birds on the coast of Louisiana. In 1940 a severe storm wiped out the Louisiana flock, and by 1945, when Robert Porter Allen began to study the species, only 18 birds were left in the Aransas flock. It was not until 1954 that the breeding grounds of the Texas flock were discovered at Wood Buffalo Park in Canada. A decade later, when the captive-breeding program for this species was established at Patuxent Wildlife Research Center, Maryland, the wild flock had increased to 32 birds, thanks primarily to an educational campaign carried out by National Audubon that discouraged shooting along the migratory route of the bird. In 1967 six eggs were removed from wild nests in Canada and brought to Patuxent. Finally, in 1975 the first egg hatched at Patuxent from an artificially inseminated crane—after eight years only one pair had bred in captivity.

Partly because of the obvious difficulty of breeding whooping cranes in captivity, attention turned to cross-fostering of whooping cranes with sandhill cranes in 1975. Between 1977 and 1978, 75 eggs were taken to Wood Buffalo Park and placed in sandhill crane nests at Gray's Lake. Of those, 45 hatched and 11 chicks lived long enough to migrate with their foster parents to their wintering grounds at Basque del Apache, New Mexico.

In the next five years breeding techniques at Patuxent improved, and eggs from the captive flock began to be taken to Gray's Lake for cross-fostering, but the cross-fostering experiment was not turning out well. Despite the placement of nearly 300 eggs at Gray's Lake for cross-fostering in sandhill crane nests, in 1991 there were only 13 whooping cranes there, nine males and four females, none of which gave any indications of breeding on their own.

In the meantime, the wild flock at Aransas had grown steadily. In the winter of 1990–91 there were 136 whooping cranes at Aransas, all reared by their parents in the wild without human intervention. Obviously, the captive-breeding program had done little to help the species, and the birds had improved the situation on their own. Clearly, this case is not one that fits the no-alternative scenario.

In other cases, in which the requirement of no alternative is met, reintroduction programs appear futile because the programs are undertaken without a plan for correcting the original problem that threatened the species in the wild. If we just do the same thing over, why do we expect the results to differ? In my opinion, the attempts to reintroduce the red wolf into the wild are almost certain to fail despite the enormous amount of effort expended on the program and even though it is often presented as a paradigm of a successful captive-breeding program. The reason that the red wolf became endangered was only partly loss of habitat and persecution by humans. The major factor was hybridization with coyotes. In 1975 the decision was made to let the red wolf go extinct in the wild and capture all the pure wolves remaining in the wild for the captive-breeding program. Of the 400 animals trapped from the wild as a result of this decision, only 40 were judged to be pure enough to admit to the captive-breeding program. Even some of those were later judged to be hybrids, so in the end the future of the species rested on just 17 animals that were judged to be pure red wolves.[9]

The program was successful and red wolves were introduced at five sites. Generally about half of the wolves died within the first year after release, but with plenty of wolves in captivity to replace the casualties, a population can be sustained in the wild. That is, until they encounter coyotes.

You will recall that the reason the red wolves were brought into the captive-breeding program in the first place was that they were breeding themselves out of existence by hybridizing with coyotes. The theory is that with so few pure wolves left in the wild, the social structure of the species was disrupted to the extent that the wolves began to mate with the coyotes, which had been invading their range in the east for some decades. It seems obvious that with far fewer animals in the wild today than when hybridization started, the red wolves will have to be watched day and night to guard against hybridization. In short, the program is premised on continual human intervention forever. On coyote-free islands, hybridization will not be a problem, but these islands are small and family groups will quickly become inbred. Therefore, genetic shuffling of new bloodlines on and off the islands will be necessary forever.

Another program that has been widely acclaimed as an example of a successful program is the case of the black-footed ferret. Yet when one looks at the causes of endangerment, it becomes difficult to see why. The immediate cause of the demise of the ferret was epidemic disease in the ferrets and in the population they preyed upon. Breeding ferrets in captivity is unlikely to establish populations in perpetuity as long as the animals are susceptible to epidemics.

Despite the frightful difficulties plaguing captive-breeding programs, some have been successful by any standards. Of the programs I am familiar with the most successful has been the reintroduction of the Arabian oryx. In the 1930s Arabian nobility began hunting the animals from motor vehicles. By 1960 there were fewer than 100 in the wild, all at the southern end of the Arabian peninsula. Then a hunting party from Qatar killed 48 animals, about half the wild population. At that point the Faunal Preservation Society of Great Britain became interested in the species. The next year (1962) another 16 animals were killed by hunters, leaving fewer than 30 Oryx remaining in the wild. Major Grimwood captured 3, which were sent to Kenya and then to Phoenix, Arizona, where they were joined by 7 more from the London Zoo and from the private collections of noblemen from Kuwait and Saudi Arabia. The first calf was born in captivity in 1963. By 1971 there were 30 animals in captivity in zoos, and the next year the last 6 wild oryx were killed by hunters. In 1974 Sultan Qaboos of Oman became interested in the project and offered to support it financially. In 1980 there were some 80 to 100 animals held by wealthy Arabs, approximately the same number in the captive breeding program in zoos. In 1981, 14 oryx were released at Yalooni in Oman, and the next year 10 more were released after a period of acclimatization in large pens. A key element in the success of this program was involving the local population, a nomadic group known as the Harasis, in the program, hiring them as rangers and cultivating good relations with them. In 1984 a second herd of 11 animals was released. The period 1984–86 was very dry, so the oryx were fed supplementary rations. In 1987 and 1989 two more herds were released to enhance the genetic diversity of the population. By 1990,

99 animals were ranging in the wild, they were reproducing well in the wild, the survival rate for released animals was very good, and the hunting problem had been controlled. The target of the program is 300 animals in the wild.

The cause of the demise of the Arabian oryx was sport hunting. Human behavior is much easier to control than that of any other animal (if one has the necessary resolve). Therefore, once this cause of mortality was brought under control, the reintroduction became relatively easy. There was plenty of habitat, no pandemic diseases, and enough genetic diversity for the species to make it. Hoofed stock generally breeds well in captivity, so the program worked.

Given this pattern of successes mixed with failures, it is obviously difficult to generalize regarding the necessity and likely success of captive-breeding programs. An observer might get the impression that the institution of a captive-breeding program depends more on accidental features of a situation—the existence of a dedicated advocate for a particular species well placed in a modern zoo with captive-breeding facilities, for example—than on a careful evaluation of the program based on its unavoidability and likelihood of success. An important priority for the future must therefore be to develop and apply natural criteria for determining when a captive-breeding program is justified. Hopefully, the various arguments listed above in this chapter will provide guidance for such an understanding.

Reintroduction, Training, and the Welfare of Released Animals

I represent the land-ethic position in environmental ethics, and in this role I will examine captive breeding programs from that viewpoint. In contrast to the animal-rights position defended by Tom Regan,[10] the land-ethic position emphasizes the well-being of ecosystems and stems from Aldo Leopold. J. Baird Callicott, the ablest defender of the land ethic, articulates Leopold's ethic as placing the locus of value in entire ecosystems rather than in individual nonhuman animals, or even in species.[11] Individuals and entire species have a role to play in an ecosystem and are valued for that functional role, not for themselves. Thus the land ethic can tolerate actions that are detrimental to individual animals but draws the line at actions detrimental to the system as a whole. This philosophy is based in large part on a certain understanding of the science of ecology wherein individuals are dispensable so long as there are enough other individuals to play their roles in the system. This is why our detractors, with deliberate abusiveness, refer to the land-ethic position as "environmental fascism."

Thus the land-ethic position in environmental ethics generally attributes more value to a member of an endangered species than to an individual of a common species, even if the endangered organisms are at the same or lower level of cognitive or sensory development. Thus the land ethic can value an endangered plant more than an introduced rabbit that is eating it, even though the rabbit is much more highly sentient and has an incomparably higher level of cognition. The rabbit may well be in the wrong place at the wrong time ecologically and is thus judged to be disrupting the system. If this is the case, then human action to the detriment of the rabbit in favor of the plant is justified.

According to the land ethic, each species has a role to play as an energy manager within the ecological system and therefore has a functional relationship within it, because other species in the system depend on efficient energy management for their continued flourishing. Accordingly, those particular members of a species that is threatened with extirpation take on special value as essential to the maintenance of energy flows in the larger system. The land-ethic position therefore differs sharply from the animal-rights position, because the latter cannot attribute more value to an individual of an endangered species than to a member of a common species, other things being equal. Thus a California condor individual is due exactly the same moral consideration as a European starling, no more and no less.[12] It is therefore difficult for animal rights advocates to justify any special efforts to save species by affording special treatment to members of remnant populations. This difficulty makes the animal-rights position unattractive to most committed conservationists.

It is also a widely recognized shortcoming of the animal-rights position, from the viewpoint of conservationists, that it cannot attribute intrinsic or inherent value to plants and those animals so low on the chain of cognitive and sensory development that they cannot be considered subjects of a life. Thus plants can have instrumental value at most. On the other hand, the land-ethic has no difficulty extending moral consideration to plants, microorganisms, insects, or even abiotic components of an ecological system, not for their own sake but for the sake of the system as a whole—thus the holistic thrust of the land-ethic position.

This philosophical difference makes a profound practical difference when it comes to the question of management of wild animals. According to animal rights advocates, "With regard to wild animals, the general policy recommended by the rights view is let them be."[13] In contrast to the hands-off posture intrinsic to the animal-rights position, the land ethic can accept management of natural systems if the goal is to restore the system to a facsimile of its original condition or to maintain the system in a semblance of its natural condition. Thus the land ethic can accept captive breeding if and only if the goal is restoration. It cannot accept captive breeding if the goal is to keep the species on ice, that is, to keep the species in captivity forever.

When I first began to think about captive breeding as a moral issue, I was under the naive impression that the goal of the species survival plans (SSPs) was to reintroduce endangered species into the wild and phase them out in captivity. I soon realized that this is not what is being contemplated. The plan is to reintroduce some individuals back into the wild and perhaps build self-sustaining populations out there, but also to keep a breeding population of the species in captivity forever. Zoos are fond of calling themselves modern versions of Noah's ark, but there is one important difference—the animals got off and left the empty ark behind once it had fulfilled its purpose. Is that what the SSPs have in mind, or are SSPs to be perpetual arks? With the red wolf, to take just one example, the plan calls for 330 animals in captivity forever and 220 in the wild.

Most SSPs include no provision for reintroduction. In most cases it remains a pious hope. What is being contemplated in most cases is keeping the species alive in captivity for 200 years. Arguing from the land ethic, therefore, I assert that only those plans that include reintroductions are deserving of social support. The goal should be to get the species out of the cages and into the wild where they belong.

The reason that most SSPs do not include reintroduction as more than a pious hope is that there is simply no habitat available. Even if we can breed the species in captivity, we will end up with an animal all dressed up and no place to go. Zoos therefore ought to devote more attention to preserving whole intact ecosystems than they do. When the Cleveland Zoo expends some $24 million on a new rainforest exhibit, one has to wonder whether it would have been wiser, from a conservationist viewpoint, to preserve a significant amount of existing intact rain forest in Brazil or Belize rather than to build one from scratch in Cleveland.

Captive breeding is the extreme example of autecological (single-species) management. Single-species management is, quite simply, biologically unsound. It is also philosophically unsound, it can be argued, in that an individual or a species has little value apart from its context. As David Brower has put it, " A [California condor] is only 5 percent bone and feathers. Ninety-five percent of condor is place." When an animal is removed from its context, the ecosystem, it is degraded immediately—most of its value is lost. Therefore, captive breeding ought to be only a temporary measure to perpetuate species until they can be restored to the wild.

For this reason, there is an important conceptual distinction between a zoo and a captive-breeding facility. The difference is in purpose. The purpose of a captive-breeding facility is, or ought to be, to restore ecosystems by reintroducing species into the ecological niches they once occupied. The purpose of a zoo is to retain animals in captivity for the purpose of display, including entertainment, education, and economic development (chiefly to promote tourism). Although these two activities are often carried on by the same corporate entity, thus blurring the distinction, conceptually the two are quite distinct—a zoo is not a captive-breeding facility.

Reintroduction is frightfully difficult and can be very hard on individual animals. Survival rates are often low because pen-reared animals sometimes lack the skills needed to survive in the wild. Nevertheless, we must reject the argument that reintroduction is too cruel to be morally justified in any case. We must reject the outdated philosophy of Heini Hediger, who argued that humans are actually doing wild animals a favor by taking them into captivity where they will not be subjected to the hardships and hazards of the wild.[14] Hediger argued that wild animals are not really free, since they are bound by instincts into behavior patterns that are inflexible. He also pointed out that some animals live longer in zoos than in the wild as evidence that the ones in captivity are the lucky ones. That has been challenged empirically in recent decades, but even if Hediger was right, even if some animals do live longer in zoos, their lives, and ours, are correspondingly impoverished.[15]

Therefore, we are morally obligated to attempt to condition individuals as thoroughly as we can before release so that their chances of survival in the wild are maximized, even if this entails inflicting suffering on individual animals. Beck has suggested that a part of the conditioning of golden lion tamarins for release should include the sudden introduction of predators into their cages to condition the monkeys to avoid predators in the wild.[16] Any monkeys killed by successful predators would be eaten by the predator in full view of the surviving members of the group. This is harsh, but it is no harsher than what these monkeys have to face daily in the wild. If it can be shown that this kind of conditioning actually works, if it improves predator avoidance in the wild, these harsh measures and others equally harsh are justified.

For how long should the zoos monitor the reintroduced animals? The answer, apparently, is "forever." I'm not sure humans realize the implications of that. One of the most important aspects of intact natural systems is that they are self-sustaining—they take care of themselves. But once the system is degraded by human action to the point that we decide to take the species in out of the cold and put it on ice, we have made a permanent commitment to that wild species that is irrevocable. We have assumed the obligation to protect it from extinction, and extinction is forever. Hence, survival is forever as well. Not two hundred years, not fifty thousand years, not one hundred thousand years, but forever. If you are not awed by that prospect, you should be.

For eight years, I built nest boxes for a threatened subspecies of American kestrel in Florida. American kestrels are common and in no need of human assistance throughout most of their range, but the Florida breeding population is classified as threatened. Many scientists believe nest cavities are the limiting factor, and have suggested that nest boxes erected and maintained by humans in suitable habitat would augment the population. And they do. The kestrels move right in and fledge young if the box is in the right place. But after eight years, I asked myself the question, who is going to be doing this one thousand years from now? This brought me some clarity, because I realized my efforts were futile unless I was in a position to make a permanent commitment on the part of future humans to maintain boxes for kestrels. If I could not make that commitment, I was not doing the kestrels any favors. Once they become dependent on boxes provided by humans for their continued survival, they become clients.

We have done that with one wild North American bird, the purple martin. The practice originated in colonial times because martins will mob raptors, thus a flourishing martin colony on one's farm provided significant protection for one's chickens. Today the purple martin is an industry, and in the western United States, they seldom if ever nest anywhere except in structures provided by humans. I decided that I would prefer not to put the kestrel on the same path to dependency. Better to try to preserve some remnant of kestrel habitat and let the birds make it on their own.

Hence, I believe that animals should be taken out of the wild for captive-breeding programs only if there is no alternative to extinction. Although there really is no alternative in at least some cases, the priorities should be on maintaining the species intact in the wild and correcting the causes of endangerment in the first place. Captive breeding ought to be only a last-ditch, desperation effort.

Many will disagree with this conclusion because once a species gets to the point of desperation it becomes very difficult to save. Beck has pointed out that out of 145 reintroduction projects only 16, about 11 percent, have made any contribution to maintaining a self-sustaining wild population.[17] But this rather underwhelming result does not necessarily lead to the conclusion that reintroduction is so difficult and unlikely to help that it ought not to be attempted. The alternatives to captive breeding and release are extinction or keeping the species on ice (that is, in zoos) forever. Neither alternative is acceptable from the point of view of the land ethic. We are morally obligated to attempt reintroduction and to do our best to make it work whether we succeed or not. In short, we must try.

NOTES

1. D. Jamieson, "Wildlife Conservation and Individual Animal Welfare," in *Ethics on the Ark: Zoos, Animal Welfare, and Wildlife Conservation*, ed. Bryan G. Norton et al. (Washington, D.C.: Smithsonian Press: 1995).

2. J. DeBlieu, *Meant to Be Wild: The Struggle to Save Endangered Species through Captive Breeding* (Golden, Colo.: Fulcrum, 1991).

3. Ibid.

4. L. Durrell and J. Mallinson, "Reintroduction as a Political and Educational Tool for Conservation," *Dodo: The Journal of the Jersey Wildlife Preservation Trust* 24 (1987): 6–19.

5. M. W. Fox, *Inhumane Society: The American Way of Exploiting Animals* (New York: St. Martin's Press, 1990).

6. J. R. Luoma, *A Crowded Ark* (Boston: Houghton Mifflin, 1987).

7. J. Cherfas, *Zoo 2000: A Look Behind the Bars* (London: British Broadcasting Corp., 1984).

8. J. R. Luoma, "A Wealth of Species on the Forest Floor," *New York Times*, July 2, 1991.

9. De Blieu, *Meant to Be Wild*.

10. T. Regan, *The Case for Animal Rights* (Berkeley: University of California Press, 1983); T. Regan, "Are Zoos Morally Defensible?" in *Ethics on the Ark*.

11. J. B. Callicott, *In Defense of the Land Ethic* (Albany: State University of New York Press, 1989).

12. P. Singer, *Animal Liberation* (New York: Avon Books, 1975).

13. Regan, *The Case for Animal Rights*.

14. H. Hediger, *Wild Animals in Captivity* (New York: Dover Publications, 1964).

15. D. Jamieson, "Wildlife Conservation and Individual Animal Welfare," in *Ethics on the Ark*.

16. B. Beck et al., unpublished manuscript.

17. B. Beck, "Reintroduction, Zoos, Conservation, and Animal Welfare," in *Ethics on the Ark*.

References

Beck, B. "Reintroduction, Zoos, Conservation, and Animal Welfare." In *Ethics on the Ark: Zoos, Animal Welfare, and Wildlife Conservation*, edited by Bryan G. Norton, Michael Hutchins, Elizabeth F. Stevens, and Terry Maples. Washington, D.C.: Smithsonian Press, 1995.

Callicott, J. B. *In Defense of the Land Ethic*. Albany: State University of New York Press, 1989.

Cherfas, J. *Zoo 2000: A Look Behind the Bars*. London: British Broadcasting Corp., 1984.

DeBlieu, J. *Meant to Be Wild: The Struggle to Save Endangered Species through Captive Breeding*. Golden, Colo.: Fulcrum, 1991.

Durrell, L., and J. Mallinson. "Reintroduction as a Political and Educational Tool for Conservation." *Dodo: The Journal of the Jersey Wildlife Preservation Trust* 24 (1987): 6–19.

Fox, M. W. *Inhumane Society: The American Way of Exploiting Animals*. New York: St. Martin's Press, 1990.

Hediger, H. *Wild Animals in Captivity*. New York: Dover Publications, 1964.

Jamieson, D. "Against Zoos." In *In Defense of Animals*, edited by P. Singer. New York: Harper & Row, 1985.

———. "Wildlife Conservation and Individual Animal Welfare." In *Ethics on the Ark: Zoos, Animal Welfare, and Wildlife Conservation*, edited by Bryan G. Norton, Michael Hutchins, Elizabeth F. Stevens, and Terry Maples. Washington, D.C.: Smithsonian Press, 1995.

Luoma, J. R. *A Crowded Ark*. Boston: Houghton Mifflin, 1987.

———. "A Wealth of Species on the Forest Floor." *New York Times*, July 2, 1991.

Regan, T. *The Case for Animal Rights*. Berkeley: University of California Press, 1983.

——— "Are Zoos Morally Defensible?" In *Ethics on the Ark: Zoos, Animal Welfare, and Wildlife Conservation*, edited by Bryan G. Norton, Michael Hutchins, Elizabeth F. Stevens, and Terry Maples. Washington, D.C.: Smithsonian Press, 1995.

Singer, P. *Animal Liberation*. New York: Avon Books, 1975.

8

HELPING A SPECIES GO EXTINCT:
THE SUMATRAN RHINO IN BORNEO

Alan Rabinowitz

INTRODUCTION

It is no small miracle that rhinos still walk the face of the earth. No other group of animals has been so highly prized for so long yet managed to survive human onslaught. The focus of our obsession with this animal has revolved around the protuberance of hardened hair on the rhino's head known as rhino horn. Rhino horn played an important role in medieval Chinese medicine, a role that it continues to play in traditional Chinese practices of today.

The use and trade in rhino horn is recorded from China as early as 2600 B.C.E.,[1] spreading in later years to Western Asia and the Roman Empire.[2] But what was once a familiar animal throughout much of China was already considered a rarity "by the time of the ages illuminated by books."[3] By the T'ang Dynasty (600–900 C.E.), large quantities of horn were being imported to China. With the opening of new trade routes, horns were brought to China from northern Somalia, the Arab states,[4] and the southeast Asian areas of modern day Vietnam, Java, Sumatra,[5] the Malay Peninsula,[6] Borneo,[7] Cambodia,[8] Laos,[9] and Thailand.[10] The near extinction of the Javan and Sumatran rhinos in modern times has been largely attributed to the trade during the T'ang Dynasty.[11]

The preparation of rhino horn for particular ailments is often cited from the *Divine Peasant's Herbal*, written in the first century B.C.E.,[12] and from the *Pen Ts'ao Kang Mu*, a well-known sixteenth-century Chinese medical text. Although there have been modifications and revisions to the Chinese medical

---

This article originally appeared in *Conservation Biology* 9 (June 1995): 482–88. Reprinted by permission of Blackwell Science, Inc. Copyright © 1995 Blackwell Science, Inc. All rights reserved.

pharmacopoeia since those times, modern medical and popular books contain both old and new applications for rhino horn.[13] Many licensed doctors and pharmacists in Taiwan continue to sell or prescribe rhino horn for their patients.[14] In mainland China, an increase in the availability of rhino horn and an increased demand by the pharmacies is of growing concern.[15]

The rhino family, containing five living species, once ranged widely throughout the more open habitats of Africa and the tropical and subtropical habitats of eastern Asia, including Sumatra, Java, and Borneo. Today rhinos survive only in small, disjunct populations. The Sumatran rhinoceros, the smallest of the rhino species, was once found throughout Assam, Myanmar, Thailand, Indo-China, the Malay peninsula, Sumatra, and Borneo. Today, breeding populations of this species are thought to exist only in Sumatra, the Malay peninsula, and northeast Borneo.

The survival of all five rhino species into the twentieth century can be attributed to a number of factors: legal protection of the species, an increase in the number of protected areas where they survive, the ability for certain rhino species to live in rugged and isolated forested areas, and political and socioeconomic factors that have closed down many of the historic trade routes for rhino horn. The traditional use of rhino horn has not faded with time, however, and with the present Chinese economy growing at an unprecedented rate, these products are becoming ever more affordable to the new consumer class.

During the 1970s, rising prosperity in parts of Asia created a resurgence in demand for rhino parts, and this demand, coupled with escalating prices, encouraged greater hunting of the rhino. Between 1970 and 1987, an estimated 85 percent of the world's remaining rhino population was lost.[16] Many small, fragmented populations were wiped out. As millions of dollars were spent on efforts to reverse this trend, most rhino populations continued to decline.

I examined the case of the Sumatran rhino in Sabah, Malaysian Borneo, where at least two important populations of this species might still survive. First I discuss how, for the last two decades, highly publicized efforts to save the Sumatran rhino have been concerned more with high-profile, technical issues than with the more difficult job of protection and management in the field. Then I will show how the decline of this species in Borneo has been watched and documented for more than a century, while efforts to remedy this situation have fallen terribly short of what is needed.

INTERNATIONAL AND REGIONAL EFFORTS TO SAVE THE SUMATRAN RHINO

In response to continued concern for the decline of Asian rhino species, the Asian Rhino Specialist Group (ARSG) was created by the Species Survival Commission of the World Conservation Union. The first meeting of group, convened in Thailand in 1979, emphasized the need for data collection, research and monitoring efforts, protection of rhino habitats, reduction of poaching, and strict control of trade in rhino products. A second meeting of the ARSG, held in Malaysia in 1982, analyzed Asian rhino distribution patterns, estimated numbers of animals, and put forth conservation requirements. By the third meeting in Singapore in 1984, the ARSG decided to launch a program to capture "doomed" Sumatran rhinos for breeding in captivity in Asian, European, and North American zoos. Doomed rhinos were loosely defined as animals whose lives were in immediate danger due to the clearing or conversion of forest for other uses.

The Sumatran Rhino Trust (SRT), an organization spawned from the American Association of Zoological Parks and Aquariums, initially worked out an agreement with Malaysia for the export of animals to the United States with the aim of establishing a captive-breeding program. But protests over the shipping of Malaysian rhinos to Western zoos resulted in the dissolution of the proposed agreement and the establishment of a separate Malaysian captive-breeding program. Political differences between the state of Sabah and the national government then led to the creation of two separate Malaysian breeding programs, one in Peninsular Malaysia and one organized by the newly formed Sabah Rhino and Wildlife Conservation Committee, each to be funded and coordinated individually.

Because of the lack of cooperation between the different countries in the region, the fourth and fifth meetings of the Asian Rhino Specialist Group in Indonesia (1986) and Malaysia (1987), respectively, were held to design a comprehensive conservation action plan for all Asian rhino species. The subsequent plan concluded that there was still time to reverse the rapid decline of the Sumatran rhino.[17] The creation of captive populations was deemed an important component of any Sumatran rhino conservation plan. While recognizing the importance of in situ protection and management of wild populations, this plan clearly emphasized ex situ management of captive rhino populations by the ARSG.

In 1987, the SRT signed an agreement with the Indonesian government. It continued to acknowledge that protection and management in situ was a top priority for Sumatran rhino conservation, but the agreement stipulated the following:

(1) A donation of $60,000 per rhino would be paid to the newly established Indonesia Rhino Foundation once rhinos were received in SRT facilities in North America.

(2) In the event of death during transport to the zoos and for a period of one year, an indemnity of $25,000 per rhino would be paid by SRT to the Indonesia Rhino Foundation.

(3) In the event of death during capture, $5000 per rhino would be paid by SRT to the Indonesia Rhino Foundation.

(4) All expenses for the survey, capture, and transport of rhinos would be covered by SRT.

(5) SRT would contribute $20,000 per year for the duration of this agreement for improving protection and management for rhinos in National Parks.

In 1993, the SRT was dissolved after five years and a cost of more than $2.5

million. Virtually none of the money went to improving the protection and management of wild rhinos in existing protected areas. This program, along with the similar efforts in Sabah and Peninsular Malaysia to catch doomed rhinos for breeding, were expensive failures resulting in the capture of thirty-five rhinos and the deaths of twelve rhinos between 1984 and 1993.[18] The failure was partly a result of the skewed sex ratio of captured animals. Still, as of 1993, the surviving twenty-three rhinos (fourteen females, nine males) were being held in ten separate areas in Indonesia, Peninsular Malaysia, Sabah, the United Kingdom, and the United States. Other than one facility in Peninsular Malaysia with five rhinos, no more than three rhinos were at any of the other facilities.[19] Because adult males and females were never together in the same place for a significant amount of time, there have been no births from captive Sumatran rhinos to date, except for one female who was pregnant when captured.

THE SUMATRAN RHINO IN BORNEO

Although Borneo was once home to both the Javan and the Sumatran rhino, the Javan rhino was thought to have disappeared due to natural causes about 12,000 years ago.[20] The Sumatran rhino, described as a distinct subspecies on Borneo,[21] was still considered relatively common into the early twentieth century.[22] The harvesting and sale of rhino horn, regarded by the government as simply another forest product, was encouraged throughout the early 1900s.[23]

By the turn of the century, the alarm was already being sounded about the rhino's decline, because hunting for the highly prized horn continued unabated to support a primarily Chinese market.[24] By the 1950s it was reported that the Sumatran rhino has been hunted to near extinction in Borneo,[25] partially due to the hunting skills of the indigenous people.[26] This did little to dampen trade however, as countries such as Singapore continued to obtain rhino horn from Borneo.[27]

By the 1960s Harrisson estimated that there were no more than two rhinos left in Sarawak, possibly five in Kalimantan, and between eleven and thirteen in Sabah.[28] The Fauna Conservation Ordinance of 1963 in Sabah and the Wild Life Protection Ordinance of 1958 in Sarawak protected rhinos on paper but did little to deter poaching or to ensure the prosecution of offenders. Ten years later there was still virtually nothing known of existing rhino numbers.[29] In 1982, Davies and Payne estimated that between fifteen and thirty rhinos remained in Sabah and recommended protected status for two areas that still contained numbers of rhinos: Silabukan and Danum Valley.[30] Shortly thereafter a summary of reports compiled by van Strien indicated that rhinos were virtually gone from Sarawak and most of Kalimantan.[31] At this point, Sabah contained the most important populations of Sumatran rhino outside of Sumatra and Peninsular Malaysia.

Efforts to Protect the Sumatran Rhino in Sabah

Between 1979 and 1987, as Sabah became the focus of attention for Sumatran rhinos in Borneo, some positive steps were taken by the Game Branch of the Sabah Forest Department and subsequently by the newly formed Sabah Wildlife Department to protect the areas where these last populations existed.

Danum Valley was long considered one of the most pristine lowland areas left in Borneo. Free of human habitation and known to contain a rich diversity of wildlife, the area was assumed to be relatively undisturbed because of its ruggedness and inaccessibility.[32] When the presence of rhinos was first suspected in this area in 1976, the Danum Valley was proposed as a national park;[33] it was later recommended for protection as a game sanctuary.[34] However, the state-run Sabah Foundation, which maintained a long-term timber concession in the area, did not want to relinquish its rights to the land. Instead, in 1982 a 438-square-kilometer area was designated as "Danum Valley Conservation Area" in which logging would be prohibited but control would remain under the Sabah Foundation. Soon thereafter, buildings for research and visitor accommodations were constructed at the site.[35] Research conducted at the site in the late 1980s verified that at least one population of rhinos was declining in numbers.[36] By 1989 a traverse through the area recorded only a single set of rhino footprints.[37]

A second area, the Silabukan Forest Reserve, had been commercially logged since the 1960s, even while it was thought to contain one of the largest remaining concentration of rhinos in Sabah. In the early 1980s, Davies and Payne verified the presence of a breeding population of Sumatran rhinos in this lowland forest and pushed for protection of the area.[38] Finally, in 1984, 1220 square kilometers were gazetted by the Sabah government as the Tabin Wildlife Reserve, primarily for the protection of rhinos.[39] But selective logging in the reserve continued under license through 1986[40] and "unofficially" through the early 1990s.

Six walk-through surveys in Tabin conducted by the Wildlife Department between 1980 and 1991 indicated a minimum of three to seven rhinos in the area, with steady declines in rhino sign between the 1982 and 1991 surveys.[41] Noticeable shifts in rhino distributions between surveys caused enough alarm for the recommendation of urgent follow-up research to investigate the possibility of declining rhino numbers.[42] No such research was ever conducted. The first management plan for the sanctuary listed rhino poaching as the most serious threat to the value of Tabin.[43]

In the Asian Rhino Action Plan, Tabin Wildlife Reserve and Danum Valley were singled out as the two main areas where viable populations were likely still to exist in Sabah.[44] The plan cited estimates of twenty and ten individuals, respectively, although no definitive surveys had been carried out at either site. Specific activities recommended by the plan for protecting rhinos in Sabah included the following:

(1) strengthening the staffing, funding, and logistical support of the Sabah

Wildlife Department to allow for effective protection and research of wild rhino populations;

(2) stricter legislation against rhino poaching;

(3) review of the size and protected status of Danum Valley Conservation Area and Tabin Wildlife Reserve;

(4) surveys in Danum and Tabin to determine the true status of the rhinos there; and

(5) capture of isolated or threatened rhinos for captive breeding or translocation.

These recommendations, while appropriate, did little more than rephrase similar recommendations made during the first meeting of the Asian Rhino Specialist Group in 1979. The fact that there had been little progress on these issues, ten years after they had first been discussed, was not mentioned. As of 1992, there were still no reliable estimates of rhino densities for any part of Sabah. Of the five activities recommended by the Action Plan, only the capture of doomed rhinos was carried out with any serious intent.

In September 1992, I organized a rhino survey, at the request of the Sabah Foundation and the Sabah Wildlife Department, to assess rhino abundance and to standardize a methodology for future rhino surveys and monitoring in the area. The survey was also intended to provide data to the Sabah Wildlife Department for use in upgrading the Greater Danum Valley Conservation Area into a park or wildlife reserve.

Using methodology developed by Borner and van Strien,[45] two small groups of rhinos, each consisting of two to three individuals, were found through intense surveying of areas totalling 80 square kilometers.[46] Assuming that other rhinos might be similarly distributed, an estimate of thirteen to twenty-three rhinos was made for the 1,000-square-kilometer Greater Danum Valley Conservation Area. While this estimate was more than twice that speculated by the Asian Rhino Action Plan,[47] this survey put to rest the assumption that much of the area was undisturbed and protected by virtue of its ruggedness and isolation.

Only two out of seven teams found recent evidence of rhino presence. Five teams encountered only old rhino sign, along with old hunting camps. This included an area where rhinos had been studied in 1986[48] but were now no longer present. Of the two teams that discovered fresh rhino sign, one was located adjacent to the field station and tourist accommodations, an area with regular human activity but no hunting. The second team, which was dropped by helicopter into the most remote section of the study area, encountered an ongoing rhino-poaching expedition. The hunters fled along a well-used trail peppered with old campsites, indicating a history of poaching in the area.

Despite the serious and unexpected nature of these findings, there was no attempt by the Wildlife Department to look into the situation. The following year there were still no patrols sent into the area nor any effort to check or monitor the recent rhino sign that had been detected. Because no immediate action was taken to change the protected status of the Danum Valley despite the survey, the Wildlife Department did not feel compelled to pursue further surveys or management activities in the area.

In Tabin Wildlife Reserve, meanwhile, other activities were underway. As part of an environmental management project funded by the United Nations Development Program in the early 1990s, a wildlife specialist was hired as a consultant to the Sabah Wildlife Department, and a New Zealand consulting firm was contracted to provide a manager for the Tabin Reserve. A second Tabin Management Plan was produced that did little more than restate the initial 1986 plan.[49] Illegal logging and poaching were still identified as the major threats to the reserve.

Despite new infrastructure, the assignment of a full-time staff, and the presence of foreign consultants assigned to Tabin Reserve, there were still no systematic patrols or surveying of the area when I visited and trained staff there in 1992. During a 1992 elephant census in Tabin, spoor of only one rhino was encountered in 118 kilometers of transects.[50] Later that year, rhino tracks were sighted close to the Tabin ranger station in an area frequented by visitors and researchers but with virtually no hunting pressures. Although the implication of these track locations, which were similar to some of the track locations in the earlier Danum rhino survey, were of potential management importance, there was never any follow-up to the reports. At the time of this writing, there has not been a single reliable estimate of the number of rhinos that might still survive in Tabin, nor has any systematic management been carried out for the species.

With encouragement from the foreign wildlife specialist, the Sabah Wildlife Department shifted most of its emphasis to the capture of doomed Sumatran rhinos—this, despite the fact that organized patrols in the field were not being encouraged, proper surveys were not being carried out, and the foreign consultants themselves were insufficiently trained to handle wild-caught rhinos. Furthermore, the definition of doomed rhinos had now been expanded to include any rhino found or captured outside of an already existing protected area, which did little to encourage new rhino surveys or the protection of remaining forest areas where rhinos still survived.

Of two new rhinos captured since 1992, both in the forests of an area proposed for protection along the Kinabatangan River, one died in captivity under the care of a foreign veterinarian sponsored by the United Nations Development Program and another was radio-collared by the program's wildlife specialist and put in an enclosure in Tabin. The rhino immediately broke free of the enclosure and went into the forest. Despite the collar, the animal was never followed after its escape. Under the same management, efforts to capture, collar, and relocate additional rhinos were continued.

Discussion

Despite protective legislation and the creation of protected areas where rhinos survive, Sumatran rhino populations continue to decline. Within the last two decades, the international community has stepped in to assist in the protection of this species. During that time, every report, management strategy, and action plan has come to the same conclusion: The decrease in rhino populations is due to poaching carried out primarily to collect the horn, and to habitat loss as land is converted to other uses.

The problem, however, has been that once the causes of decline of the Sumatran rhino were recognized, the actions needed to remove or neutralize these causes were never fully implemented. Both Malaysia and Indonesia acceded to the Convention on International Trade of Endangered Species of Wild Fauna and Flora (CITES), in 1978 and 1979 respectively, which effectively banned the legal trade in rhino products. Yet the legislation needed to fully implement CITES was never enacted in either country.[51] Furthermore, even the existing legislation relating to wildlife protection in Malaysia and Indonesia was rarely used to discourage trade in rhino parts or to prosecute offenders.

In Sabah, as elsewhere, the easiest, most palatable, and most visible steps toward Sumatran rhino conservation were taken first. Rhino habitat was better secured through the creation of protected areas that were not controversial and that caused minimal interference with ongoing logging activities and agricultural development plans. Tabin Wildlife Reserve, for example, gained full protection only after most of the valuable timber had been taken out, and Danum Valley remains protected only at the discretion of the Sabah Foundation, the state's largest timber concessionaire. Other management activities, such as antipoaching patrols, education campaigns, and surveys to assess the adequacy of reserve size, were increasingly discussed but never implemented because they were more difficult, time consuming, and sometimes controversial if they conflicted with existing land-use policies.

Emphasis in time, money, and effort has been placed on the capture and breeding of rhinos, despite the fact that such activities alone, even if successful, would not solve the problem or remove the causal factors of rhino decline in the wild. Although such activities involve known techniques and provide a high-profile outlet for government spending and international funding, the implication that captive breeding can save the Sumatran rhino makes the failure of in situ conservation seem less serious. This, in turn, helps create a self-fulfilling prophecy that wild populations have a low probability of survival.

Caughley distinguishes two advancing fronts in the field of conservation biology.[52] The first, which he calls the declining-population paradigm, is concerned with the external causes that drive populations toward extinction. Research efforts are aimed at determining why populations are declining and how to neutralize the causes. The second, called the small-population paradigm, deals with the risk of extinction as a consequence of small population size. Here one deals with the genetics and dynamics of a small, finite population. While the former paradigm is mostly empirical and lacks scientific rigor, the latter is mostly theoretical and thus more attractive by virtue of its seemingly "hard" scientific approach.

The small-population paradigm dominated much of the science of conservation biology in the 1980s,[53] but it is almost completely removed from the real world.[54] The proponents of this approach, using terms such as extinction vortices, minimum viable populations, population and habitat viability analysis, inbreeding depression, and metapopulation analysis, do their field work in the laboratory, in captive-holding facilities, and at the computer. They acknowledge the need for in situ protection of wild populations, but their results almost always point to the same conclusion: Declining populations in the wild will eventually become extinct, and thus captive breeding is needed to save the species.

Using decision analysis, Maguire et al. predicted the probability of Sumatran rhino extinction if certain actions were or were not taken by Indonesia and Malaysia.[55] The choice of possible actions included increased control on poaching, new and/or expanded protected areas, fencing of existing protected areas, translocation, and captive breeding. Not surprisingly, the capture and breeding of wild rhinos were viewed as the most promising means of saving the species.

But as with other attempts at linking theory with management applications, the actual attempts to establish a captive Sumatran rhino herd that would help repopulate the wild herd fell far short of expectations. Not only was the sex ratio of captured Sumatran rhinos highly skewed, but those in captivity proved extremely difficult to breed. Furthermore, the international and regional captive-breeding programs were subjected to the same political and economic realities that caused Maguire et al. to so easily discard other conservation actions.

While some of the blame for the decline of the Sumatran rhino must be placed on the Indonesian and Malaysian governments, the rest of it falls squarely in the lap of international funding and conservation organizations. The international community, with its funding and expertise, has played a major role in directing the course of rhino conservation over the last quarter century. Unfortunately, it has tried to avoid dirtying its hands with controversial and difficult issues such as poaching, protected-area staff training and wages, and the establishment of new reserves in areas where local communities, government agencies, or entrepreneurs wish to alter or use the land for other purposes. Foreign advisers and nongovernment conservation organizations have all too often avoided such issues because of the risk of becoming an unwelcomed guest.

While political, cultural, and socioeconomic issues in Indonesia and Malaysia continue to interfere with Sumatran rhino protection, these difficulties have never been insurmountable. The rhino simply has not been considered important enough for governments and large funding agencies to tackle these realities. Only when a firm commitment is made to save the Sumatran rhino will the species stand a chance of survival. Regrettably, our years of accumulated failures and avoidance of issues have not moved us closer to this kind of a commitment. The 1993 report of the Asian Rhino Specialist Group to the United Nations Environment Program Conference for Rhinoceros Range States, Consumer States, and Donors, estimated a new three-year cost for rhino conservation in Indonesia and Malaysia at approximately $14 million. As part of this cost, a $2-million program by the Global Environmental Facility is already underway to establish yet another conservation strategy for southeast Asian rhinos in Indonesia and Malaysia. This "new" strategy, based primarily on viable population theory, entails the following components: wild population protection sanctuary management, captive propagation, and gene-bank technologies. The strategy ignores the fact that the only means likely to save the rhino in the wild involves intensive, on-the-ground protection and management activities.

Meanwhile, the decline of the Sumatran rhino continues. In August 1994, twelve more Sumatran rhino horns were confiscated in Taiwan that had been smuggled on a fishing boat from Malaysia.[56] In Sabah, the Wildlife Department continues to capture doomed rhinos from areas that have not been adequately surveyed nor even considered for protected status. After all these years, do we know how many Sumatran rhinos we are dealing with? No, but soon we might have a nice round figure.

Notes

1. K. Nowell, Chyi Wei-Lin, and Pei Chia-Jai, *The Horns of Dilemma: The Market for Rhino Horn in Taiwan* (Cambridge, England: Traffic International, 1992).

2. F. Hirth and W. W. Rockhill, *Chau Ju-Kua: His Work on the Chinese and Arab Trade in the Twelfth and Thirteenth Centuries, Entitled "Chu-fan-chi"* (St. Petersburg, Russia: Imperial Academy of Sciences, 1911); and E. H. Schafer, *The Golden Peaches of Samarkand: A Study of T'ang Exotics* (Berkeley: University of California Press, 1963).

3. Schafer, *The Golden Peaches of Samarkand*.

4. Hirth and Rockhill, *Chau Ju-Kua*.

5. J. V. G. Mills, *Ying-Yai Sheng-Lan of Ma Huan (1433)* (Cambridge, England: The Hakluyt Society, 1970).

6. Hirth and Rockhill, *Chau Ju-Kua*.

7. E. Mjöberg, *Forest Life and Adventures in the Malay Archipelago* (London: George Allen & Unwin, 1930).

8. Chou Ta-Kuan, *The Customs of Cambodia: 1296–1297* (Bangkok: Siam Society, 1993).

9. G. van Wusthof, "Voyage Loitain aux Royaumes de Cambodge et Laouwen," *Bulletin de la Societe de Geographie* 6, no. 10 (1871).

10. N. Gervaise, *The Natural and Political History of the Kingdom of Siam*, trans. John Villiers (Bangkok: White Lotus Press, 1989); J. Bowring, *The Kingdom and People of Siam* (Kuala Lumpur: Oxford University Press, 1969); C. Bock, *Temples and Elephants: The Narrative of a Journey of Exploration Through Upper Siam and Lao* (London: Sampson Lew, Marston, Searle and Rivington, 1884).

11. Schafer, *The Golden Peaches of Samarkand*.

12. Nowell, Wei-Lin, and Chia-Jai, *The Horns of Dilemma*.

13. B. E. Read, *Chinese Medica Materia: Animal Drugs* (Taipei: SMC Publishing, 1982); Kun-Ying Yen, *The Illustrated Chinese Materia Medica: Crude and Prepared* (Taipei: SMC Publishing, 1992).

14. Nowell, Wei-Lin, and Chia-Jai, *The Horns of Dilemma*; J. Loh and K. Loh, "Rhino Horn in Taipei, Taiwan," *Traffic Bulletin* 14, no. 2 (1994): 55–58.

15. J. Loh and K. Loh, "A Spot Check on the Availability of Rhino Products in Guangzhou and Shanghai, China," *Traffic Bulletin* 14, no. 2 (1994): 79-80.

16. S. Fitzgerald, *International Wildlife Trade: Whose Business Is It?* (Washington, D.C.: World Wildlife Fund U.S.A., 1989).

17. M. Khan, *Asian Rhinos: An Action Plan for Their Conservation* (Gland, Switzerland: World Conservation Union, 1989).

18. T. J. Foose and Z. Z. Zainuddin, *International Studbook for the Sumatran Rhino* [Dicerorhinus sumatrensis] (Columbus, Ohio: The Wilds, 1993).

19. Ibid.

20. G. G. Cranbook, *The Riches of the Wild: Land Mammals of Southeast Asia* (Singapore: Oxford University Press, 1987).

21. C. P. Groves, "Description of a New Subspecies of Rhinoceros, from Borneo, *Didermocerus sumatrensis harrisoni*," *Saugtierkundliche Mitteilungen* 13, no. 3 (1965): 128–31.

22. W. C. M. Weedon, "A Journey Across British North Borneo Together With a brief Account of Some Further Explorations in the Interior of That State," unpublished report, Sabah State Archives, Malaysia; Mjöberg, *Forest Life and Adventures in the Malay Archipelago*.

23. J. Payne, *The Distribution and Status of the Asian Two-Horned Rhinoceros (Dicerorhinus sumatrensis harrisoni) in Sabah Malaysia*, Project 3935 (World Wildlife Fund–Malaysia, 1990).

24. R. W. C. Shelford, *A Naturalist in Borneo* (London: T. Fischer Unwin, 1916); T. Harrison, ed., *Borneo Jungle: An Account of the Oxford University Expedition of 1932* (Oxford: Oxford University Press, 1988).

25. T. Harrison, "Borneo Fauna Anxieties," *Oryx* 3, no. 3 (1955): 134–37; T. Harrison, "Rhinoceros in Borneo and Traded to China," *The Sarawak Museum Journal* 7, no. 8 (1956).

26. N. van Strien, *The Sumatran Rhinoceros Dicerorhinus sumatrensis (Fischer, 1814) in the Gunung Leuser National Park Sumatra, Indonesia* (Hamburg: Verlag Paul Parey, 1986).

27. L. Talbot, "A Look at Threatened Species," *Oryx* 5, no. 4, 5 (1960): 153–295.

28. T. Harrison, "A Future for Borneo's Wildlife," *Oryx* 8, no. 2 (1965): 99–104.

29. L. C. Rookmaker, "The Distribution and Status of the Rhinoceros *Dicerorhinus sumatrensis* in Borneo—A Review," *Bijdragen Tot de Dierkunde* 47, no. 2 (1977): 197–204.

30. G. Davies and J. Payne, *A Faunal Survey of Sabah: Report to the Game Branch, Sabah Forestry Department, Kuala Lumpur* (World Wildlife Fund–Malaysia, 1982).

31. van Strien, *The Sumatran Rhinoceros Dicerorhinus sumatrensis*.

32. C. W. Marsh and A. G. Greer "Forest Land-Use in Sabah, Malaysia: An Introduction to Danum Valley," *Philosophical Transactions of the Royal Society of London B* 335 (1992): 331–39.

33. B. H. Kiew, *A Survey of the Proposed Danum Valley National Park, Sabah* (Kuala Lumpur: World Wildlife Fund–Malaysia, 1976).

34. Davies and Payne, *A Faunal Survey of Sabah*.

35. P. Andau, "Conservation of the Sumatran rhinoceros in Sabah, Malaysia," *Rimba Indonesia* 21, no. 1 (1987): 39–46.

36. Hamid Bin Ahman, Abd, "A Study of Current Abundance and Several Aspects of the Feeding Ecology of Sumatran Rhinoceros (*Dicerorhinus sumatrensis harrisoni* Groves

1965) in the Danum Valley Field Centre Area, Lahad Datu, Sabah," B.S. thesis, University Kebangsaan Malaysia, Sabah, 1991.

37. J. Payne, *Report on a Survey of the Distribution of Rhinoceros in Southeastern Sabah, November–December 1989* (World Wildlife Fund–Malaysia, Sabah Department, 1990).

38. Davies and Payne, *A Faunal Survey of Sabah*.

39. Andau, "Conservation of the Sumatran Rhinoceros in Sabah, Malaysia."

40. J. Payne, *Tabin Wildlife Reserve, Sabah: A Preliminary Management Plan* (Report to Sabah Forest Department, World Wildlife Fund–Malaysia, Sabah Department, 1986).

41. C. Jomitin, *A Report on a Survey of Rhinoceros in Tabin Wildlife Reserve Sabah 16–28 July 1991* (Report to the Sabah Wildlife Department, Malaysia, 1991).

42. M. N. Shukor, L. N. Ambu, and A. Tuuga, "Rhino Survey in Tabin Wildlife reserve, Lahad Datu," *Sabah Museum Monograph* 3 (1989): 77–81.

43. Payne, *Tabin Wildlife Reserve, Sabah: A Preliminary Management Plan*.

44. Khan, *Asian Rhinos*.

45. M. Borner, *A Field Study of the Sumatran Rhinoceros* Dicerorhinus sumatrensis: *Ecology and Behavior Conservation Situation in Sumatra* (Zurich: Juris Durck & Verlag, 1979); van Strien, *The Sumatran Rhinoceros* Dicerorhinus sumatrensis.

46. Alan Rabinowitz, *Rhino Survey in the Greater Danum Conservation Area* (New York: Wildlife Conservation Society, 1992).

47. Khan, *Asian Rhinos*.

48. Hamid Bin Ahman, Abd, "A Study of Current Abundance and Several Aspects of the Feeding Ecology of Sumatran Rhinoceros."

49. Payne, *Tabin Wildlife Reserve, Sabah: A preliminary Management Plan*.

50. S. Dawson, *Estimating Elephant Numbers in Tabin Wildlife Reserve, Sabah* (Report to the Sabah Wildlife Department, World Wildlife Fund–Malaysia, 1991).

51. D. G. Nichols et al., *Wildlife Trade Laws of Asia and Oceania* (Washington, D.C.: Traffic US, 1991).

52. G. Caughley, "Directions in Conservation Biology," *Journal of Animal Ecology* 63 (1994): 215–44.

53. See M. E. Soulé and B. K. Wilcox, eds., *Conservation Biology: An Evolutionary-Ecological Perspective* (Sunderland, Mass.: Sinauer Associates, 1980); O. H. Franklin and M. E. Soulé, *Conservation and Evolution* (Cambridge: Cambridge University Press, 1981); M. E. Soulé, *Conservation Biology: The Science of Scarcity and Diversity* (Sunderland, Mass.: Sinauer Associates, 1986); and Soulé, ed. *Viable Populations for Conservation* (Cambridge: Cambridge University Press, 1981).

54. Caughley, "Directions in Conservation Biology."

55. L. A. Maguire, U. S. Seal, and P. F. Brussard, "Managing Critically Endangered Species: The Sumatran Rhino as a Case Study," in *Viable Populations for Conservation*.

56. *The Jakarta Post*, August 9, 1994.

References

Ahmad, Abd, Hamid Bin. "A Study of Current Abundance and Several Aspects on the Feeding Ecology of Sumatran Rhinoceros (*Dicerorhinus sumatrensis harrisoni* Groves 1965) in the Danum Valley Field Centre Area Lahad Datu, Sabah." B.S. thesis, University Kebangsaan Malaysia, Sabah, 1991.

Andau, P. "Conservation of the Sumatran Rhinoceros in Sabah, Malaysia." *Rimba Indonesia* 21, no. 1 (1987): 39–46.
ANZEC Limited Consultants. *Sabah Environmental Management Plan for Tabin and Kulamba Wildlife Reserves*. Report to United Nations Development Program. Kota Kinabalu, Malaysia, 1992.
Bock, C. *Temples and Elephants: The Narrative of a Journey of Exploration Through Upper Siam and Lao*. London: Searle & Rivington, 1884.
Borner, M. *A Field Study of the Sumatran Rhinoceros* Dicerorhinus sumatrensis: *Ecology and Behaviour Conservation Situation in Sumatra*. Zurich: Juris Durck & Verlag, 1979.
Bowring, J. *The Kingdom and People of Siam*. Kuala Lumpur: Oxford University Press, 1969.
Caughley, G. "Directions in Conservation Biology." *Journal of Animal Ecology* 63 (1994): 215–44.
Cranbrook, G. G. *The Riches of the Wild: Land Mammals of Southeast Asia*. Singapore: Oxford University Press, 1987.
Davies, G., and J. Payne. *A Faunal Survey of Sabah*. Report to the Game Branch, Sabah Forestry Department. Kuala Lumpur: World Wildlife Fund–Malaysia, 1982.
Dawson, S. *Estimating Elephant Numbers in Tabin Wildlife Reserve, Sabah*. Report to Sabah Wildlife Department. Kuala Lumpur: World Wildlife Fund–Malaysia, 1992.
Fitzgerald, S. *International Wildlife Trade: Whose Business Is It?* Washington, D.C.: World Wildlife Fund–USA, 1989.
Foose, T. J. and Z. Z. Zainuddin. *International Studbook for the Sumatran rhino* (Dicerorhinus sumatrensis). Columbus, Ohio: The Wilds, 1993.
Franklin, O. H. and M. E. Soulé. *Conservation and Evolution*. Cambridge: Cambridge University Press, 1981.
Gervaise, N. *The Natural and Political History of the Kingdom of Siam*. Translated by John Villiers. Bangkok: White Lotus Press, 1989.
Groves, C. P. "Description of a New Subspecies of Rhinoceros, from Borneo, *Didermocerus sumatrensis harrisoni*." *Saugetierkundliche Mitteilungen* 13, no. 3 (1965): 128–31.
Harrison, T. "Borneo Fauna Anxieties." *Oryx* 3, no. 3 (1955): 134–37.
———. "Rhinoceros in Borneo and Traded to China." *Sarawak Museum Journal* 7, no. 8, (1956).
———. "A Future for Borneo's Wildlife." *Oryx* 8, no. 2 (1965): 99–104.
Harrison, T., ed. *Borneo Jungle: An Account of the Oxford University Expedition of 1932*. Oxford: Oxford University Press, 1988.
Hirth, F., and W. W. Rockhill. *Chau Ju-kua: His Work on the Chinese and Arab Trade in the Twelfth and Thirteenth Centuries, Entitled Chu-fan-chi*. St. Petersburg, Russia: Imperial Academy of Sciences, 1911.
Jomitin, C. *A Report on a Survey of Rhinoceros in Tabin Wildlife Reserve, Sabah 16–28 July 1991*. Report to the Sabah Wildlife Department. Kota Kinabalu, Malaysia, 1991.
Khan, M. *Asian Rhinos: An Action Plan for Their Conservation*. Gland, Switzerland: World Conservation Union, 1989.
Kiew, B. H. *A Survey of the Proposed Danum Valley National Park, Sabah*. Kuala Lumpur: World Wildlife Fund–Malaysia, 1976.
Loh, J., and K. Loh. "Rhino Horn in Taipei, Taiwan." *Traffic Bulletin* 14, no. 2 (1994): 55–58.
———. "A Spot Check on the Availability of Rhino Products in Guangzhou and Shanghai, China." *Traffic Bulletin* 14, no. 2 (1994): 79–80.
Maguire, L. A., U. S. Seal, and P. F. Brussard. "Managing Critically Endangered Species:

The Sumatran Rhino as a Case Study." In *Viable Populations for Conservation*, edited by M. E. Soulé. Cambridge: Cambridge University Press, 1987.

Marsh, C. W., and A. G. Greer. "Forest Land-Use in Sabah, Malaysia: An Introduction to Danum Valley." *Philosophical Transactions of the Royal Society of London* B 335 (1992): 331–39.

Mills, J. V. G. *Ying-Yai Sheng-Lan of Ma Huan*. Cambridge: Hakluyt Society, 1970.

Mjöberg, E. *Forest Life and Adventures in the Malay Archipelago*. London: George Allen & Unwin, 1930.

Nichols, D. G., K. S. Fuller, E. McShane-Caluzi, and E. Klerner-Eckenrode. *Wildlife Trade Laws of Asia and Oceania*. Washington, D.C.: Traffic U.S.A., 1991.

Nowell, K., Chyi Wei-Lien, and Pei Chia-Jai. *The Horns of a Dilemma: The Market for Rhino Horn in Taiwan*. Cambridge, England: Traffic International, 1992.

Payne, J. *Tabin Wildlife Reserve, Sabah: A Preliminary Management Plan*. Report to Sabah Forest Department. Kuala Lumpur: World Wildlife Fund–Malaysia, 1986.

———. *The Distribution and Status of the Asian Two-horned Rhinoceros* (Dicerorhinus sumatrensis harrisoni) *in Sabah Malaysia*. Project 3935. Kuala Lumpur: World Wildlife Fund–Malaysia, 1990.

———. *Report On a Survey of the Distribution of Rhinoceros in Southeastern Sabah, November-December 1989*. Kota Kinabalu, Malaysia: World Wildlife Fund–Malaysia, 1990.

Rabinowitz, A. *Rhino Survey in The Greater Danum Conservation Area*. New York: Wildlife Conservation Society, 1992.

Read, B. E. *Chinese Materia Medica: Animal Drugs*. Taipei: Southern Materials Center, 1982.

Rookmaker, L. C. "The Distribution and Status of the Rhinoceros, *Dicerorhinus sumatrensis* in Borneo—A Review." *Bijdragen Tot de Dierkunde* 47, no. 2 (1977): 197–204.

Schafer, E. H. *The Golden Peaches of Samarkand: A Study of T'ang Exotics*. Berkeley: University of California Press, 1963.

Shelford, R. W. C. *A Naturalist in Borneo*. London: T. Fischer Unwin, 1916.

Shukor M. N., L. N. Ambu, and A. Tuuga. "Rhino Survey in Tabin Wildlife Reserve Lahad Datu." *Sabah Museum Monograph* 3 (1989): 77–81.

Soulé, M. E. *Conservation Biology: The Science of Scarcity and Diversity*. Sunderland, Mass.: Sinauer Associates, 1986.

———. *Viable Populations for Conservation*. Cambridge: Cambridge University Press, 1987.

Soulé, M. E., and B. K. Wilcox, eds. *Conservation Biology: An Evolutionary-Ecological Perspective*. Sunderland, Mass.: Sinauer Associates, 1980.

Ta-Kuan, Chou. *The Customs of Cambodia (1296–1297)*. Translated by J. Gilman d'Arcy Paul. Bangkok: The Siam Society, 1993.

Talbot, L. "A Look at Threatened Species." *Oryx* 5, no. 4, 5 (1960): 153–295.

van Strien, N. *The Sumatran Rhinoceros* Dicerorhinus sumatrensis *(Fischer, 1814) in the Gunung Leuser National Park Sumatra, Indonesia*. Hamburg: Verlag Paul Parey, 1986.

van Wusthof, G. "Voyage Loitain aux Royaumes de Cambodge et Laouwen." *Bulletin de la Societe de Geographie* 6, no. 10 (1871).

Weedon, W. C. M. *A Journey Across British North Borneo Together with a Brief Account of Some Further Explorations in the Interior of that State*. Sabah State Archives. Kota Kinabalu, Malaysia, 1906. Unpublished.

Yen, Kun-Ying. *The Illustrated Chinese Materia Medica: Crude and Prepared*. Taipei: SMC Publishing, 1992.

Part III

People, Politics, and Wildlife

9

APPROACHES TO CONSERVING VULNERABLE WILDLIFE IN CHINA:
DOES THE COLOR OF CAT MATTER—IF IT CATCHES MICE?

Richard B. Harris

INTRODUCTION

How do Chinese value wildlife, and how might (oft-times divergent) Western values affect our efforts to assist the Chinese in its conservation? Do we really understand what we mean when we talk to Chinese of "conservation?"[1] For that matter, what do we mean by "wildlife?" What is the desired state of some piece of land, if something other than the status quo? And what mechanisms do we use to achieve our objectives, if a change in the behavior of a group of people is necessary for success?

My purpose here is to address broadly-defined "strategies" for conservation of wildlife in China, and to suggest that attitudes we in the Western world hold inevitably influence the choice of strategies we will conclude to be effective. Although I address approaches currently and potentially taken by the Chinese themselves, my primary audience is the concerned Westerner. While conservation of Chinese wildlife is ultimately China's responsibility, knowledgable Chinese are frank about needing help, and thus Westerners can legitimately claim an interest. Here, I argue that we in the West have often encouraged ineffective strategies for wildlife conservation in China because we have inappropriately applied our own terms of reference, and that, all too often, Chinese policy makers have made the mistake of taking our advice.

This article originally appeared in *Environmental Values* 5 (1996): 303–34. Reprinted by permission. Copyright © 1996 The White Horse Press. All rights reserved.

Of Cats, Mice, and Attainable Objectives

My subtitle comes from an ancient Chinese *chengyu* (proverb) about practicality. The old saying was dusted off a few years back by China's paramount leader, Deng Xiaoping, and now has become part of the contemporary Chinese lexicon. On hearing complaints about the resemblance between his "Socialism with Chinese Characteristics" to capitalism, Deng is said to have reminded his listeners, "It doesn't matter if the cat is black or white, so long as it catches mice."

For a number of reasons, it is an apt metaphor to keep in mind while examining strategies for wildlife conservation in China. It reminds us of the changes currently occurring in everyday Chinese life, particularly in the economic sphere to which Deng referred. Where once throngs dressed in drab Mao suits and caps, today's city dwellers spend much of their free time shopping for colorful and fashionable clothing. Where once "capitalist roaders" were reviled, today's roads are filled with free markets selling everything from vegetables to Rambo posters. Where once a bicycle was a high point on the consumer's wish list, today color televisions abound, portable cassette stereo players are almost ubiquitous, refrigerator sales are on the march, and one can only guess when air conditioners will replace fans and electric skillets will outsell iron woks.

The "cats and mice" metaphor is also appropriate because it reminds us that the dominant attitude most Chinese hold toward animals is utilitarian. Why have a cat? To catch mice, of course. If the Chinese appreciation of cats is limited to their mouse-catching ability, at least that ability provides a reason to value and conserve cats. (It hardly bears pointing out the obvious limitations of this attitude as the sole basis for conservation, particularly from the viewpoint of mice).

Viewed in one light, the "cats and mice" proverb would seem to be the logical equivalent to "the ends justify the means," a position that may be morally abhorrent. I don't intend for it to be interpreted that way. Rather, I see it as a clarion call to abandon orthodoxy, to loosen ourselves of whatever ideological bent we come armed with. We need not renounce passion, convictions, or persistence, but we would be well advised to recall that our actions and attitudes can frequently be counterproductive when applied to other countries if we fail to understand the cultural basis of local beliefs and the social constraints on notions we'd take for granted at home.

The noted African ecologist Richard Bell has summarized well the situation we find ourselves in: "Conservation is based on a conflict of value systems. If everyone was a conservationist, there would be no need for conservation." Further, that

> there are two main avenues towards resolution of these conflicts: one involves bringing our ideals more closely into line with physical, biological or economic realities, at the cost of compromising our conservation ideals; the other is to bring physical, biological and economic events[2] into line with our ideals with costs, in terms of management, enforcement and public relations, corresponding to the degree of the discrepancy between ideals and events.

Bell concludes that "What we need to do is to map out a strategy of the attainable."[3]

I don't pretend to know what is attainable for wildlife in China. But we can at least examine the possible strategies to conserve wildlife that one might encourage. I broadly lump these strategies into the three categories *legal*, *educational*, and *economic*, and argue that, while aspects of all three are ultimately necessary for successful conservation, Westerners have tended to overemphasize the first, have unwarranted faith in the second, and underutilize the potential of the third.

Further, I suggest that Westerners have focused unproductive criticism on Chinese utilitarian views of wildlife, which may not, in themselves, be detrimental to conservation. Meanwhile, we have remained silent on the most fundamental deficiency in current Chinese efforts to conserve, namely their inability to develop new, or foster existing, social institutions that can cope with protecting wildlife in the face of increasing demand for use and competition for space. I shall defend these contentions through the use of some examples and a general discussion on contemporary Chinese culture.

Before describing potential conservation strategies and their prospects for success in China, however, two quick overviews are needed: first, of the attitudes Chinese generally display toward wildlife; and second, of social factors operating in China that presently make conservation difficult, and indirectly constrain what is attainable. A review of the state of Chinese wildlife is not attempted here; suffice it to say that in the eastern, agricultural and urban section of China wildlife has been reduced to those few species able to survive in habitats dramatically altered by mankind, while in the arid, mostly pastoral areas, most indigenous species remain, but are generally reduced greatly in numbers.[4]

CHINESE ATTITUDES TOWARD WILDLIFE

Within the Chinese language are suggestions that, traditionally, wildlife and wilderness have been viewed negatively. The word *yeshengdongwu* ("wildlife") currently carries no great emotional baggage. However, it is suggestive that the adjective *ye* ("wild") also occurs in such derogatory terms as *yexing* ("unruliness") and *yexin* ("ambition," in the sense of "overweening," rather than "noble," ambition). Similarly, the word used for wilderness, *huangdi*, is equally accurately translated as "wasteland" or "place of desolation."

However, Chinese culture is hardly unique here. Roderick Nash has discussed at length how the traditionally Western values of wildlife and wilderness also were predominantly negative. He summarises, "If paradise was early [Western] man's greatest good, wilderness, as its antipode, was his greatest evil."[5] The Chinese language is considerably older and more resistant to change than European languages, so one would not be surprised to find ancient concepts embodied in current usage.

There have been no scholarly studies of contemporary Chinese attitudes toward wildlife,[6] yet a broad view of the literature and of language used in Chinese publications suggests that they are predominantly "utilitarian," and secondarily "dominionistic" and "aesthetic."[7] Simply put, most Chinese traditionally view wildlife in terms of its impact on human life and livelihood, and secondarily as objects of beauty, but only when under the control of man.

Utilitarian View

Others have already noted the predominance of the utilitarian view of wildlife among most Chinese. Writers in English have generally been explicit about this, Chinese less so. For example, Greer and Doughty concluded that

> Current trends in the utilisation of wildlife in China continue a tradition of satisfying material needs for meat, apparel, and medicinal and other products. . . . Decisions about conserving or protecting animals are therefore based largely upon utilitarian premises.[8]

Shen and colleagues noted that while "conservation figures importantly in the national development plan . . . in practice it is being promoted for utilitarian reasons," and later, "China's cultural heritage values wildlife . . . but the people have always adopted a utilitarian attitude."[9] In addition to the utilitarian attributes of wildlife that Westerners are familiar with (e.g., uses such as meat and fur), Chinese culture includes an elaborate relationship with wildlife for medicinal uses, for which there is no clear Western analogy.

But perhaps an even more convincing indication of the depth and breadth that the utilitarian view of wildlife holds on the Chinese mind comes from reading works in Chinese, including those not directly related to the issue of wildlife value. Sheng provides an overview of contemporary Chinese use of mammalian fauna; the list is long.[10] Almost invariably, when rationalizing research on one of China's many *zhengui* ("precious and valuable") species, the words *weile baohu* . . . ("in order to protect . . .") are followed immediately by . . . *yu heli liyong* (". . . and rationally use").[11] Wildlife is also more frequently referred to as a "resource" (*ziyuan*) in the Chinese technical literature than we are used to seeing in the West.

Evidence of the utilitarian view exists in the nonscientific literature as well. A short educational reader intended to simultaneously provide reading practice for elementary school students and indoctrinate them with the proper attitude toward nature and the environment serves as a good example.[12] The book follows the experiences of a group of youngsters as they are educated about environmental issues at the hands of the father of one of them, Mr. Lin, an environmental engineer. It features chapters focusing on water and air pollution, solid waste, protection of greenbelts, and other fashionable topics with environmental themes. Of course, there is also a chapter focusing on wildlife.

After hearing the children exclaim that such creatures as bears, snakes, and wild boars can be dangerous to people, Mr. Lin sagely interrupts them: "These animals can be dangerous, but they are also beneficial! Take the tiger, for example. People call it 'King of the Mountain,' but one could also say it's quite a treasure." The children protest, "But tigers threaten people!" Wanting to appear reasonable, Mr. Lin responds, "Yes, that's true, but the benefits to people from tigers are also great." What argument does he use to convince the youngsters? "The entire body of a tiger is a treasure! Why, one could say that the tiger is a drug store capable of curing 100 ills!"[13] In other words, not only do tigers threaten people, but they are useful to people, at least when dead.

Going on, Mr. Lin asks the children "Has any of you ever seen a snake?" Needless to say, the response of the children to the notion of a snake is less than positive; one wonders why the authors bring up this particular example if not to attempt to provide a more balanced view to the prevailing one of snakes as evil. But once again, it's the snake's "usefulness" that is called upon to justify its protection. Mr. Lin explains, "Snakes can cure diseases with the medicine produced from them, and they can also catch rats (which in turn, saves grain for human consumption). In southern China, we also like eating snake meat."[14] Nowhere in this lesson are messages of ecosystem integrity attempted or nonanthropocentric values suggested.

Dominionistic/Aesthetic

An additional attitude toward wildlife can be discerned among typical Chinese that is somewhere between Kellert's "dominionistic" and "aesthetic" types. Here, the notion is of nature the beautiful, but always in its tame state, as in a symmetrically ordered garden. Schafer reviewed the long history of captive rearing of animals from a diverse array of taxonomic groups in China. "Throughout, the controlled environment (e.g., zoo, garden) is seen as the preferable state to the natural."[15]

Some examples from translated literature[16] serve to illustrate this view of nature. From an essay describing the same forest in which the wolf-Buddhist pilgrim incident[17] occurred, comes this romantic view:

> The ... Forest is often compared to a vast natural zoo and arboretum combined. It is more than that. It is really a colorful, fantastic fairyland! It is grown with towering spruces and Chinese pines [sic] and other groves where scores of different shrubs were blossoming.... Oh, how luscious and attracting the berries are? And how gorgeous are the azaleas in this flowering season! During the ... journey on horseback ... we were always surrounded by hospitable animals. Some of them acted as our guides ... some as followers.... The small number of Kangbaren [Kham speaking people] ... who live here ... are kind at heart and never dream of harming the 'residents' in the forest, excepting, of course, vermin. This area is also a paradise for birds. Tibetan pheasants ... looked very much like bouquets of flowers.[18]

Note that throughout, in addition to the inability of the author to resist the reference to "bad" animals (vermin), nature is compared favorably to a garden or zoo, rather than vice versa. Another reference, here to the newly established

Qiangtang Reserve in Tibet, again uses the garden analogy, stretched as it might seem given what precedes it:

> With an average annual temperature between minus 3 degrees and zero Celsius and a yearly precipitation between 100 and 300 mm, Qingtang [sic] is too cold and dry to accommodate human souls, but it is what Chinese and American zoologists call "a rare animal garden."[19]

Similarly, a semipopular account of China's nature reserve system refers to such reserves (ziran baohuqu) set aside specifically for protection of endangered wildlife as "natural zoos."[20] In a recent study of attitudes in Japan, Kellert found a similar tendency to value wildlife most when it accords with preexisting visions of beauty and harmony.[21]

In recent years there have been expressions of an expanded set of values for wildlife, particularly among scientists. The same book on Chinese nature reserves includes a good many rationalizations for the existence of nature reserves that could easily come from western scientists, e.g., integrity of ecosystems, "balance of nature," and so on.[22] Another recent, somewhat more scholarly, Chinese book on nature reserves also includes a variety of rationales, including those that are recognizably ecological, scientific, and naturalistic.[23] Nonetheless, the tendency for utilitarian, aesthetic, and dominionistic values to dominate in most Chinese writing is clear.

Attitudes and Interpretation of Behaviors

It bears pointing out that the prevalence of utilitarian and dominionistic and/or aesthetic views are hardly uniquely Chinese, nor do they, in themselves, prevent good land management. In particular, the utilitarian outlook was recently found to be among the most common ones in a large-scale survey of the American public,[24] and even ranked number one when data were obtained from references in newspapers, rather than from random surveys of individuals.[25] Further, the prevalence of both the utilitarian and dominionistic attitudes expressed in the American press was found to increase as one looked further back in time toward the year 1900.[26] To the degree that China is still not as "modern" as America, one should therefore not be surprised to find these attitudes still strong.

What is worth noting, however, is that the utilitarian and dominionistic attitudes expressed in China come not merely from the general public, but also from the educated and interested segment of society that deals with wildlife, directly or indirectly.[27] Here, current American attitudes clearly diverge from Chinese. Among college-educated Americans, utilitarian and dominionistic attitudes ranked last, and among bird watchers, trappers, hunters, and those active in environmental/wildlife organizations, naturalistic and ecological attitudes ranked high.[28] Returning to the story at the beginning of this essay, one would not be surprised to find either a "typical" Chinese or Westerner hold that some animals are "good" and others "bad"; it would, however, be rare indeed to find such an attitude expressed by a contemporary Western zoologist closely associated with wildlife conservation.

We need not condone practices toward wildlife in contemporary China, but we only criticize them blindly if we fail to recall their genesis in prevailing attitudes. Thus, this review of Chinese attitudes toward wildlife is presented not as criticism, but rather because understanding attitudes may be useful in interpreting some of the actions in China that have been viewed by Westerners as greedy or malicious. For example, much is made of the willingness the Chinese express to engage in consumptive use of wildlife (i.e., hunting and trapping), even if populations are not known to be stable.[29] The notion that the Chinese government has been "using" pandas, by renting them out to foreign zoos (from which they receive both dollars and goodwill) is seen as simply immoral.[30] Controversy similarly surrounds Chinese attempts to satisfy demands for the traditional medicinal properties of bear gall bladders by raising fistulated bears in cages. One need not necessarily defend these types of Chinese activities; even from a strictly mechanistic point of view, we rarely have enough data to assess whether or not they threaten population viability or continuous functioning of ecosystems. But if viewed from the Chinese basis that animals are primarily to be used, then Chinese instincts to allow hunting when in doubt, or to earn money on "precious" species, are more understandable. At the least, such an understanding allows one to see these actions as arising from fundamental cultural attitudes, rather than solely from "greed, corruption, and stupidity."[31]

Another criticism of Chinese wildlife policy, one that finds resonance in the academic community in addition to readers of the popular press, is that excessive attention has been paid to captive breeding at the expense of habitat protection.[32] But should we be surprised to find this in a culture with a long history of domestication? Indeed, Chinese culture is deeply rooted in the intensive agriculture of the Yellow River basin, the "cradle" of Chinese civilization, which is considered to have advanced only after the great river was "controlled" by the emperor Yu.[33] With the land needing cultivation, ideas of mastery and manipulation over nature come easily. It is but a small transition to imagine that, like crops, wildlife can be better off under the care and kindness of educated and benevolent mankind.[34]

A Typology of Conservation Strategies

Potential strategies for wildlife conservation clearly overlap in their fundamental assumptions and suggested courses of actions. Classification systems are necessarily Procrustean; nevertheless, they may help to clarify our thinking. Lester Ross categorized what he termed "modes of implementation" into three he named "bureaucratic-authoritarian," "campaign-exhortation," and "market exchange."[35] He then analyzed how well each performed in forestry, water conservation, pollution control, and other aspects of environmental policy. Very briefly, the first can be equated with centralized control and directive; the second with the mass campaigns that typified the Cultural Revolution (such as planting trees in the "four arounds," and doing away with the "four pestilences"); and the third with policies that reward economically efficient practices, such as under the responsibility system.

Here, I similarly categorize possible wildlife conservation strategies very broadly into three: *legal, educational,* and *economic*. This typology is similar, although not identical to Ross's, with legal corresponding to his "bureaucratic-authoritarian," educational to his "campaign-exhortation," and economic, very broadly to his "market exchange." I note below some distinctions from Ross's approach.

Legal Strategies

This category includes most of what goes on under the label of wildlife protection in the West: regulations and proscriptions against killing wildlife and legal protection for land, including establishment of reserves. I contend that this strategy, obvious as it seems to us, has already proven its weakness, as seen by the fact that almost every species in China for which overexploitation is threatening oblivion is fully protected by law.[36] A wealth of nature reserves and protected areas have been set up;[37] additional areas are legally given status as provincially-recognized "no-hunting" areas. These legal actions are trumpeted both in Chinese and English, and appear to represent attempts to adopt perceived Western models.[38] Yet we know that poaching is ubiquitous and that it not only occurs in protected areas, but is often carried out by the very people hired for protection. I'll illustrate this point with three examples, two from Qinghai, one from Yunnan.

My musk deer study in Qinghai's Baizha Provincial Forest[39] was initially designed under the assumption, gained through incomplete interviews and a short visit to the area, that musk deer in Baizha were legally harvested in some, but not other, regions within the forest. In fact, the entire Baizha forest was off-limits to hunting, and musk deer were listed as Class 2 Protected Species.[40] However, in practice, protection—not only of wildlife but of all natural resources—was ineffectual. Timber harvest was indeed controlled, but only during the timber harvest season. As soon as the season ended, the "unofficial" season began, operated at a profit by forest guards, and the harvest during the latter seemed roughly equal to the former. As one Chinese friend there put it, when the official season ended, they "closed the front door and opened the back door."

A similar situation existed with regard to poaching wildlife. Musk deer poaching in the vicinity of the forest guard station was uncommon, but most participants were the forest guards themselves. The mountain directly back of the guard station also contained a small population of blue sheep; it was kept small by the constant pressure put on it by the well-armed forest guards. Our study team learned this from personal experience: Suspicions about the source of gunfire heard earlier in the day were confirmed one evening by a dinner of blue sheep meat—courtesy of the forest guards.[41]

However, one place in the forest benefited from virtually complete protection: the area surrounding the local Tibetan Buddhist monastery. It is now well documented that Tibetan Buddhism has suffered greatly under the rule of the Chinese since liberation, particularly during the Cultural Revolution. But since about 1980, religion has been allowed to flourish in Tibetan society again, and has rapidly re-emerged as a major component of Tibetan life. In Qinghai in particular, Tibetan monasteries have been restored in recent years, and young people are once again entering monastic life.[42] Within monasteries, individuals who are considered to be reincarnated deities, known locally as *akkas*[43] are revered, and they generally make clear that there is to be no hunting in the vicinity of their monasteries.

The effectiveness of this kind of unwritten protection was brought home one day during summer 1989, when a group of hunters from nearby Tibet-proper happened onto monastery land while hunting for deer. Despite being armed with military weapons, they were apprehended, arrested, and their guns confiscated by a group of unarmed monks. Their case was brought before the local *akka*, who then handed them over to the civil authorities for punishment. Meanwhile, the same civil authorities routinely turned a blind eye to poaching that occurred within the forest in areas further away from the monastery.

In reality then, there were three levels of protection given to wildlife within the so-called no-hunting area of Baizha Forest. In general, protection was quite limited, probably not materially different from that on any legally unprotected lands. Near the forest guard station, outsiders were deterred, but the forest guards themselves did most of the poaching. Near the monastery, protection was virtually complete. Notice, however, that according to law, these three areas should have been treated identically, but that economic (in the case of the guard station) and cultural/religious (in the case of the monastery) factors dictated the operative level of protection.

Another illustrative example was provided in a high valley in western Qinghai, locally termed Yeniugou ("wild yak valley"), where the indigenous high Tibetan plateau fauna still held on in moderate numbers.[44] Illegal hunting of all protected species occurred, the most damaging from outsiders, not local pastoralists. During summer, the valley became a travel route for itinerant gold-miners traveling with little food but lots of ammunition. Not only was their poaching illegal, but their entry into the valley itself was proscribed by law. A police checkpoint was set up during summer 1991 to stop the miners; it slowed them down, but didn't stop them. By September, the checkpoint had been removed and miners traveled freely. The legal means to prohibit entry existed; it simply wasn't effective.

In Yunnan, illegal hunting had, by the late 1980s, become such a problem that authorities ordered a complete, province-wide ban on hunting for three years, beginning on March 1, 1991.⁴⁵ While no statistics are available with which to assess compliance, experience in the mountainous area of western Yunnan (which, unlike much of central and eastern Yunnan, still had some wildlife of interest to subsistence hunters) suggested that the ban was not merely ineffective, it was completely *unknown*, both to hill-tribe hunters, and to the county and township officials entrusted with the responsibility of enforcing it.⁴⁶ Everywhere we traveled in hill country, local people with strong hunting traditions were continuing to hunt and always traveled armed. Government and forestry officials could hardly intervene even if they wished to, as they were usually full participants. Thus, it appeared that this dramatic gesture, an absolute ban on hunting, had effect only in places with no wildlife (i.e., cities), but was a nonstarter precisely in those places it was intended to apply.

These examples serve to illustrate that the peripheral concept of law in China, as discussed earlier, applies equally well to protection of wildlife and natural habitats. Wildlife finds itself outside the concerns of *li*, but not yet included within the slowly growing realm wherein *fa* is truly effective. Thus, I argue that when Westerners propose more laws, more legally protected lands, or simply better enforcement of current laws as a solution to China's loss of wildlife, they inappropriately apply concepts from our frame to theirs, failing to appreciate important differences between China's legal heritage and our own. An analysis of current patterns of Chinese wildlife conservation suggests that, put in terms of Deng's analogy, the legal system simply isn't a cat that catches many mice.

Educational Strategies

In admitting the deficiencies of a purely legal strategy, many—both Chinese and Western—argue that wildlife's salvation ultimately depends on education. But confusion lies in wait here, because education to Westerners implies formal schooling, while to Chinese it implies propaganda.

Schools. One problem with formal schooling is that many of the people we might be interested in reaching through the school system get only a rudimentary education, and some don't have access to schools at all. Among hill tribes in western Yunnan, while most children have access to elementary schools, their curriculum is largely limited to the stereotypical "reading, writing, and arithmetic." Educational materials are poor, and less than 2 percent of children move beyond the fifth grade level.⁴⁷ In the Tibetan area of Qinghai (where musk deer are being poached), young children who are being educated are not in government-run schools at all, but rather are in monasteries (and those not entering monasteries simply stay illiterate until adulthood).⁴⁸ I've already argued that monks are, on balance, a positive force for conservation, but it is difficult to imagine a Western-inspired environmental education curriculum having any relevance in this realm. In poor, rural China, school systems become weaker (to the point of nonexistence) as one travels closer to the communities that actually live and interact with the species about which they presumably need to be educated.

Secondly, the Chinese educational system, based largely on rote memorization and blind obedience to the teacher-authority, is particularly conservative and resistant to change. Curriculum reforms face considerable institutional and cultural obstacles.[49] Thus waiting for the school system to enhance typical rural Chinese appreciation for nature may take a long time. Further, even elevating "average" awareness about wildlife, (i.e., educating the "typical" person) is not sufficient because a very small minority, engaged in unrestricted wildlife exploitation, may be all it takes to cause tremendous damage.

Propaganda. When most Chinese use the word education in this context they are actually thinking more of the word "*xuanchuan*," which is translated as "propaganda," but doesn't carry with it nearly the negative connotation it does in English, being rather closer to propaganda's root, meaning simply "to propagate, or disseminate" information.[50] For example, despite efforts at disassociation with official government policy, a pilot educational program we attempted in Yunnan for hill tribesmen (and that, in truth, was primarily functioning as community organization) was routinely referred to by Chinese participants as *xuanchuan*.[51] The elementary children's reader referred to earlier provides another illustration.[52] In the first chapter, one of the young children announces with great pride that she has become a "propagandist" (*xuanchuanyuan*), and later she and her siblings march happily out to join the local neighborhood propaganda team instilling environmental responsibility in their fellow citizens. This is written with complete seriousness and none of the sense of embarrassment Westerners usually associate with the word "propaganda."

In fact, a fair amount of "propagating" basic wildlife conservation messages has already been done. While the era of "mass campaigns" may be over, one cannot travel far in China without encountering posters and billboards exhorting environmental responsibility. In Baizha, where musk deer are poached and timber illegally removed, a large sign in front of the field headquarters details the illegality of these acts. The highway leading to Yeniugou, where gold miners and others poach yaks and antelope, is posted in numerous places with large warnings against both poaching and illegal mining. In Kunming, where one can easily buy endangered species in free markets, neighborhood Party committees frequently use their public blackboards to display a summary of the national wildlife law. In cities, colorful banners extolling the virtues of environmental protection hang from doorways black with coal soot, while in villages near forests subject to unsustainable levels of logging, large characters painted on the adobe walls proclaim the importance of following the rules about conserving forests and soils (whatever they are).

Unfortunately, social propaganda in China has been overdone by so many, for so long, and toward so many issues, that it has lost any power it may have once had to influence public thought. Most Chinese are simply inured to the messages, having been anesthetized by the blaring loudspeakers from decades of failed campaigns. People routinely smoke underneath large "smoking strictly forbidden" signs, throw trash in places marked "absolutely no littering," and practice small-scale capitalism beside remonstrations to "uphold the socialist path."

Might the relative success enjoyed by Chinese family-planning programs constitute a counterargument? Propaganda to convince young couples of the virtue of stopping at one is ubiquitous in China. However, the history of family-planning programs in China argues *against* the effectiveness of propaganda, and rather for the effectiveness of economic incentives and institutional arrangements to alter social behavior. Propaganda notwithstanding, preferred family size continues to be substantially higher than sanctioned family size.[53] Family planning policy has been markedly less successful in rural areas than in the big cities,[54] and this is due not to differences in levels of propaganda, but rather to the degree of control that local governments have over people in the two areas. Family-planning policy is backed up by stiff economic sanctions on violators: Parents may incur substantial tax penalties or even lose their jobs, while extra children are denied the social benefits essential in Chinese society.[55] In cities, where most people work for a central work unit (*danwei*), these economic sanctions are relatively easily enforced. In rural areas, where farmers—substantially freed by the responsibility system—are more self-sufficient, it is easier to find ways around these penalties.

To summarize, one can hardly argue against the benefits of "education." As Chinese society changes, attitudes and values will also, no doubt, evolve in parallel. But Westerners are naive to believe that education alone will save species currently under threat.

Economic Strategies

Economic incentives are not be equated here with simple reliance on market forces. In fact, unrestrained market forces are often destructive of natural resources in whatever country they operate in,[56] and one should have no illusions that, operating on their own, they do any differently in China. Some economists have argued that, in an ideal state, markets can act to conserve natural resources, but even they have acknowledged the frequent occurrence of market failure in natural-resource management. Market failure, very briefly, is the failure of the commodity-exchange system to accurately measure the true value of a resource, or the failure of accurate information regarding its value to percolate through the system. Market failures occur with great regularity in Western economies with their relatively sophisticated web of oversight and regulatory functions,[57] so it strains credulity to believe that market failure doesn't also happen in China, which is struggling with markets after four decades of state control.

Thus I distinguish a very broad category of "economic" strategies from Ross's "market exchange implementation." The latter would "reaffirm the basic superiority of a system under which self-interested individuals strive for a maximum return."[58] Under such a strategy, "Prices . . . serve the same communication function for the environment as they do for raw materials, energy or labour." While not denying the pervasiveness of self-interest or the potential for pricing to drive conservation, I would submit that wildlife can rarely be privatized, and even private-property mimicking structures can be difficult to arrange. Rather, my intention in delineating economic strategies of wildlife conservation is to emphasize the importance of integrating wildlife into daily material life in its various forms. Such economic use may involve markets and prices, but may also involve subsistence use, religious use, and so on. An "economic" value for wildlife is that it be seen as part of the family household, rather than set apart, because it will be conserved most effectively when it is most integral to the lives of the people on whose survival it depends.

The quest for economic improvement currently pervades all aspects of Chinese life.[59] Economic motives for conservation, in the broadest sense but not excluding market-exchange mechanisms, are thus coincident with the daily lives of most Chinese people and likely to find favor with little need of disrupting other agendas. Further, while Chinese public policies call for protection of wildlife, governments allocate precious little funding to implement them.[60] Programs that can provide economic incentives, that perhaps make money rather than cost money, will have the best chance to be taken seriously by Chinese government officials.

However, even to bring up economics is to raise a red flag (pun notwithstanding) for much of the Western environmental movement. The patron saint of the American conservation movement, Aldo Leopold, cautioned against using economics as a rationale for conservation, arguing that it can never serve as a substitute for a land ethic.[61] While Leopold's argument is persuasive when applied in our culture, it may not be elsewhere. Below, I outline four potential economically based vehicles through which wildlife conservation might be more effectively realized in China, and a caveat underlying them all.

Generating foreign currency through ecotourism. Nature tourism, or "ecotourism," has often been touted as an effective way to make the protection of an area's natural resources economically attractive.[62] Such tourism is in its infancy in China, although each year sees more and more trips oriented toward scenic and cultural attractions. Little of this effort has gone toward promoting wildlife per se as a tourist attraction.[63]

Nature tourism no doubt has potential to make a greater contribution to wildlife conservation in China than it currently does. However, it's likely that its contribution will be limited, for a number of reasons. First, it is important to make sure that when we speak of ecotourism we are speaking clearly about the situation as it pertains in today's China, with the opportunities for wildlife viewing and photography as they actually are. Too often the only models for international nature travel are those of Kenya, Thailand, or Costa Rica, which have made a success of drawing in foreigners to national parks.[64] But even other countries in Africa and Central America have been unable to emulate the success of Kenya or Costa Rica in this regard, and many areas in China that could benefit from tourist revenues have neither the spectacular wildlife and ability to withstand the large numbers of people that characterize Kenyan parks, nor the relative accessibility to foreigners of Thai or Costa Rican parks.

A second problem with ecotourism is that the link between attracting tourists and wildlife protection is often blurred. For example, Qinghai province currently has one wildlife-oriented tourism attraction, Qinghai Lake's Bird Island (*niao dao*), an island long known for its rookery for geese, gulls, cormorants, and other waterfowl and shorebirds. Tourism has been promoted here for years, but thus far the income derived has been underwhelming, not even paying for the upkeep of the small hotel that has been built to accommodate the few tourists who do arrive. Further, it's not clear that whatever economic incentives thus far produced have resulted in effective conservation. Domestic livestock continue to trespass in the reserve (despite, of course, legal prohibitions), and water levels in Qinghai Lake decline yearly—probably due to nearby agricultural practices—thus potentially threatening the security of the birds' nesting areas by turning Bird Island into a bird peninsula.[65] The forces behind this threat have nothing to do with tourism, of course, but neither has the economic power of tourism had any effect in reversing the course of this environmental change.

Further, it is fairly well documented that most money generated by foreign tourism stays in the county of origin, and relatively little is available for use in the destination.[66] To obtain some data on the potential for nature tourism to contribute economically to conservation of Yeniugou, I conducted a survey of North American nature tourism operators offering nature tours in Asia, asking them to indicate the disposition of funds received for their tours. Results suggested that only a relatively small amount was available for direct conservation activities.[67]

It may be that, in a perfect market, simply raising the value of a pristine landscape or of the survival of a species is enough to ensure its perpetuation. But such a simple linkage is made at one's peril in China, where cost, value, and resulting actions often have little to do with one another. Thus, in addition to simply "making the resource valuable" in a generic, undefined sense, one must generate specifically earmarked funds for well-designed conservation activities. Only then would such a program make significant local contributions to wildlife conservation.

Generating foreign currency through international hunting. The mere mention of trophy hunting usually produces an audible gasp among many concerned with wildlife conservation. One study found that 80 percent of Americans surveyed disapproved of trophy hunting.[68] As well, current trends in environmental conservation emphasize biodiversity, as differentiated from the single-species management implied by trophy hunting programs.[69]

But whether or not we approve on moral grounds, it is difficult to deny that enthusiasm can often be generated for the conservation of wildlife, both on the part of local people and—perhaps equally importantly—local officials, from this perhaps somewhat mercenary activity.[70] Such hunting may potentially have the ability to provide funding not just for low-wage jobs, but for vehicles and gasoline for guards, direct compensation payments for habitat protection, and even what the Chinese call *xiao fei* (literally "small fee"; more prosaically, "bribe") for local administrators, which might also be built in to keep their enthusiasm high.[71] Under trophy-hunting schemes, harvest rates are kept low, typically to about 1 or 2 percent of the population.[72] It is therefore unlikely for such activity, by itself, to cause population declines.

Of course, trophy hunting contributes only additional mortality if its presence does not create an incentive for better conservation. However, there exists conceptual in addition to empirical reason for believing that trophy hunting programs often do provide a basis for precisely this incentive. In a system premised on maintaining a huntable number of older animals, the link between the health of the habitat producing them and the profit earned by harvesting them should be clear. Particularly in arid western China, in which most of the limited opportunities for trophy hunting remain, pastoralists are intimately familiar with the basic concept of sustained yield. They know that killing too many causes long-term loss and that healthy habitats are ultimately the source of production. Few of them retain a pure subsistence economy, however; most use money and enjoy having more of it. Thus, international hunting, rather than merely recreation for wealthy foreigners, can legitimately be viewed as the basis for an incentive-based conservation system.

Subsistence hunting. Western conservation groups often support a policy of reducing the dependence of local people on nearby wild resources,[73] and this fits well with Chinese paradigms of development that view progress and urbanization as virtually synonymous. Yet it is reasonable to question whether attempts to abolish, rather than to reform, bush-meat traditions are productive in the long term. Outside of nature reserves, lands that can support wildlife are under great pressure for conversion to higher-efficiency uses, and it will be increasingly difficult to thwart this tide if the bush meat produced by them is no longer valued. Thus I would argue for a policy of supporting, rather than suppressing, bush-meat traditions, but additionally, for assisting in the necessary transition toward effective means of monitoring and control in the face of increased populations and more efficient weaponry.

Using local demand: cottage industries. In many cases, wildlife can provide the basis of a successful cottage industry to enhance the living standards of relatively poor rural people. Chinese already do this with many species, albeit with essentially no monitoring or control, but tend to run afoul of international sensitivities when export of such wildlife products are considered.[74] In some cases, a financial connection might be enough to tip the balance away from an open-access mentality in favor of habitat protection and harvest control. For example, Tibetan villagers living in Baizha might be more successful in fending off the incursions of outsiders who poach local musk deer if they themselves were allowed to use the deer for local use or sale. These villagers appeared to possess the requisite social feedback mechanisms to be able to use, rather than abuse, a renewable resource such as musk.[75]

There are, or course, stumbling blocks to the expansion of these types of cottage industries. As in any economic enterprise, the profit earned must be worth the effort, and more than competing enterprises might earn for a similar amount of effort. Some of the Baizha pastoralists already participate in a small cottage industry based on the white-lipped deer. A factory in Germany that makes speciality buttons for clothing buys the dried and shed antlers of the deer, thus creating a market for them. Theoretically, this provides an incentive for live over dead deer, because live deer produce a new crop of shed and dried antlers each year. In actual fact, it is unclear how well the incentive is working. In 1990, shed hard antlers could be sold to the local government store for 80 yuan per kilogram; velvet antlers of the same species (but from animals poached during spring) would fetch up to 2,000 yuan per kilogram on the black market.[76]

Local Control of Resources: The Critical Caveat

The larger obstacle to any economic strategy however, be it ecotourism, international hunting, subsistence hunting, or some form of cottage industry, is that to succeed, local people must have secure access and control of the resource. For local activities involving "key" species to become established and expand, higher government sanction of local autonomy is ultimately required. For example, in Zimbabwe, the acclaimed CAMPFIRE program was made possible only after the central government passed a statute specifically devolving power to local councils, deliberately limiting its own power to control these resources.[77] Zimbabwean officials have gone so far as to formally embrace a policy of benefits from wildlife "*biased* towards the communities and areas that generate them."[78]

Issues of local control are often particularly cloudy in rural western China, where local people are usually different ethnically, linguistically, religiously, and even racially from the dominant Han majority. One can hardly begin to deal seriously with local control issues without asking sensitive questions that bear on Han chauvinism at best and racism at worst. As well, a history of oppression and civil war have often uprooted local people, and political events have sometimes caused wholesale movements of ethnic groups. For example, Yeniugou has changed hands twice during the past forty years, from Tibetans, to Kazakhs, and then to Mongols.[79] It is difficult to promote local control as an effective conservation policy when it's not clear who the local people are.

But even more problematic is that underlying any such discussion of allocation of benefits is the assumption that groups of people will naturally differ in their views, and often have competing claims to wildlife. This is a particularly difficult issue in China where the traditional Confucian view of society, reinforced by the Maoist legacy of exhortations to "unity" (*tuanjie*), is that society is fundamentally harmonious, and that conflicts arise only from lack of education (in the Confucian view), or from exploitation of the ruling classes (in the Communist view).[80] Even to admit that there exist competing claims requiring adjudication is to question this underlying philosophy.

Conclusion: An Approach for Westerners

I have argued from two premises: (1) That the prevailing view of Chinese toward wildlife is utilitarian, and (2) that numerous characteristics of contemporary Chinese society make implementation of agreed-upon social goals, such as conservation of wildlife, particularly difficult. These forces, together with increasing population and resource demands spurred by greater affluence, have reduced wildlife populations even in the relatively sparsely populated western part of the country, and threaten to cause additional extinctions. Given these premises, and the additional one that further reductions in wildlife populations are undesirable, it seems that one can propose changes along one of two lines: One can attack the utilitarian attitude and press for a transition to more humanistic, ecologistic, or other attitudes that would reduce demand for wildlife; or, alternatively, one can encourage the development and fostering of social institutions which can cope effectively with the increased demand.

Neither strategy would be easy to implement. But I submit that while the second strategy would be daunting, the first qualifies as nearly impossible. There is little reason to believe that fundamental values toward wildlife, formed and kept over millennia, can change quickly enough to avert the loss of many species. Admittedly, China will also find it difficult to develop rapidly social institutions that can more effectively conserve wildlife. However, the mechanistic requirements of doing so are more in keeping with current Chinese realities than are the philosophical requirements of a wholesale change in environmental values. As well, among the largely non-Han communities of rural western China, there already exist social institutions (in the form of religious or traditional strictures, communal use of land, and village-based taboos) that, with appropriate adjustments to new markets and increased access, could be mobilized to deal with the modern threats to wildlife.

Yet many Western observers have been unable to overcome an aversion to wildlife use, seemingly insisting not only that China conserve its wildlife, but that it do so for the correct reasons. The recent citation of China under the U.S. government's Pelly Amendment, and near invocation of trade sanctions, dealt nominally with the issue of Chinese trade in endangered species originating in other countries. However, lurking not far from the surface were issues of values: Not merely was China subverting CITES, but what were those old-fashioned Chinese doing consuming tiger and rhinoceros products in the first place?[81]

Some writers seem to have recommended that Chinese simply alter their way of valuing wildlife. For example, Shen and colleagues write that "The single most important factor *hampering* wildlife conservation in China is the traditional use of wild animals for medicinal purposes, meat and skins."[82] A later report commissioned by a prominent nature conservation organization, referring specifically to bears, concluded that "No campaign to slow or stop the bear trade will ever succeed without *understanding* Asian attitudes," but later suggested that "educational efforts should *promote the value* of bears as wild animals and *important members of the world's ecological community.*"[83] The authors thus legitimately suggest that we (presumably non-Asians) better understand Asian attitudes, but then seem to propose that we do so in order to show them the error of their ways.

Yet there is no inherent reason that utilitarianism cannot be consistent with conservation, at least as narrowly defined. If members of society use and value a renewable resource, they generally have reason to conserve it so as to be able to continue using it. Some Westerners argue that the utilitarian view of wildlife makes it more valuable dead than alive. But a similar argument could be made about the wildlife conservation system for many species in western North America, where the premise is largely that a broad constituency, much of which values wildlife primarily for food or sport, nevertheless provides the political muscle to conserve that wildlife's natural habitat. Such a system has not resulted in, for example, deer being worth more dead than alive.

The rub, of course, comes in institutionalizing incentives for players at all levels of the system to prioritize long-term benefits over short-term profits. It is worth pondering if this is possible in a society with so little tradition of power coming from the bottom, spread reasonably evenly over individuals assumed to be equal in prerogatives and responsibilities. Yet such does not seem totally unrealistic if management units are focused on local rural areas sufficiently small that they retain an internal sense of "*guanxi*," and particularly if management policy attempts to harmonize with the tradition of viewing wildlife primarily in terms of its material value. This challenge, in any case, is the real one facing China's vulnerable wildlife. But what has the West done? Although the Pelly confrontation (and Chinese response) resulted in some limited offers of official U.S. government training assistance to Chinese law enforcement, the greater need for outside technical help in developing local institutions that encourage sustainability of wildlife use remains unmet, having been virtually ignored by Western governments and NGOs alike.

Our view of Chinese attitudes toward wildlife suggest the strategies we will embrace. While advocating economic strategies, I am not arguing that legal and educational strategies be abandoned, if for no other reason than there will continue to exist conservation problems for which economic strategies, no matter how broadly interpreted, offer no solution.[84] Some species can be saved only by a commitment by Chinese society to restrain destructive behavior, codified in laws and maintained by supportive attitudes. For another, even in China, attitudes do change, and in time, it may be appropriate to focus on those wildlife values currently more in vogue in the West.

However, laws and education are not synonymous with conservation. Discouraging wildlife use is likely to have the effect of further alienating local rural people—already partly disenfranchised under the current system—from the wildlife whose fate lies in their hands. There are in China, as in Africa, "many paths to wildlife conservation."[85] It is important that we in the West look realistically at those paths that might work in present-day China. We may not feel totally comfortable with the ethic implied by Deng's "doesn't-matter-what-kind of-cat" sentiment, but that may be a small price to pay for ensuring that the "mice" in China—mice that are slowly but steadily destroying its priceless natural heritage—are not allowed free reign simply because we weren't satisfied with the color of the cat supplied to control them.

NOTES

1. There exists no Chinese term that carries with it the flavor of the word "conservation" as I use it here, i.e., benevolent use, or limitations of use patterns by current generations, in favor of nature. The most often seen term is simply *baohu*, which translates more nearly to the considerably more restrictive "protection." Occasionally, one also encounters *baohu* followed by *guanli* ("management"), which seems to imply that protection and use are not mutually exclusive.
2. To which I would add "cultural" events.
3. R. H. V. Bell, "Problems in Achieving Conservation Goals," in *Conservation and Wildlife Management in Africa*, ed. R. H. V. Bell and E. McShane-Caluzi (Washington, D.C.: U.S. Peace Corps, 1984).
4. For general reviews of the status of Chinese wildlife in English (which vary considerably in outlook), the reader is referred to C. E. Greer and R. W. Doughty, "Wildlife Utilization in China," *Environmental Conservation* 3 (1976): 200–208; G. P. Qu, "Wildlife Conservation in China," *Mazingira* 6 (1982): 52–61; G. B. Schaller, "Saving China's Wildlife," *International Wildlife* (Jan/Feb 1990): 30–41; G. B. Schaller, "In Search of the Kylin: The Endangered Wildlife of China," in *The Last Panda* (Chicago: University of Chicago Press, 1993); G. Rowell, "China's Wildlife Lament," *International Wildlife* 13 (1983): 4–11; S. Shen, E. D. Ables, and Q. Z. Xiao, "The Chinese View of Wildlife," *Oryx* 16 (1982): 340–47; Y. C. Wang, "Wildlife Protection in Northwest China," *New China Quarterly* 9 (1988): 108–10; W. H. Li and X. Y. Zhao, *China's Nature Reserves* (Beijing: Foreign Languages Press, 1989); and H. F. Xu and R. H. Giles Jr., "A View of Wildlife Management in China," *Wildlife Society Bulletin* 23 (1995): 18–25. More general treatments of China's environmental problems, which tend to focus little attention on conservation of native flora and fauna, include B. Boxer, "China's Environmental Prospects," *Asian Survey* 24, no. 7 (1989): 669–86; B. C. He, *China on the Edge: The Crisis of Ecology and Development* (San Francisco: China Books and Periodicals, 1991); and V. Smil, *China's Environmental Crisis: An Inquiry into the Limits of National Development* (Armonk, N.Y.: M. E. Sharpe, 1993).
5. R. Nash, *Wilderness and the American Mind*, rev. ed. (New Haven: Yale University Press, 1973), p. 9. Nash also contrasted this attitude with what he believed to be a more positive one on the part of traditional "Eastern" cultures, including China's. (Ibid., p. 20) However, others (for example, R. Guha, "Radical American Environmentalism and Wildlife Preservation: A Third World Critique," *Environmental Ethics* 11 [1989]: 71–83) have argued that Nash and other similar thinkers have used a selective reading of Eastern religions in their attempts to show that a biocentric view of nature is universal and has precedent.
6. S. R. Kellert, "Americans' Attitudes and Knowledge of Animals," *North American Wildlife and Natural Resources Conference* 45 (1980): 111–24.
7. Here and throughout, I use the ten-category typology of nonexclusive attitudes

toward wildlife developed by Kellert. These are termed "naturalistic" (interest in wildlife and the outdoors), "ecologistic," (interest in systems and interrelationships), "humanistic" (interest in individual animals, primarily pets), "moralistic" (concern for animal welfare), "scientistic" (interest in biological functioning of animals), "aesthetic" (interest in artistic or symbolic qualities), "utilitarian" (concern for practical, material value of animals), "dominionistic" (interest in mastery and control), "negativistic" (avoidance or dislike of animals), and "neutralistic" (indifference). See Kellert, "America's Attitudes and Knowledge of Animals."

8. Greer and Doughty, "Wildlife Utilization in China."

9. Shen, Ables, and Xiao, "The Chinese View of Wildlife."

10. H. L. Sheng, "Perspectives and Use of Chinese Mammal Resources" (in Chinese), *Chinese Wildlife* (January 1988): 3–5.

11. Examples include H. L. Sheng, H. F. Xu, and H. J. Lu, "Home Range and Habitat Selectivity of Musk Deer" (in Chinese), *Journal of the East China Normal University*, Mammalian Ecology Supplement (1990): 14–20; and Q. S. Yang, J. C. Hu, and J. T. Peng, "A Study on the Population Ecology of Forest Musk Deer in the Northern Hengduan Mountains" (in Chinese), *Acta Theriologica Sinica* 10 (1990): 255–62.

12. P. J. Qin and D. X. Qin, *Unusual Cases from the Countryside* (in Chinese) (Harbin: Heilongjiang People's Publishing House, 1985).

13. Ibid.

14. Ibid.

15. E. H. Schafer, "Hunting Parks and Animal Enclosures in Ancient China," *Journal of Economic and Social History of the Orient* 2 (1968): 318–43.

16. Here I use only material published originally in English to avoid any biases that might arise from my own translations.

17. See R. B. Harris, "Conservation Prospects for Musk Deer and Other Wildlife in Southern Qinghai, China," *Mountain Research and Development* 11 (1991): 353–58.

18. Q. Zhu, "In the Nangqen Primitive Forest," in *China's Majestic and Richly Endowed Qingliai Plateau*, trans. Jin Yushi (Xining: Qinghai People's Publishing House, 1987).

19. H. B. Li, "Tibet to Establish World's No. 1 Preserve," *Beijing Review*, April 9–15, 1990: 11–12.

20. Li and Zhao, *China's Nature Reserves*.

21. S. R. Kellert, "Japanese Perceptions of Wildlife," *Conservation Biology* 5 (1991): 297–308.

22. Li and Zhao, *China's Nature Reserves*.

23. X. B. Wang et al., *Theory and Practice of Nature Reserves* (in Chinese) (Beijing: Chinal Environmental Science Press, 1989).

24. Kellert, "Americans' Attitudes and Knowledge of Animals."

25. S. R. Kellert and M. O. Westervelt, "Historical Trends in American Animal Use and Perception," *North American Wildlife and Natural Resources Conference* 47 (1982): 649–64.

26. Ibid., p. 656.

27. In addition to the published examples previously cited, personal observations include a researcher who killed a snake upon finding it—even though he knew it to be a harmless species—because he "didn't like snakes," and a mammalian taxonomist who, while acknowledging the near extinction of tigers in China, allowed that he would himself use tiger bone for medicinal purposes, given the chance.

28. Kellert, "Americans' Attitudes and Knowledge of Animals"; S. R. Kellert,

"Affective, Cognitive, and Evaluative Perceptions of Animals," in *Behavior and the Natural Environment*, vol. 6: Human Behavior and Environment, ed. I. Alterman and J. F. Wohlwill (New York: Plenum Publishers, 1983).

29. D. Wade, "U.S. Hunters Invited to Kill Tibet's Rare Wildlife," *South China Morning Post*, September 1, 1992.

30. S. Begley, "Killed by Kindness," *Newsweek*, April 12, 1993.

31. Ibid.

32. Shaller, "Saving China's Wildlife"; Schaller, "In Search of the Kylin."

33. On this, as well as on the erroneous viewpoint of some Westerners that Chinese environmental actions were traditionally in accord with Taoist or Buddhist teaching, see Y. F. Tuan, "Discrepancies Between Environmental Attitude and Behavior: Examples from Europe and China," *Canadian Geographer* 12 (1968): 176–91.

34. This description of Han civilization can be contrasted with the largely pastoral societies living to the north and west. While far too much is often made of these distinctions, particularly in the case of Tibet, which is usually (and erroneously) taken to be a culture in which absolutely no use of wildlife occurs (see e.g., G. Rowell, "The Agony of Tibet," *Greenpeace* 15 (1990): 6–11), there is a fundamental difference. In the pastoral society, a livelihood is made by the use and husbandry of a naturally occurring resource, i.e., native vegetation. Wholesale modification is impossible; instead, pastoralists must adjust their use patterns to biological constraints (e.g., productivity, yearly fluctuations in water availability). Pastoralism thus provides a closer analogy to Western notions of "responsible wildlife use" than does intensive cultivation of crops.

35. L. Ross, *Environmental Policy in China* (Bloomington: Indiana University Press, 1998).

36. Most of the extant legal protection for wildlife nationally is condensed into a single piece of legislation, termed simply the "National Wildlife Law" (1998, *yeshengdongwu baohu fa*), which has been published in Chinese by Zhongguo linye chubanshe (Chinese Forestry Publishing House), Beijing. The main import of the law is to identify species of concern and/or interest, and to classify them as "key" species of either first or second class. The principal difference between these two classes is that permits to take specimens of first-class species must be obtained directly from central government representatives in Beijing, whereas permits for second-class species are issued by respective provinicial authoritites. In both cases, permits are to be issued only for limited scientific and educational reasons, and are never given to local people for subsistence or sport hunting (although they are occasionally issued to foreigners participating in sanctioned hunts). In effect, permits are reserved for those few with the ability to deal with bureaucracy in central cities; for rural people, all "key" species are effectively endangered species. Species not listed as "key" are not treated in any way. Provinces generally have parallel legislation, sometimes including additional areas specified as off-limits to hunting, and similarly ignoring species not listed as "key."

37. The exact number is difficult to determine because sources differ in what they include under the definition of a nature reserve. Wang et al, *Theory and Practice of Nature Reserves*, list 514 reserves, but a recently updated database at the Kunming Institute of Zoology (Chinese Academy of Sciences) includes 786 areas. (Zhu Jianguo, personal communication with the author, 1994)

38. X. Yan, "Status, Problems, and Development of China's Nature Preserves," *Journal of Northeast Forestry University* 17 (1989): 79–83; J. Zhu, "Nature Conservation in China," *Journal of Applied Ecology* 26 (1989): 825–33; Xu and Giles, "A View of Wildlife Management in China."

39. Harris, "Conservation Prospects for Musk Deer and Other Wildlife in Southern Qinghai, China."

40. See n. 50; also Qinghai Province, *Qinghai Province Wildlife Resources Protection and Management Regulations* (in Chinese), Report of the Standing Committee of the Seventh People's Congress of Qinghai Province, April 25, 1988.

41. This particular poaching incident may have set a record for the greatest number of simultaneous infractions: A fully protected species was killed by a forest guard in a no-hunting area with an illegal (military) weapon during a time of year in which no hunting of any kind was allowed, even outside the no-hunting zone.

42. I acknowledge that resistance to the full flowering of Tibetan Buddhism, both official and unofficial, still exists in China, and that the religion is nowhere near the powerful force it was prior to the 1950s. In particular, the large monasteries near Lhasa house only a fraction of their historic numbers of monks. As well, Chinese writing on the issue of Tibet continues to be strongly influenced by political concerns, and Tibetans generally suffer from oppressive practices. However, the popularly accepted Western notion that rehabilitated monasteries are merely facades to mislead tourists does not hold in Qinghai, where tourists are virtually absent. Pu's account of monasteries in Qinghai appears accurate, based on the evidence that his descriptions match my own obsercations for every one of the eight monasteries I visited, seven of them located in areas closed to tourists. See W. C. Pu, ed., *Tibetan Buddhist Monasteries in Qinghai and Gansu* (in Chinese) (Xining: Qinghai People's Publishing House, 1990).

43. Meaning "living Buddha," or *houfo*, in Chinese.

44. R. B. Harris, "Wildlife Conservation in Yeniugou, Qinghai Province, China," Ph.D. diss., University of Montana, Missoula, 1993; R. B. Harris and D. J. Miller, "Overlap in Summer Habitats and Diets of Tibetan Plateau Ungulates," *Mammalia* 59, no. 2 (1995): 197–212. Species included wild yak (*Poephagus mutus*), as the name suggests, but also Tibetan gaxelle (*Procapra picticaudata*), Tibetan antelope (*Pantholops hodgsoni*), Tibetan wild ass (*Equus kiang*), white-lipped deer (*Cervus albirostris*), blue sheep (*Pseudois nayaur*), argali (*Ovis ammon*), wolves (*Canis lupus*), and brown bear (*Ursus arctos*).

45. Yunnan Linyeting, "Notice on Protecting Wildlife Resources in Yunnan Province" (in Chinese), *Kunming* (March 1992): 226.

46. S. L. Ma et al., "Faunal Resources of the Gaoligongshan Region of Yunnan, China: Diverse and Threatened," *Environmental Conservation* 22 (1995): 250–58.

47. Ibid.

48. An admission of falling elementary school enrollment rates in Tibetan areas of Qinghai is found in W. Zhang, "Do a Good Job with Family Planning in Pastoral Regions, Improve the Competence Level of the Minority Population" (in Chinese) *Inquiries Into the Problems of Minority Populations in Qinghai* (Xining: Qinghain People's Publishing House: 1984).

49. An illustration is provided by comments made to me by a junior-high teacher in Kunming, Yunnan, while attending a training session on implementing an innovative environmental-education curriculum originally developed by a Western-based conservation organization. After about two weeks of enthusiastically participating in the session, I asked her for her reaction to the new ideas. She responded that she liked the concepts very much and was enjoying the training. However, she added that she didn't see how any of it could actually be incorporated into the city school system, because none of it related to standardized tests that were the focus of both teachers' and students' attention. She had not yet brought this to the attention of the Western trainers, and did not intend to.

50. On this point in a different context, see also P. West, "The Korean War and the Criteria of Significance in Chinese Popular Culture," *Journal of American–East Asian Relations* 1 (Winter 1992): 383–408.

51. R. B. Harris and S. L. Ma, "Initiating a Hunting Ethic in Lisu Villages, Western Yunnan," *Zoological Research* (Kunming), in press.

52. Qin and Qin, *Tibetan Buddhist Monasteries in Qinghai and Gansu*.

53. M. K. Whyte and S. Z. Gu, "Popular Response to China's Fertility Transition," *Population and Development Review* 13 (1987): 3.

54. S. R. Conley and S. L. Camp, *China's Family Planning Program: Challenging the Myths*, Country Study Series 1 (Washington: The Population Crisis Committee, 1992).

55. Conley and Camp report that "Rewards for couples who adhere to official policy can reportedly be substantial in the wealthier communities," while "small fines may be imposed for failure to use a contraceptive. Steeper financial penalties, denial of free social services, demotion, and other administrative punishments may be administered for continuing an unapproved pregnancy." (*China's Family Planning Program*)

56. C. W. Clark, "The Economics of Overexploitation," *Science* 181 (1973): 630–34.

57. R. M. Rasker, V. Martin, and R. L. Johnson, "Economics: Theory versus Practice in Wildlife Management," *Conservation Biology* 6 (1992): 338–49.

58. Ross, *Environmental Policy in China*.

59. O. Schell, *Discos and Democracy: China in the Throes of Reform* (New York: Pantheon Books, 1988); P. Link, "China's 'Core' Problem," in *China in Transformation*, ed. W. M. Wu (Cambridge: Harvard University, 1994); and N. D. Kristoff and S. WuDunn, *China Wakes* (New York: Times Books, 1994).

60. Unpublished data, interviews with regional staff at Gaoligongshan, Tonbiguan, and Niaodao Nature Reserves, and forestry officials in Yunnan and Qinghai.

61. A. Leopold, *A Sand County Almanac, With Essays on Conservation from Round River* (New York: Ballantine Books, 1966).

62. T. Whelan, *Nature Tourism: Managing for the Environment* (Washington, D.C.: Island Press, 1991).

63. Harris, "Wildlife Conservation in Yeniugou, Qinghai Province, China."

64. J. A. Dixon and P. B. Sherman, *Economics of Protected Areas: A New Look at Benefits and Costs* (Washington, D.C.: Island Press, 1990).

65. S. L. Bian, "Qinghai Lake, How Long Can You Endure It?" (in Chinese), *Zhongguo huanjing bao* (*China Environmental News*), December 6, 1990.

66. K. Lindberg, *Policies for Maximizing Nature Tourism's Ecological and Economic Benefits* (Washington, D.C.: World Resources Institute, 1991).

67. R. B. Harris, "Ecotourism versus Trophy-Hunting; Incentives Toward Conservation in Yeniugou, Tibetan Plateau, China," in *Integrating People and Wildlife for a Sustainable Future*, ed. J. A. Bissonette and P. R. Krausman (Bethesda, Md.: The Wildlife Society, 1995).

68. In contrast, 82 percent approved of traditional native subsistence hunting, and 85 percent approved of hunting for meat. See Kellert, "Americans' Attitudes and Knowledge of Animals."

69. For a critique of the single-species approach to conservation, see R. L. Hutto, S. Reel, and P. B. Landres, "A Critical Evaluation of the Species Approach to Biological Conservation," *Endangered Species Update* 4 (1987): 1–4. A counterargument is that highly priced species may act as proxies for others whose value cannot be appropriated, and thus

foster incentives to invest in natural habitats needed by both. (T. M. Swanson, "The Role of Wildlife Utilization and Other Policies for Diversity Conservation," in *Economics of the Wilds*, ed. T. M. Swanson and E. B. Barbier [Washington, D.C.: Island Press, 1992])

70. The Chinese Wildlife Conservation Association (CWCA), a quasi-governmental arm of the Ministry of Forestry, now devotes the majority of its time and efforts to the fledgling international hunting program. (Wang Wei, personal communication with the author, 1993).

71. The notion that hunting-based money-making schemes can raise consciousness about conservation was given additional support when an American businessman, operating completely independently of our research, arrived on Qinghai, suggesting a business proposition to expand international trophy hunting. Within a few weeks, the local authorities, who had heretofore shown only perfunctory interest in our wildlife research, were in high gear, eager to see our results. While their official mandate calls for protection of wildlife, they had no funds, no personnel, and most importantly, no great interest in doing so. Now, given the chance to provide funding to their agency (and perhaps themselves) through the proposed venture, they quickly became enthusiastic about the subject.

72. R. H. V. Bell, "Carrying Capacity and Off-Take Quotas," in *Conservation and Wildlife Management in Africa*; and B. Child, "Notes On the Safari Hunting Industry," in *Living With Wildlife: Wildlife Resource Management with Local Participation in Africa*, ed. A. Kiss, World Bank Technical Paper No. 130, Washington, D.C., 1990.

73. A European adviser to a survey team working among hill-tribesmen in Yunnan, representing a large international conservation organization, objected to his Chinese colleagues' collecting specimens for scientific purposes, on the grounds that it provided a poor model for the local residents. In his report back to the organization's headquarters, he wrote that killing animals was "the very behavior we're trying to halt."

74. Sheng, "Perspectives and Use of Chinese Mammal Resources."

75. Harris, "Conservation Prospects for Musk Deer and Other Wildlife in Southern Qinghai, China."

76. Ibid., p. 356.

77. Zimbabwe Trust, *CAMPFIRE Approach to Rural Development in Zimbabwe* (Harare: Zimbabwe Trust, 1990).

78. Ibid., emphasis added.

79. Harris, "Wildlife Conservation in Yeniugou, Qinghai Province, China."

80. See A. Nathan, *Chinese Democracy* (Berkeley: University of California Press, 1985), for an explanation of Chinese concepts of "democracy." Democracy in traditional Chinese thought is viewed as a means to use more efficiently the ideas and talents of people who are assumed to fundamentally agree with one another. By contrast, the Western conception of democracy is as a means by which policy can be made in the face of conflicts of interest that are assumed to characterize society.

81. Although the petition to invoke the Pelly Amendment over CITES infractions involving tiger and rhinoceros (species China had already nearly or completely extirpated) became widely publicized, it was soon followed by a pair of less-publicized requests to invoke Pelly with respect to species that China both possesses and views as useful resources: musk deer and a collection of species including bears, leopards, and pangolins. It seems, therefore, that some groups were attempting to use Pelly not merely to pressure Chinese compliance with CITES but to censure domestic use of these species as well. I don't argue that China has managed any of these species well: quite to the contrary. However, while CITES itself takes no position on the ethics of consumptive use of any species,

it seems that Pelly certification provided credibility to attempts to use CITES as a ruse to batter China on its environmental ethics, in addition to its enforcement of trade laws.

82. Shen et al., "The Chinese View of Wildlife," emphasis added.

83. J. A. Mills and C. Servheen, *The Asian Trade in Bears and Bear Parts* (Washington, D.C.: World Wildlife Fund–U.S., 1991), emphasis added.

84. For example, it is difficult to imagine their application to such species as tigers (*Panthera tigris*) and giant pandas (*Ailuropoda melanoleuca*), whose populations and rate of reproduction are too low to allow consumptive use, and which are nearly impossible to view or photograph in the wild.

85. T. M. Caro, "The Many Paths to Wildlife Conservation in Africa," *Oryx* 20 (1986): 221–29.

References

Begley, S. "Killed by Kindness." *Newsweek*. April 12, 1993.
Bell, R. H. V. "Problems in Achieving Conservation Goals." In *Conservation and Wildlife Management in Africa*, edited by R. H. V. Bell and E. McShane-Caluzi. Washington, D.C.: U.S. Peace Corps, 1984.
———. "Carrying Capacity and Off-Take Quotas." In *Conservation and Wildlife Management in Africa*, edited by R. H. V. Bell and E. McShane-Caluzi. Washington, D.C.: U.S. Peace Corps, 1984
Bian, S. L. "Qinghai Lake, How Long Can You Endure It?" (in Chinese). *Zhongguo huanjing bao (China Environmental News)*, December 6, 1990.
Boxer, B. "China's Environmental Prospects." *Asian Survey* 24. no. 7 (1989): 669–86.
Caro, T. M. "The Many Paths to Wildlife Conservation in Africa." *Oryx* 20 (1986): 221–29.
Child, B. "Notes On the Safari Hunting Industry." *Living With Wildlife: Wildlife Resource Management with Local Participation in Africa*, edited by A. Kiss. World Bank Technical Paper No. 130, Washington, D.C., 1990.
Clark, C. W. "The Economics of Overexploitation." *Science* 181 (1973): 630–34.
Conley, S. R., and S. L. Camp. *China's Family Planning Program: Challenging the Myths*. Country Study Series 1. Washington, D.C.: The Population Crisis Committee, 1992.
Dixon, J. A., and P. B. Sherman. *Economics of Protected Areas: A New Look at Benefits and Costs*. Washington, D.C.: Island Press, 1990.
Greer, C. E., and R. W. Doughty. "Wildlife Utilization in China." *Environmental Conservation* 3 (1976): 200–208.
Guha, R. "Radical American Environmentalism and Wildlife Preservation: A Third World Critique." *Environmental Ethics* 11 (1989): 71–83.
Harris, R. B. "Conservation Prospects for Musk Deer and Other Wildlife in Southern Qinghai, China." *Mountain Research and Development* 11 (1991): 353–58.
———. "Wildlife Conservation in Yeniugou, Qinghai Province, China." Ph.D. diss., University of Montana, Missoula, 1993.
———. "Ecotourism versus Trophy-Hunting; Incentives Toward Conservation in Yeniugou, Tibetan Plateau, China." In *Integrating People and Wildlife for a Sustainable Future*, edited by J. A. Bissonette and P. R. Krausman. Bethesda, Md.: The Wildlife Society, 1995.

Harris, R. B., and D. J. Miller. "Overlap in Summer Habitats and Diets of Tibetan Plateau Ungulates." *Mammalia* 59, no. 2 (1995): 197–212.
Harris, R. B., and S. L. Ma. "Initiating a Hunting Ethic in Lisu Villages, Western Yunnan." *Zoological Research (Kunming)*, in press.
He, B. C. *China on the Edge: The Crisis of Ecology and Development*. San Francisco: China Books and Periodicals, 1991.
Hutto, R. L., S. Reel, and P. B. Landres. "A Critical Evaluation of the Species Approach to Biological Conservation." *Endangered Species Update* 4 (1987): 1–4.
Jianguo, Z. Personal communication with the author, 1994.
Kellert, S. R. "Americans' Attitudes and Knowledge of Animals." *North American Wildlife and Natural Resources Conference* 45 (1980): 111–24.
———. "Affective, Cognitive, and Evaluative Perceptions of Animals." In *Behavior and the Natural Environment*. Vol. 6: Human Behavior and Environment, edited by I. Alterman and J. F. Wohlwill. New York: Plenum Publishers, 1983.
———. "Japanese Perceptions of Wildlife." *Conservation Biology* 5 (1991): 297–308.
Kellert, S. R., and M. O. Westervelt. "Historical Trends in American Animal Use and Perception." *North American Wildlife and Natural Resources Conference* 47 (1982): 649–64.
Kristoff, N. D., and S. WuDunn. *China Wakes*. New York: Times Books, 1994.
Leopold, A. *A Sand County Almanac, With Essays on Conservation from Round River*. New York: Ballantine Books, 1966.
Li, H. B. "Tibet to Establish World's No. 1 Preserve." *Beijing Review* April 9–15, 1990: 11–12.
Li, W. H., and X. Y. Zhao. *China's Nature Reserves*. Beijing: Foreign Languages Press, 1989.
Lindberg, K. *Policies for Maximizing Nature Tourism's Ecological and Economic Benefits*. Washington, D.C.: World Resources Institute, Washington, 1991.
Link, P. "China's 'Core' Problem." In *China in Transformation*, edited by W. M. Tu. Cambridge: Harvard University Press, 1994.
Ma, S. L., L. X. Han, D. Y. Lan, W. Z. Ji, and R. B. Harris. "Faunal Resources of the Gaoligongshan Region of Yunnan, China: Diverse and Threatened." *Environmental Conservation* 22 (1995): 250–58.
Mills, J. A., and C. Servheen. *The Asian Trade in Bears and Bear Parts*. Washington, D.C.: World Wildlife Fund–U.S., 1991.
Nash. R. *Wilderness and the American Mind*. Rev. ed. New Haven: Yale University Press, 1973.
Nathan. A. *Chinese Democracy*. Berkeley: University of California Press, 1985.
Pu, W. C., ed. *Tibetan Buddhist Monasteries in Qinghai and Gansu* (in Chinese). Xining: Qinghai People's Publishing House, 1990.
Qin, P. J. and D. X. Qin. *Unusual Cases from the Countryside* (in Chinese). Harbin: Heilongjiang People's Publishing House, 1985.
Qinghai Province. *Qinghai Province Wildlife Resources Protection and Management Regulations* (in Chinese). Report of the Standing Committee of the Seventh People's Congress of Qinghai Province. April 25, 1991.
Qu, G. P. "Wildlife Conservation in China." *Mazingira* 6 (1982): 52–61.
Rasker, R. M., V. Martin, and R. L. Johnson. "Economics: Theory versus Practice in Wildlife Management." *Conservation Biology* 6 (1992): 338–49.
Ross, L. *Environmental Policy in China*. Bloomington: Indiana University Press, 1988.
Rowell, G. "China's Wildlife Lament." *International Wildlife* 13 (1983): 4–11.

———. "The Agony of Tibet." *Greenpeace* 15 (1990): 6–11.
Schafer. E. H. "Hunting Parks and Animal Enclosures in Ancient China." *Journal of Economic and Social History of the Orient* 2 (1968): 318–43.
Schaller, G. B. "Saving China's Wildlife." *International Wildlife* (January/February 1990): 30–41.
———. "In Search of the Kylin: The Endangered Wildlife of China." In *The Last Panda*. Chicago: University of Chicago Press, 1993.
Schell, O. *Discos and Democracy: China in the Throes of Reform*. New York: Pantheon Books, 1988.
Shen, S., E. D. Ables, and Q. Z. Xiao. "The Chinese View of Wildlife." *Oryx* 16 (1982): 340–47.
Sheng, H. L. "Perspectives and Use of Chinese Mammal Resources" (in Chinese). *Chinese Wildlife* (January 1988): 3–5.
Sheng, H. L., H. F. Xu, and H. J. Lu. "Home Range and Habitat Selectivity of Musk Deer" (in Chinese). *Journal of the East China Normal University Mammalian Ecology Supplement* (1990): 14–20.
Smil, V. *China's Environmental Crisis: An Inquiry into the Limits of National Development*. Armonk, N.Y.: M. E. Sharpe, Inc., 1993.
Swanson, T. M. "The Role of Wildlife Utilization and Other Policies for Diversity Conservation." In *Economics for the Wilds*, edited by T. M. Swanson and E. B. Barbier. Washington, D.C.: Island Press, 1992.
Tuan, Y. F. "Discrepancies Between Environmental Attitude and Behavior: Examples from Europe and China." *Canadian Geographer* 12 (1968): 176–91.
Wade, D. "U.S. Hunters Invited to Kill Tibet's Rare Wildlife." *South China Morning Post*, September 1, 1992.
Wang, Y. C. "Wildlife Protection in Northwest China." *New China Quarterly* 9 (1988): 108–10.
Wang, X. B., J. M. Jin, L. B. Wang, and J. S. Yang, eds. *Theory and Practice of Nature Reserves* (in Chinese). Beijing: China Environmental Science Press, 1989.
Wei, W. Personal communication with the author, 1993.
Wells, M. and K. Brandon. *People and Parks: Linking Protected Area Management with Local Communities*. Washington, D.C.: The World Bank/The World Wildlife Fund/U.S. Agency for International Development, 1992.
West, P. "The Korean War and the Criteria of Significance in Chinese Popular Culture." *Journal of American–East Asian Relations* 1 (winter 1992): 383–408.
Whelan, T., ed. *Nature Tourism: Managing for the Environment*. Washington, D.C.: Island Press, 1991.
Whyte, M. K., and S. Z. Gu. "Popular Response to China's Fertility Transition." *Population and Development Review* 13 (1987): 3.
Xu, H. F., and R. H. Giles Jr. "A View of Wildlife Management in China." *Wildlife Society Bulletin* 23 (1995): 18–25.
Yan, X. "Status, Problems, and Development of China's Nature Reserves" (in Chinese). *Journal of Northeast Forestry University* 17 (1989): 79–83.
Yang, Q.S., J.C. Hu, and J.T. Peng. "A Study on the Population Ecology of Forest Musk Deer in the Northern Hengduan Mountains" (in Chinese). *Acta Theriologica Sinica* 10 (1990): 255–62.
Yunnan Linyeting. "Notice on Protecting Wildlife Resources in Yunnan Province" (in Chinese). *Kunming* (March 1992): 226.
Zhang, W. "Do a Good Job with Family Planning in Pastoral Regions, Improve the Com-

petence Level of the Minority Population" (in Chinese). In *Inquiries Into the Problems of Minority Populations in Qinghai*. Xining: Qinghai People's Publishing House, 1984.

Zhu, J. "Nature Conservation in China." *Journal of Applied Ecology* 26 (1989): 825–33.

Zhu, Q. "In the Nangqen Primitive Forest." In *China's Majestic and Richly Endowed Qingliai Plateau*, translated by Jin Yushi. Xining: Qinghai People's Publishing House, 1987.

Zimbabwe Trust. *CAMPFIRE Approach to Rural Development in Zimbabwe*. Harare: Zimbabwe Trust, 1990.

10

LOCAL CHALLENGES TO GLOBAL AGENDAS:
CONSERVATION, ECONOMIC LIBERALIZATION, AND THE PASTORALISTS' RIGHTS MOVEMENT IN TANZANIA

Roderick P. Neumann

I do not want to spend my holidays watching crocodiles. Nevertheless, I am entirely in favor of their survival. I believe that after diamonds and sisal, wild animals will provide Tanganyika with its greatest source of income.
Julius K. Nyerere, ca. 1961[1]

Apartheid is [the] superiority of one human being over another human being. In the Ngorongoro Crater case [it] is [the] superiority of wild animals over human being[s].
Tipilit ole-Saitoti, 1994[2]

INTRODUCTION

The history of wildlife conservation and the emergence and growth of a mass tourism industry are interwoven in Tanzania. Tourism was first established under colonial rule as an industry catering to the needs of Westerners coming to observe or hunt exotic animals. It later became a key industry in the development plans of the independent government. Politically, this plan carried potential hazards, since many Africans saw tourism as exhibiting the worst aspects of the old colonial order of European dominance and African subservience.[3] The inherent conflicts are exemplified by the epigraph above, wherein one pastoralist organizer goes so far as to compare wildlife conservation in present-day Tanzania's Ngorongoro Conservation Area to apartheid. Yet, in the preceding epigraph, we find the "father of African socialism," former Tan-

---

This article originally appeared in *Antipode* 27, no. 4 (1995): 363–82. Reprinted by permission. Copyright © 1995 Blackwell Publishers. All rights reserved.

zanian President Julius Nyerere, wholeheartedly embracing wildlife tourism as a major source of foreign exchange.

Recent changes in Tanzania's political economy have sharpened the apparent contradictions among wildlife conservation, revenue generation, and social justice revealed in these quotes. Economic "liberalization," exemplified by the new investment code, has opened the floodgates for foreign investment in lodge development in the country's parks and reserves. Privatization of the tourism industry and the private titling of land are providing the basis for a new phase of rapid accumulation, with the bulk of profits captured by foreign-owned companies. Simultaneously, "democratization" is offering new opportunities for people to voice their discontent over this process. Recent years have witnessed an unprecedented rise in the number of political parties and nongovernmental organizations (NGOs),[4] many of which have focused their agendas on issues of cultural survival and economic and social justice. One of the most urgent narratives is produced by an increasingly organized and vocal pastoralist society which has suffered the loss of prime grazing lands to wildlife conservation setasides, only to watch the resulting flow of tourism profits bypass their devastated communities.

I argue in this paper that Tanzania's ongoing political-economic transition has the potential to significantly reconfigure the terrain of power in the decades-old conflict between local land rights and state-directed wildlife conservation. To demonstrate precisely how this reconfiguration is unfolding, I begin with a historical synopsis of the politics of wildlife conservation in Tanzania. Next, I present two sections on "liberalization" and "democratization," giving emphasis in each to the relevance for wildlife tourism and conservation. These are followed by a section which uses the new pastoralists' rights organizations as examples for analyzing political activism in Tanzanian society, giving particular attention to their rhetorical challenges to state wildlife conservation policies and tourism. In the closing section, I argue that the space of nature—national parks and related protected areas—currently provides the center stage for political action in much of rural Tanzania. Upon these sites, evolving debates over nature protection, privatization, economic justice, ethnic politics, and cultural survival converge and are struggled over, often with contradictory outcomes.

Conservation Without Representation

The British took control of Tanzania (then Tanganyika) as a League of Nations Mandated Territory following World War I. Almost immediately, they regazetted the game and forest reserves that their German predecessors had established and further elaborated the game and forest laws. Wildlife conservation in the form of reserves and hunting laws were thus part of British rule from its inception. During the years of British control, wildlife conservation was characterized by top-down, often coercive policies that displaced African settlement and land use.[5] From about 1930 well into the 1950s, these policies were, to a significant degree, the product of intervention by a group of international conservationists headquartered in London.[6]

One effect of state conservation policies throughout the colonial period was to outlaw African rights of access to land and natural resources.[7] Often, the Africans who were most affected were the last to be informed about the transfer of control over access. For example, a colonial officer commenting on Sukuma lands in Serengeti National Park observed that the Sukuma "were arbitrarily deprived of [their lands] years ago, without any prior consultation with their local representatives."[8] The lack of due process and the repressive conditions of colonial rule limited the potential for organized political responses by peasant farmers and pastoralists who objected to losing ancestral lands to national parks and reserves. Thus, "everyday forms of resistance"[9] such as illegal hunting, grazing trespass, and fuel wood theft prevailed in colonial protected areas.[10] As Scott argues, these types of actions are aimed not at reforming the legal order, but at "undoing its application in practice."[11] While this conceptualization of resistance tends "to lump together many actions with different intentions and outcomes,"[12] it nevertheless advances our understanding of African responses by recognizing explicitly that wildlife conservation is political.[13] As Blaikie explains: "Land-using practices are economic activities, in which political-economic relations may be crucial. . . . Conservation policies *must* affect these relations."[14] Viewed from this perspective, violations of wildlife conservation laws are not merely the result of ignorance or population pressure, but are often highly politicized acts of protest aimed at policies which threaten rural livelihoods.[15]

Another characteristic of colonial policies was that they limited African involvement in state wildlife conservation activities to employment as game scouts and reserve guards. This aspect, combined with the history of local resistance to conservation laws, limited the ability and desire of the postcolonial government to take over the administration of wildlife conservation programs. Thus, as independence approached, it was clear to international conservationists that the new government would need persuasion and assistance. International environmental NGOs, such as the International Union for the Conservation of Nature and Natural Resources (IUCN; now the World Conservation Union) and the African Wildlife Foundation (AWF), began to aid the decolonized nations in planning and managing their national parks and conservation programs. Nyerere embraced their ideas of wildlife conservation in a 1961 speech known as the Arusha Manifesto, which reads, in part:

> In accepting the trusteeship of our wildlife we solemnly declare that we will do everything in our power to make sure that our children's grandchildren will be able to enjoy this rich and precious inheritance.

This passage is followed immediately by an invitation to outside intervention:

The conservation of wildlife and wild places calls for specialist knowledge, trained manpower and money and we look to other nations to cooperate in this important task.

Conservationists were delighted by the manifesto and it continues to be cited in their documents and publications as a positive example of African government interest and cooperation in protecting wildlife. Usually unmentioned, however, is the fact that it was written for Nyerere's speech by members of Western conservation organizations.[16]

The manifesto was delivered as part of the initiation of the IUCN's African Special Project. This campaign helped to maintain and expand the conservation programs of the newly independent African states. The IUCN and other international organizations focused much of their attention on Tanzania, funneling money and technical support for protected areas establishment and management. In the years following independence in 1961, these NGOs aimed a great deal of effort toward training government bureaucrats and technicians. As a result, they helped to create in Tanzania an elite class of conservation bureaucrats, trained in Western ideologies and practices of natural resource conservation.[17] During the first two decades of independence, this new class of conservation officials was confronted by the continued resistance of peasants and pastoralists. In the early 1980s, conservationists responded by increasingly emphasizing "local participation" and "community development" as key to wildlife protection.[18] By that time, however, an African administrative system which essentially replicated the top-down, repressive practices of colonial rule had become firmly entrenched.

To elaborate on the last point, conservation goals in Tanzania have often been realized through mass relocations of African populations from areas designated as parks or reserves.[19] Pastoralists, principally Maasai, have perhaps been the most severely affected by protected area establishment, as grazing lands which once overlapped with prime wildlife habitat are now off-limits.[20] As an illustration of how this practice continues, in 1988 Maasai pastoralists were forced out of the Mkomazi Game Reserve after refusing to obey a government eviction order.[21] The original 1951 order which created the reserve legally guaranteed the residents of the area the right to continue living and grazing livestock there. By the 1970s, wildlife officials were claiming that the reserve had become overrun by pastoralists and their cattle, although this is strongly disputed.[22] In 1976 the reserve manager informed the pastoralists they must move out. Resident pastoralist efforts to get the eviction decision reversed were answered with threats from armed Game Department agents.[23] Finally in 1987 a directive from the Ministry of Natural Resources and Tourism canceled all previous permits for grazing and residency.[24] In July of 1988, over five thousand pastoralists were evicted from Mkomazi, many by force after refusing to leave voluntarily.

The potential to monopolize revenue from tourism helps to explain why the government did not pursue a more participatory and cooperative approach to conservation. Tanzania National Parks' (TANAPA) first postcolonial director recognized the role of wildlife in generating revenue, and under his guidance the agency sought to "encourage by every means the growth of a tourist industry."[25] For the government, park tours became an exportable commodity with great potential to fuel state accumulation. To this point, an official wrote that the purpose of the parks "is the earning of foreign exchange, in the same way that one looks upon the exports of coffee, sisal, cotton, tea or diamonds."[26] As the opening epigraph shows, Nyerere himself was quite unambiguous in his position on the role of wildlife in bringing in foreign exchange.

There has been, in sum, a distinct continuity between the colonial and postcolonial situations. The independent government has not significantly altered park and wildlife laws, boundaries are mostly unchanged, forced relocations persist, African personnel are trained in practices developed in the West, and international conservation organizations still play a critical role. As a consequence of the coercive nature of wildlife conservation, "everyday forms of resistance" to national park establishment and management policies have persisted throughout the period of independence. Despite the prevalence of resistance, there has been no organized and coordinated "movement" to reclaim control of pastures and forests from the state. As we will see below, this may be about to change.

TANZANIA'S CHANGING POLITICAL ECONOMY

Much has been written on Tanzania's attempt at development through a uniquely African style of socialism, labeled by former President Julius Nyerere as *Ujamaa*.[27] This form of "statism" was partly characterized by the expansion of the state sector (including nationalizations of banking, agriculture, and industry) combined with the monopolized control of political organizing by *Chama Cha Mapanduzi* (CCM, Party of the Revolution).[28] During most of the independence period, CCM (or its predecessor, TANU) was the sole legal party and all mass political organizations (women, workers, youth) were under its leadership. Essentially, the mass of peasants and workers were denied the right to organize independently, and thus could not counter the formation of a bureaucratic ruling class. The party leadership and state bureaucrats (often interchangeable) facilitated their control of popular politics through coercive practices reinforced by a legitimating populist discourse which represented Tanzania as a classless, egalitarian society.[29]

Today, Tanzania's political landscape is reconfigured as the country moves toward a multiparty political system and subjects itself to the International Monetary Fund's (IMF) conditionalities. The seeds of recent change can be traced to 1985 when Nyerere resigned from the presidency and his successor, Ali Hassan Mwinyi, agreed to adopt the IMF's structural adjustment program. Pressure from foreign aid donors and within CCM to retreat from single-party rule increased in the late 1980s, and in May 1992 the Tanzanian Parliament approved a multiparty system. I want to examine the linked processes of "liberalization" and "democratization" in greater detail and investigate the implications for the transformation of ecological politics in Tanzania.

Liberalization

When President Mwinyi succeeded Nyerere, he initiated a series of economic "reforms" demanded by Tanzania's agreement with the IMF. The resulting new investment code offers foreign investors exemptions from import duties, tax-free periods, and no restrictions on the repatriation of capital gains. The consequences for increased foreign direct investment (FDI) and market deregulation are typical of those experienced by Third World countries across the globe.[30] In Tanzania, however, there are particular ecological ramifications because of the emphasis on wildlife tourism as the cornerstone of economic development. The IMF reforms have already helped to dramatically restructure the tourism industry.

For twenty years, the tourist industry had been controlled by a government parastatal, the Tanzania Tourist Corporation (TTC), which owned and operated fifteen hotels and lodges in the national parks and elsewhere. In his push to privatize the sector, President Mwinyi fired the general manager of the TTC and demoted other officials in September of 1989.[31] Finally, in May of 1993, the TTC was dissolved and reconstituted as the Tanzania Tourist Board (TTB), a promotional office for private investors. The dissolution of the TTC combined with the new investment code has resulted in a massive infusion of foreign capital in tourism. Even before the parastatal was eliminated, the government had brought in the French hotel chain Accor/Novotel as partners in the management of seven of the TTC's hotels and lodges, including the main facilities in Ngorongoro Conservation Area and Lake Manyara and Serengeti National Parks.[32] Management and renovation of the lodges is supported by a $16.5 million loan from the German Investment Bank, the Swiss International Finance for Development in Africa, and the East African Development Bank.[33] Another $35 million in financial backing is coming from the European Investment Bank and the African Development Bank.

In the fall of 1993, the Investment Finance Corporation (IFC, a member of World Bank), the Aga Khan Fund for Economic Development (AKFED), and the United Kingdom's Commonwealth Development Corporation (CDC) agreed to collectively invest $33 million in tourism development.[34] The project involves the construction of three lodges in Serengeti and Lake Manyara National Parks and Ngorongoro Conservation Area and a tented camp in Serengeti. An AKFED affiliate, Tourism Promotion Services Tanzania, will construct the lodges and Serena Promotion Services of Switzerland will manage them as part of their chain of four- and five-star lodges in East Africa.

Tanzania's wildlife is now closer to becoming the profit-generator it was always hoped to be. A recent report based on information from the United States Embassy in Dar Es Salaam tells American business that "tourism is perhaps Tanzania's most attractive investment opportunity, with several of the world's largest game reserves, a variety of resorts, hotels and ecotourism."[35] Tourism is currently Tanzania's fastest growing industry, expanding by 600 percent from 1985 to 1990.[36] Revenues rose by nearly half again in 1991 when the government received $94.7 million from tourism.[37] Figures from Tanzania's Ministry of Tourism, Natural Resources, and Environment put 1992–93 financial year earnings at $129 million. The government's goal is to generate $500 million annually, which is about 60 percent of what the country currently receives in foreign aid.

With the reduction in state expenditures under IMF-imposed conditionality, the importance of international conservation NGOs in developing and funding projects will likely increase. These groups have also joined in celebrating the triumph of the market, reversing their traditional position in which state regulation is seen as essential for reining in the destructive forces of capitalism[38]. Increasingly, the formulation of conservation projects by these groups reflects the discourse of "market idolatry"[39] which now prevails among First World development institutions. Conservationists interested in Third World environmental problems are basing the future hope of conservation on the commodification of resources.[40] In 1989 in Tanzania, the IUCN conducted a study which estimated the gross annual value of wildlife at over $120 million and encouraged further state and private investment in this sector as a way of making conservation profitable. Thus, the frenzy of privatization prompted by the IMF meshes well with conservationists' vision for wildlife in Tanzania.

Not everyone, however, is pleased with the results of structural adjustment, or what Tanzanian scholar Issa Shivji calls "the third great rape" of Tanzania (following slavery and colonialism).[41] The privatization of land ownership, in particular, is now a major source of rural conflict. A recent government report on land matters points out that a major result of "liberalization is the acquisition of land by foreign investors, a situation that is clouded by lack of land records."[42] Politicians have been accused of granting ownership transfers of large tracts of land without following legal procedures.[43] Recently, for example, an unauthorized leasing of Loliondo Game Controlled Area in Ngorongoro District to an army officer from the United Arab Emirates for use as a private hunting reserve became a national scandal.[44] Land scams like "Loliondogate" (the local media's label for the scandal) have reverberated throughout rural areas and have significantly influenced the structure of political organizing in "democratizing" Tanzania, particularly among displaced pastoralists.

Democratization

As with numerous countries across the African continent since 1989, Tanzania has begun the transition from single-party rule to multiparty politics. On May 7, 1992, the Tanzanian Parliament approved amendments to the constitution to legalize a multiparty system and registered political parties quickly multiplied. By the end of 1993, eleven had been fully registered, giving CCM its first legal opposition in nearly three decades. By mid-1994, there had been four parliamentary by-elections, all won by CCM. Of more direct relevance here, these changes in electoral politics have been accompanied by a loosening of the party's grip over mass political organizations.[45]

The increased latitude for grassroots organizing has resulted in the establishment of new interest groups which have forged important positions in the arena of ecological politics. There has been a dramatic increase in the number of domestic, environmentally oriented NGOs in Tanzania. Organization and coordination is increasing, as evidenced by the founding in December 1992 of the Tanzania Environmental NGOs Networking (TANEN), an assemblage of twenty individual NGOs. The most well organized and funded of these is the Wildlife Conservation Society of Tanzania (WCST), established in 1988. WCST publishes a periodical titled *Miombo* covering national wildlife issues. While the organization is technically independent of the state, it could hardly be described as "grassroots." Its members include business leaders and government conservation officials, with President Mwinyi serving as the WCST patron.

Pastoralists whose property rights have been adversely affected by national park establishment and related tourist developments are also beginning to organize. Groups are now registering with the government as NGOs and establishing contacts with other NGOs and institutions, domestically and internationally. In 1992, the Pastoral Network of Tanzania was established by an alliance of pastoralists, NGO researchers, and donor representatives. The first and most visible example is the Korongoro Integrated Peoples Oriented to Conservation (KIPOC). The acronym is a Maasai word that translates as "we shall recover."[46] KIPOC was organized principally by former Member of the Tanzanian Parliament, Moringe Parkipuny, a Maasai who represented the Ngorongoro District. The organization is concerned primarily with the defense of the culture and rights of "indigenous minority peoples"[47] in the Ngorongoro District and their self-directed economic development. Specific attention is given to initiatives to "restore legal and political respect to community ancestral lands."[48] Part of the aim of KIPOC is to promote among the pastoralists of the Ngorongoro District a transition "from subsistence economy to a long-term sustainable economic system," accomplished by "integrating community development with nature conservation."[49]

A second pastoralist organization developed out of the First Maasai Conference on Culture and Development, sponsored by various bilateral development organizations in December 1991. Inyuat e-Maa, translated as "Maa efforts in development," was established to promote self-directed "economic and cultural development" of Maa peoples in Tanzania.[50] Like KIPOC, Inyuat e-Maa emphasizes "integrating community development with nature conservation."[51] Also similar to KIPOC, the organization recognizes the past and present threat of protected areas and agricultural expansion to their economy and culture. It seeks to have all remaining Maa lands legally registered.[52]

These are the most visible of the new pastoralist organizations, but others are springing up seemingly with every new land conflict. Barabaig pastoralists whose lands have been illegally appropriated for a large-scale wheat scheme have formed a KIPOC chapter. At Mkomazi Game Reserve, a group has emerged among evicted pastoralists which focuses on lost rights,[53] while at Ngorongoro a group calling itself the Ngorongoro Conservation Peoples Saving Trust has organized to rectify perceived injustices by the Ngorongoro Conservation Area Authority.[54] The common interest binding all of these groups is a concern with locally directed development based on respect for customary land rights and cultural practices.

While the process of "democratization" has allowed for increasing mass involvement in political life and the public expression of opposition by marginalized people, the coincidental process of economic liberalization has encouraged political destabilization and new forms of state repression.[55] Tanzania, like many African nations undergoing similar political transformations, is experiencing new or reenergized tensions among various social groups which threaten democratic progress. As Newbury points out, "Perhaps the most salient feature of democratic openings in Africa is their fragility." The state has sometimes responded to pastoralists' political organizing with a heavy hand. The organizers of the first Maasai conference complained that they "were seen as and labeled agitators, troublemakers and forces of tribalism. We were interrogated and threatened that legal action would be taken against anyone who worked for the Conference."[56] In the context of growing economic inequities, the new openings for political participation have in part resulted in an ominous escalation of racial, ethnic, and religious politics.[57] Among pastoralists, political discontent focuses on land transfers, an issue so charged that activists' rhetoric at times borders on incendiary. "I am now blowing an alarm that unless something is done as soon as possible this land alienation will lead to social disruption. How long can one tolerate being treated as a non citizen of the area in the land of their birth?"[58] As often as not, the land alienation that has marginalized and impoverished pastoralists in Tanzania has been the result of state wildlife conservation policies.

Whose Sustainable Development?

Local Critiques of Global Agendas

The proliferation of domestic NGOs, the dramatic increase in foreign investment in tourism, and the continued involvement of international conservation NGOs produces a complicated matrix of often conflicting social processes. Many of the underlying social contradictions of wildlife conservation are masked by the rhetoric of sustainable development. In fact, as I will explore below, it has become de rigueur for all parties involved with wildlife conservation, from hotel chain magnates to dislocated pastoralists, to speak of the simultaneous need for community development and environmental stabilization. The notion of "sustainable development," however, has very different political meanings and implications for the various parties involved in Tanzania.

Some parallels with the current situation can be found in the case of the colonial soil conservation program of the 1940s and 1950s in the Shambaa kingdom of Tanzania. Feierman demonstrated how Shambaa peasants co-opted Western rhetoric of democracy and individual human rights and incorporated it into a counterdiscourse of protest over colonial soil conservation policies.[59] Following Feierman, I want to explore how pastoralist activists in the 1990s adopt the language of sustainable development to offer a localized critique of internationally constructed notions of economic development and nature protection. Inyuat e-Maa, for example, presents its goals in a manner which challenges the exclusion of "locals" from the management of and profits from wildlife, and does so within the context of protection of land rights and cultural survival:

> We shall try to diversify our economy by protecting the abundant wildlife within our range resources and making use of it. We could create multiple land use units (livestock, wildlife and tourism) in such areas and claim hunting and camping fees. . . . Much of all this, however, depends on whether we retain our land or not.[60]

More explicitly, an example from KIPOC illustrates the linkage pastoralist activists are making between social justice and environmental protection: "The required focus of action is authentic measures geared to speed up the restoration of social justice and environmental harmony in the management of the natural resources of this area."[61]

The rhetoric produced by these organizations serves to uncover the contradictions between human rights and conservation that often go unexamined in global agendas for environmental stabilization. Above all, statements such as these powerfully remind us of the limitations of analyzing ecological politics in isolation from questions of livelihood and justice.[62] Nature protection in Tanzania is muddied with sediment from the erosion of human rights, economic justice, and political accountability. Examples abound, but a quote from KIPOC provides the most compelling summation:

> Meantime, under pressure from the powerful preservation lobbies in the North extensive tracks of quality rangelands have been carved into wildlife preserves for exclusive use by wild animals and tourists, the latter from the affluent society. In that pursuit African regimes have carried on eviction of indigenous peoples and denied them access to resources vital to the viability of their flexible transhumance system of utilization of their land. These losses of land are accompanied by denial of access to critical sources of water and salt licks as well as sacred sites of worship and burial. This process of displacement, launched in the colonial era continues to date with ever increasing momentum.[63]

Clearly, activists for pastoralist rights are articulating a pointed critique of wildlife conservation practices. They offer a perspective on the history of nature protection that sharply contests the standard conservationist narrative of a morally directed mission that transcends politics.[64] Wildlife conservation, as traditionally practiced in Tanzania, represents for these activists a threat to cultural survival and serves as a central issue to rally and organize political involvement at the grassroots level. A closer examination of specific cases will illustrate how struggles over the meaning and practice of development and conservation are currently unfolding.

Conflicting Visions of Environment and Development

Investors anxious to seize the opportunities for large profits in nature tourism (enthusiastically recounted in the financial reports cited above) have presented their activities in the language of conservation and community development. The new Aga Khan-funded Serena Lodge is a case in point. The lodge is to be located at Kimba in the Ngorongoro Conservation Area on land belonging to the Maasai village of Oloirobi. The *African Economic Digest* recently reported that the new Serena site at Kimba was "chosen after extensive and satisfactory environmental studies."[65] In a press release describing the project, the IFC said the lodges would provide money for conservation, reduce poaching, and "promote the welfare of the local communities."[66] Presented in this way, the lodges seem to impeccably embody the ideals of ecodevelopment.

Maasai political organizers present the lodge at Kimba in a different light, however, one which reflects the ideals of customary rights, human rights, and legal due process. KIPOC reports that the site at Kimba has two springs which provide the only sources of reliable water during the dry season and that the lodge will be built right on the main western livestock access route in and out of the crater. They claim that the Maasai were never involved or consulted in the planning and siting process and note that the water can provide the needs of the lodge or local Maasai, but not both.[67] They also fear that, ultimately, their livestock will not be tolerated in the locality of the new lodge. They note that agreements for the sublease of village lands mandated by law have not been upheld. The claim that due process has been ignored is revealed in the accusation that "the site was given in person to Serena by the President of the United Republic of Tanzania."[68]

The highly conflictual pastoralist evictions from Mkomazi Game Reserve recounted above provide an example of contrasting interpretations of human-wildlife interactions. The conflict at Mkomazi has been detailed in *Miombo*,[69] the newsletter of the WCST. Domestic conservationists argue that the reserve is threatened by "relentless encroachment by man and his livestock"[70] and livestock keeping in general is "incompatible with wildlife."[71] Conservationists view the reserve explicitly as a national resource and drastic action to protect it is therefore justified in the national interest. Since it is "more feasible to find alternative areas for people and livestock than it is for wild animals and their habitat needs" the "pastoralists must be moved."[72] Thus, a new history for Mkomazi is invented,[73] one in which the pastoralists are no longer the principal occupants of over a century, but recent encroachers. No mention is made in the WCST account of the historic presence and legal claims of the Maasai. Rather, the area's cultural and historical importance is reduced to nineteenth-century explorer visitations and World War I bombing sites.[74]

KIPOC, on the other hand, represents Mkomazi as yet another violation of legal and human rights, calling this and other evictions "gross injustices." Pastoralists, they argue are "simply driven out of the lands of their ancestors and left to find for themselves the space to eke out a living."[75] Some have seen hidden motives in the evictions. Displaced pastoralists at Mkomazi claim "that a Wildlife Officer and a European, who allegedly 'owned' the reserve, were mining for minerals."[76] A pastoralist organizer told me in an interview, "Now you can see Pare [an ethnic group, primarily agriculturalists] settlers farming inside" the reserve, a situation which he claimed was the result of local political patronage.[77] Another activist complained of the lack of resettlement services and told how his relatives had lost most of their cattle since being evicted.[78] In these statements, pastoralists are portrayed as victims of wildlife conservation rather than its enemies.

Finally, the case of the TANAPA/AWF buffer zone project on the border of Tarangire National Park illustrates how new forms of land use control are introduced as participatory development. The idea of a buffer zone, in this project, is to encourage land uses in a ten-kilometer strip around the park that are compatible with park objectives and restrict those that are not. Faced with pastoralists' long-standing resistance and open hostility to policies which drive them from their lands, conservationists have been pushed toward political accommodation. Thus, the project would be concerned with "developing activities within the proposed buffer zone . . . for the benefit of the Maasai local communities."[79] It was hoped that this approach would defuse some of the political agitation among pastoralists and allow for a new dialogue between park managers and displaced populations.

Though presented by conservationists as participatory development, pastoralists did not enthusiastically embrace the project. Many of the Maasai in the buffer zone area had migrated into the area after being displaced from Serengeti National Park and the Ngorongoro Conservation Area. Promises concerning land rights and access in those protected areas had been made repeatedly and broken by past governments. Maasai activists learned that without legal security of tenure, agreements were meaningless. "We have suffered a lot with our land being alienated . . . to protected areas [national parks]. . . . In much of our land or areas, the whole situation is chaotic and without rules."[80] With this history in mind, local Maasai saw the buffer zone as potentially another means of outsider penetration into local land and resource control. They refused to participate in the buffer zone program until they had hired their own surveyors and legal assistance to gain title deeds to all of the village lands.[81] In a similar situation in the Loliondo area on the boundary of Serengeti National Park, KIPOC was instrumental in securing title deeds for pastoralist communities. This action was KIPOC's first major effort at community organizing and helped launch its campaign for pastoralist rights.[82]

Discussion and Conclusion

The political and economic changes underway in Tanzania are profoundly altering the politics of wildlife conservation and tourism. The level of investment in and control over the tourist industry and associated protected areas by private capital is unparalleled in the country's history. At the same time, the new democratic openings have facilitated unprecedented political mobilizations by pastoralists around land rights and cultural survival which directly challenge the very structure of conservation and tourism. The expanding role of international conservation NGOs and their new focus on "community development" has led them literally into new territory. The national parks and protected areas provide the stage upon which the political dramas produced by these phenomena are unfolding. Clearly, ecological politics in Tanzania are entering a new phase, one that is fraught with contradictions.

From the perspective of pastoralist political activists, numerous injustices have been carried out by the state in the name of wildlife conservation. The fact that pastoralist voices speaking out against conservation as usual are now heard loudly at international conferences and workshops is in itself a remarkable historical shift in Tanzania's ecological politics. Pastoralists are now armed with the potent global discourses of sustainable development and human rights. Their appropriation of this language is a significant political act which heralds a new assertiveness. Yet the emergence of ethnic politics and the rising economic stakes in Tanzania have escalated the potential for repressive government action. The willingness of the state to resort to coercion in the face of continuing economic crisis and new challenges to its authority and legitimacy is clearly evident in wildlife conservation. The 1988 paramilitary operation to evict legally resident pastoralists in Mkomazi, the questionable land transfers, and the harassment of pastoralist organizers are examples.

Furthermore, increased foreign investment adds new urgency about questions of who gains and who loses from wildlife conservation and tourist development. The idea that foreign investment in luxury lodges will lead to community development is typically based on a naive faith in a trickle-down process. Rarely are mechanisms for the redistribution of tourism profits to local communities formulated.[83] Foreign companies that invest in park lodges will inevitably have important leverage over issues of wildlife conservation which they see affecting their interests. As is demonstrated by the case of the lodge at Kimba, their interests often conflict with pastoralists' concerns for security of land tenure. There is also evidence that investors are being given lodge sites which are opposed by the conservation agencies, and that national environmental standards are being relaxed to facilitate tourism development.[84]

As for international conservation organizations, their roles are ambiguous, and democracy in Tanzania may not be helped by their involvement. While NGOs emphasize local participation, the process of international project planning and implementation is not democratic. As the leading actors in conservation and development projects, NGOs have no accountability to the people affected by their programs. Public involvement is rarely explicitly defined in the new community oriented projects and plans to implement real power-sharing between local communities and protected areas are practically nonexistent.[85] Most critically, projects are not designed to address the pastoralists' deep concern about security of land rights. In the worst cases, NGO funds for conservation can be misdirected to further the repressive activities of the state.[86] Furthermore, the specific relationship between outside NGOs and pastoralists is ambivalent. While many international NGOS, conservation oriented and otherwise, have in fact supported the efforts of pastoralist organizations in Tanzania to defend their rights, pastoralists have also at times viewed them as threats to their cultural survival.[87]

So far, pastoralists are the main social group organizing to redress the perceived injustices of wildlife conservation. Other affected groups, such as peasant farmers on other park boundaries, have not yet organized around similar issues. The potential exists, however, for a much more widespread and comprehensive political struggle over land and resource rights in protected areas, such as developed as part of the nationalist movement in the colonial period. It is clear that the selling off of state-owned properties in other sectors is already causing unrest among Tanzanians who see potential opportunities slipping into the hands of foreign investors. This, combined with shady transfers of land ownership, whether for lodges or private hunting, and the negative symbolism of tourism, create the conditions for violence in ecological politics. This does not bode well for wildlife and national parks. In other parts of the continent, for example, there are disturbing signs that as state legitimacy and power weaken, national parks become a focus of rural people's wrath.[88]

The transformation of ecological politics in Tanzania dictates that conservation as usual cannot continue. Conservationists have already been forced to reorient projects toward community development in the face of continued resistance from rural populations. Provided with new democratic openings, pastoralists are moving away from "everyday forms of resistance" and protest toward more organized and formalized forms of political action. It is difficult to predict what new structures and policies for wildlife conservation will emerge as a result of their activism. Pastoralist rights advocates have, however, made it clear that wildlife conservation issues cannot be addressed without considering broader struggles for human rights and social justice.

NOTES

1. Quoted in R. Nash, *Wilderness and the American Mind* (New Haven: Yale University Press, 1982), p. 342.
2. Ole-Saitoti is founder of the Ngorongoro Conservation Peoples Saving Trust.
3. I. Shivji, *Tourism and Socialist Development* (Dar Es Salaam: Tanzania Publishing House, 1973).

4. I use the term NGO to mean literally a nongovernmental organization. I use it as shorthand rather than as an analytical category, recognizing the problems of such an all-encompassing term.

5. See Roderick P. Neumann, "The Social Origins of Natural Resource Conflict in Arusha National Park, Tanzania," Ph.D. diss., University of California at Berkeley, 1992; Roderick P. Neumann, "Ways of Seeing Africa: Colonial Recasting of African Society and Landscape in Serengeti National Park," *Ecumene* 2 (1995): 149–69.

6. See J. MacKenzie, *The Empire Of Nature: Hunting, Conservation and British Imperialism* (Manchester, England: Manchester University Press, 1988); Neumann, "Ways of Seeing Africa."

7. See Roderick Neumann, "The Political Ecology of Wildlife Conservation in the Mount Meru Area, Northeastern Tanzania," *Land Degradation and Rehabilitation* 3 (1992): 85–98.

8. Coordinating Officer, Sukumaland Development to Provincial Commissioner, Lake Province, May 5, 1948, Tanzania National Archives, Secretariat File 34819.

9. See J. C. Scott, *Weapons of the Weak: Everyday Forms of Peasant Resistance* (New Haven: Yale University Press, 1985).

10. See Neumann, *The Social Origins of Natural Resource Conflict in Arusha National Park, Tanzania*.

11. J. C. Scott, "Resistance Without Protest and Without Organization: Peasant Opposition to the Islamic Zakat and the Christian Tithe," *Comparative Studies in Society and History* (1987): 447.

12. A. Isaacman, "Peasants and Rural Social Protest in Africa," *African Studies Review* 33 (1990): 32.

13. See, for example, D. Anderson and R. Grove, "Introduction: the Scramble For Eden: Past, Present and Future in African Conservation," in *Africa: People, Policies and Practice* (Cambridge: Cambridge University Press, 1987), pp. 1–12.

14. R. Blaikie, *The Political Economy of Soil Erosion in Developing Countries* (Essex, England: Longman, 1985), p. 78, emphasis in original.

15. L. Timberlake, *Africa in Crisis: The Causes, the Cures of Environmental Bankruptcy* (Philadelphia: New Society Publishers, 1986), pp. 163–64; Neumann, *The Social Origins of Natural Resource Conflict in Arusha National Park, Tanzania*; and Neumann, "Ways of Seeing Africa."

16. R. Bonner, *At the Hand of Man: Peril and Hope for Africa's Wildlife* (New York: Alfred A. Knopf, 1993), p. 65.

17. Similar experiences shaped other state institutions in Tanzania. See L. Fortmann, *Peasants, Officials and Participation in Rural Tanzania: Experience With Villagization and Decentralization* (Ithaca: Rural Development Committee, Cornell University, 1980); also G. Hyden, *Beyond Ujamaa in Tanzania: Underdevelopment and an Uncaptured Peasantry* (Berkeley: University of California Press, 1980).

18. M. Wells and K. Brandon, *People and Parks: Linking Protected Area Management with Local Communities*. (Washington, D.C.: World Bank, WWF, USAID, 1992).

19. Neumann, *The Social Origins of Natural Resource Conflict in Arusha National Park, Tanzania*.

20. See H. Kiekshus, *Ecology Control and Economic Development in East African History: The Case of Tanganyika 1850–1950* (Berkeley: University of California Press, 1977); K. Arhem, "Two Sides of Development: Maasai Pastoralism and Wildlife Conservation in Ngorongoro, Tanzania," *Ethnos* 49 (1984): 186–210; K. Arhem, *Pastoral Man in the*

Garden of Eden: the Maasai of the Ngorongoro Conservation Area, Tanzania. (Uppsala: University of Uppsala, 1985); and C. Diehl, "Wildlife and the Maasai: the Story of East African Parks," *Cultural Survival Quarterly* 9 (1985): 37–40.

21. In July of 1988 I visited the reserve in the company of the officer in charge during the operation to relocate the pastoralists. Although the operation had been completed just prior to my arrival, arrests were still being made of people trying to reenter the reserve and many of the former residents refused to cooperate and declined the offer of the government to transport them to their new area.

22. H. Fosbrooke, *Pastoralism and Land Tenure*, paper presented at the Workshop on Pastoralism and the Environment, April 1990, Arusha, Tanzania.

23. K. Mustafa, "Eviction of Pastoralists From the Mkomazi Game Reserve in Tanzania: A Statement," International Institute for Environment and Development, 1993, unpublished.

24. Ibid.

25. J. S. Owen, "The National Parks of Tanganyika," in *First World Conference on National Parks*, ed. A. B. Adams (Washington, D.C.: U.S. Government Printing Office, 1962), pp. 52–59.

26. Anonymous memo concerning a 1970 study on the development of the tourist industry by Arthur D. Little, Inc., TANAPA Closed Files.

27. See I. Shivji, *Class Struggles in Tanzania* (New York: Monthly Review Press, 1976); J. K. Nyerere, *The Arusha Declaration: Ten Years After* (Dar Es Salaam: Government Printer, 1977); G. Hyden, *Beyond Ujamaa in Tanzania: Underdevelopment and an Uncaptured Peasantry*; I. N. Resnick, *The Long Transition: Building Socialism in Tanzania* (New York: Monthly Review Press, 1981); J. H. Weaver and A. Kronemer, "Tanzanian and African Socialism," *World Development* 9 (1981): 839–49; P. Raikes, "The State and the Peasantry in Tanzania," in *Rural Development: Theories of Peasant Economy and Agrarian Change*, ed. J. Harriss (London: Hutchinson, 1982), pp. 350–80; and D. Bolton, *Nationalization—A Road to Socialism? The Lessons of Tanzania* (London: Zed Books, 1984).

28. A. Kiondo, "The Nature of Economic Reforms in Tanzania," in *Tanzania and the IMF: The Dynamics of Liberalization*, ed. H. Campbell and H. Stein (Boulder: Westview Press, 1992), p. 34.

29. Ibid; also see I. Shivji, *Class Struggles in Tanzania*.

30. See, for example, M. J. Watts, "Development II: The Privatization of Everything?" *Progress in Human Geography* 18 (1994): 371–84.

31. "Economist Intelligence Unit Country Report, Tanzania, 4th Quarter," EIU, 1989. According to the *African Economic Digest*, April 8, 1991, "Those critics who argued that there was no place for tourism in a socialist and self-reliant society have finally been silenced."

32. "Economist Intelligence Unit Country Report, Tanzania, 2nd Quarter," EIU, 1990.

33. Inter Press Service, April 28, 1993 (acquired from the Lexis/Nexis on-line service).

34. *African Economic Digest*, November 1, 1993 (acquired from the Lexis/Nexis on-line service).

35. National Trade Data Bank, *Market Reports*, April 14, 1993. Material in this market report was derived from a report prepared by the US Embassy, Dar Es Salaam. (acquired from the Lexis/Nexis on-line service).

36. *African Economic Digest*, April 8, 1991 (acquired from the Lexis-Nexis on-line service).

37. Inter Press Service, April 28, 1993.

38. See C. R. Koppes, "Efficiency, Equity, Esthetics: Shifting Themes in American Conservation," in *The Ends of the Earth: Perspective on Modern Environmental History*, ed. Donald Worster (Cambridge: Cambridge University Press, 1988), pp. 230–52.

39. M. J. Watts, "Visions of Excess: African Development in an Age of Market Idolatry," *Transition* 51 (1991): 125–41.

40. Two recent titles from the journal *Environment*, for example, cheerfully announce "Rain Forest Entrepreneurs: Cashing in on Conservation," (T. Carr, H. Pendersen, and S. Ramaswamy, *Environment* 35: 12–38) and "Making Biodiversity Conservation Profitable" (E. Blum, *Environment* 35: 17–45). International conservation NGOs now have product managers on their staffs. In their article about buttons manufactured from the tagua nut from the Ecuadoran rain forest, Carr, Pedersen, and Ramaswamy present information from "Robin Frank, tagua product manager at Conservation International," p. 33. See also M. Schroeder, "Contradictions Along the Commodity Road to Environmental Stabilization: Foresting Gambian Gardens," *Antipode* 27, no. 4 (1995): 325–42.

41. Quoted in "Economist Intelligence Unit Country Report, Tanzania, 2nd Quarter," EIU, 1993, p. 14.

42. *Report of the Presidential Committee of Inquiry Into Land Matters, United Republic of Tanzania*, Vol. 1 (Dar Es Salaam: Government Printers, 1992), p. 23. The study was commissioned by President Mwinyi and chaired by Issa Shivji with Solomon ole Saibul serving as vice chairman.

43. Ibid., p. 23.

44. The affair culminated in the sacking of the Minister of Tourism; Natural Resources and Environment in April, 1993.

45. Previously, for example, the only legal labor union, *Jumuiya ya Wafanyakazi Tanzania* (Juwata), was operated as a department of CCM. In late 1991, Parliament eliminated party control and replaced Juwata with the independent Organization of Tanzanian Trade Unions. The new union, despite continuing ties to the party (U.S. State Department, 1994), apparently surprised CCM by boldly organizing the first general strike against the party's wage policies in March 1994 (EIU, 1994).

46. "The Foundation Program: Program Profile and Rationale," Principal Document No. 4. (Loliondo, Tanzania: Korongoro Integrated Peoples Oriented to Conservation, 1992).

47. Ibid., p. 1.

48. Ibid., p. 7.

49. Constitution of the Korongoro Integrated Peoples Oriented to Conservation, Articles 5.1 and 5.3.

50. *Draft Constitution of Inyuat e-Maa*, Preamble.

51. Ibid., Article 7.6.

52. Saruni Oitesoi ole-Ngulay, "Inyuat e-Maa/Maa Pastoralists Development Organization: Aims and Possibilities," presented at the IWGIA-CDR Conference on the Question of Indigenous Peoples in Africa, Greve, Denmark, June 1–3, 1993, p. 5.

53. Interview with anonymous pastoralist rights activist, June 16, 1994.

54. Tipilit ole-Saitoti, "Local Perspective of Ngorongoro Conservation Area," paper presented at the Second Maa Conference on Culture and Development, May 30–June 3, 1994, Arusha, Tanzania.

55. See H. Campbell, "The Politics of Demobilization in Tanzania: Beyond Nationalism," in *Tanzania and the IMF*, pp. 85–108; Shivji, *Class Struggles in Tanzania*. Campbell

illustrates this phenomena with the example of worker protests at the Maya Sugar Factory in 1986 which were halted when the government Field Force Unit opened fire, killing four protesters. "Not even in the colonial era after the Maji Maji revolt was there an incident during which the colonial army fired on unarmed civilians," p. 100.

56. Oitesoi ole-Ngulay, "Inyuat e-Maa/Maa Pastoralists Development Organization," p. 3.

57. Reverend Christopher Mtikila, an opposition party leader whose speeches are punctuated by inflammatory anti-Asian and anti-Arab rhetoric exemplifies the trend. Mtikila claims the country is being sold to *gabacholi* (parasitic foreigners). His main following is among the black African urban workers and unemployed.

58. ole-Saitoti, "Local Perspective of Ngorongoro Conservation Area," p. 1.

59. S. Feierman, *Peasant Intellectuals: Anthropology and History in Tanzania.* (Madison: University of Wisconsin Press, 1990).

60. Oitesoi ole-Ngulay, "Inyuat e-Maa/Maa Pastoralists Development Organization," p. 5.

61. "The Foundation Program: Program Profile and Rationale," Principal Document No. 4, p. 22.

62. See S. Hecht and A. Cockburn, *Fate of the Forest: Developers, Destroyers and Defenders of the Amazon.* (New York: Harper Perennial, 1990); R. Peet and M. Watts, "Development Theory and Environment in an Age of Market Triumphalism," *Economic Geography* 69 (1993): 227–53.

63. "The Foundation Program," p. 14.

64. See Anderson and Grove, "The Scramble for Eden"; J. Caruthers, "Creating a National Park, 1910 to 1926," *Journal of Southern African Studies* 15 (1989): 188–216.

65. *African Economic Digest*, November 1, 1993.

66. Ibid., April 29, 1993.

67. "The Foundation Program," p. 18.

68. Ibid.

69. S. R. Mduma, "Mkomazi Game Reserve: Dangers and Recommended Measures For Its Survival," part 1, *Miombo* 1 (1988), part 2, *Miombo* 2 (1988).

70. Ibid., part 1, p. 19.

71. Ibid., part 2, p. 4.

72. Ibid, p. 5.

73. See E. Hobsbawm and T. Ranger, eds., *The Invention of Tradition* (Cambridge: Cambridge University Press, 1983); J. Carruthers, "Dissecting the Myth: Paul Kruger and the Kruger National Park," *Journal of Southern African Studies* 20 (1994): 263–83.

74. Mduma, "Mkomazi Game Reserve," part 1, p. 18.

75. "The Foundation Program," p. 15.

76. K. Mustafa, "Eviction of Pastoralists from the Mkomazi Game Reserve in Tanzania: A Statement," International Institute for Environment and Development, p. 18, unpublished.

77. Interview with anonymous pastoralist rights activist, June 18, 1994.

78. Interview with anonymous pastoralist rights activist, June 16, 1994.

79. From an AWF background document, 1990.

80. Oitesoi ole-Ngulay, "Inyuat e-Maa/Maa Pastoralists Development Organization," p. 5.

81. N. L. Peluso, "Whose Woods are These: Counter-Mapping Forest Territories in Kalimantan, Indonesia?" *Antipode* 27 no. 4 (1995): 383–406.

82. "The Foundation Program," p. 17.

83. A pastoralist activist told me that recently 25 million Tanzania shillings were transferred to the Ngorongoro District Council from the Ngorongoro Conservation Area Authority. The money was absorbed into the Council's general coffers and, he claimed, never reached the ground. Interview with anonymous pastoralist rights activist, June 18, 1994.

84. "Economist Intelligence Unit Country Report, Tanzania, 4th Quarter," EIU, 1992, p. 22; "Report of the Presidential Committee of Inquiry Into Land Matters, United Republic of Tanzania," vol. 1, p. 286.

85. See Wells and Brandon, *People and Parks*. In 1990, a regional IUCN official remarked to me in an interview that he was "afraid of Parkipuny's ideas about local control."

86. See R. Bonner, *At the Hand of Man: Peril and Hope for Africa's Wildlife*; N. L. Peluso, "Coercing Conservation? The Politics of State Resource Control," *Global Environmental Change* 3 (1993): 199–217; S. Ellis, "Of Elephants and Men: Politics and Nature Conservation in South Africa," *Journal of Southern African Studies* 20 (1994): 53–69.

87. See "The Foundation Program"; Oitesoi ole-Ngulay, "Inyuat e-Maa/Maa Pastoralists Development Organization."

88. See A. Lowry and T. P. Donahue, "Parks, Politics, and Pluralism: the Demise of National Parks in Togo," *Society and Natural Resources* 7 (1994): 321–29.

REFERENCES

African Economic Digest. April 8, 1991. Acquired from the Lexis/Nexis on-line service.
———. April 29, 1993. Acquired from the Lexis/Nexis on-line service.
———. November 1, 1993. Acquired from the Lexis/Nexis on-line service.
Anderson, D., and R. Grove. "The Scramble for Eden: Past, Present and Future in African Conservation." Introduction to *Conservation in Africa: People, Policies and Practice.* Cambridge: Cambridge University Press, 1987.
Arhem, K. "Two Sides of Development: Maasai Pastoralism and Wildlife Conservation in Ngorongoro, Tanzania." *Ethnos* 49 (1984): 186–210.
———. *Pastoral Man in the Garden of Eden: The Maasai of the Ngorongoro Conservation Area, Tanzania.* Uppsala: University of Uppsala, 1985.
Blaikie, R. *The Political Economy of Soil Erosion in Developing Countries.* Essex, England: Longman, 1985.
Blum, E. "Making Biodiversity Conservation Profitable: A Case Study of the Merck/INBio Agreement." *Environment* 35 (1993): 17–45.
Bolton, D. *Nationalization—A Road to Socialism? The Lessons of Tanzania.* London: Zed Books, 1984.
Bonner, R. *At the Hand of Man: Peril and Hope for Africa's Wildlife.* New York: Alfred A. Knopf, 1993.
Campbell, H. "The Politics of Demobilization in Tanzania: Beyond Nationalism." In *Tanzania and the IMF: The Dynamics of Liberalization*, edited by H. Campbell and H. Stein. Boulder: Westview Press, 1992.
Carr, T., H. Pendersen, and S. Ramaswamy. "Rain Forest Entrepreneurs: Cashing in on Conservation." *Environment* 35 (1993): 12–38.

Caruthers, J. "Creating a National Park, 1910 to 1926." *Journal of Southern African Studies* 15 (1989): 188–216.

———. "Dissecting the Myth: Paul Kruger and the Kruger National Park." *Journal of Southern African Studies* 20 (1994): 263–83.

Diehl, C. "Wildlife and the Maasai: The Story of East African Parks." *Cultural Survival Quarterly* 9 (1985): 37–40.

EIU. *Economist Intelligence Unit Country Report, Tanzania*. 4th quarter, 1989.

———. *Economist Intelligence Unit Country Report, Tanzania*. 2d quarter, 1990.

———. *Economist Intelligence Unit Country Report, Tanzania*. 4th quarter, 1992.

———. *Economist Intelligence Unit Country Report, Tanzania*. 2d quarter, 1993.

———. *Economist Intelligence Unit Country Report, Tanzania*. 2d quarter, 1994.

Ellis, S. "Of Elephants and Men: Politics and Nature Conservation in South Africa." *Journal of Southern African Studies* 20 (1994): 53–69.

Feierman, S. *Peasant Intellectuals: Anthropology and History in Tanzania*. Madison: University of Wisconsin Press, 1990.

Fortmann, L. *Peasants, Officials and Participation in Rural Tanzania: Experience with Villagization and Decentralization*. Ithaca: Rural Development Committee, Cornell University, 1980.

Fosbrooke, H. *Pastoralism and Land Tenure*. Paper presented at the Workshop on Pastoralism and the Environment, April 1990, Arusha, Tanzania.

Hecht, S., and A. Cockburn. *Fate of the Forest: Developers, Destroyers and Defenders of the Amazon*. New York: Harper Perennial, 1990.

Hobsbawm, E., and T. Ranger, eds. *The Invention of Tradition*. Cambridge: Cambridge University Press, 1983.

Hyden, G. *Beyond Ujamaa in Tanzania: Underdevelopment and an Uncaptured Peasantry*. Berkeley: University of California Press, 1980.

Inter Press Service. April 28, 1993. Acquired from the Lexis/Nexis on-line service.

Isaacman, A. "Peasants and Rural Social Protest in Africa." *African Studies Review* 33 (1990): 1–120.

Kiondo, A. "The Nature of Economic Reforms in Tanzania." In *Tanzania and the IMF: The Dynamics of Liberalization*, edited by H. Campbell and H. Stein. Boulder: Westview Press, 1992.

KIPOC. *The Foundation Program: Program Profile and Rationale*. Principal Document No. 4, Loliondo, Tanzania: Korongoro Integrated Peoples Oriented to Conservation, 1992.

Kiekshus, H. *Ecology Control and Economic Development in East African History: The Case of Tanganyika 1850-1950*. Berkeley: University of California Press, 1977.

Koppes, C. R. "Efficiency, Equity, Esthetics: Shifting Themes in American Conservation." In *The Ends of the Earth: Perspective on Modern Environmental History*, edited by D. Worster. Cambridge: Cambridge University Press, 1988.

Lowry, A., and T. P. Donahue. "Parks, Politics, and Pluralism: The Demise of National Parks in Togo." *Society and Natural Resources* 7 (1994): 321–29.

MacKenzie, J. *The Empire of Nature: Hunting, Conservation and British Imperialism*. Manchester, England: Manchester University Press, 1988.

Mduma, S. R. "Mkomazi Game Reserve: Dangers and Recommended Measures for Its Survival." Part 1. *Miombo* 1 (1988).

———. "Mkomazi Game Reserve: Dangers and Recommended Measures for Its Survival." Part 2. *Miombo* 2 (1988).

Mustafa, K. *Eviction of Pastoralists from the Mkomazi Game Reserve in Tanzania: A Statement.* International Institute for Environment and Development. Unpublished.

Nash, R. *Wilderness and the American Mind.* New Haven: Yale University Press, 1982.

Neumann, R. P. "The Social Origins of Natural Resource Conflict in Arusha National Park, Tanzania." Ph.D. diss., University of California, Berkeley, 1992.

———. "The Political Ecology of Wildlife Conservation in the Mount Meru Area, Northeastern Tanzania." *Land Degradation and Rehabilitation* 3 (1992): 85–98.

———. "Ways of Seeing Africa: Colonial Recasting of African Society and Landscape in Serengeti National Park." *Ecumene* 2 (1995): 149–69.

Nyerere, J. K. *The Arusha Declaration: Ten Years After.* Dar Es Salaam: Government Printer, 1977.

Oitesoi ole-Ngulay, Saruni. *Inyuat e-Maa/Maa Pastoralists Development Organization: Aims and Possibilities.* Paper presented at the IWGIA-CDR Conference on the Question of Indigenous Peoples in Africa, Greve, Denmark, June 1–3, 1993.

ole-Saitoti, Tipilit. *Local Perspective of Ngorongoro Conservation Area.* Paper presented at the Second Maa Conference on Culture and Development, Arusha, Tanzania, May 30–June 3, 1994.

Owen, J. S. "The National Parks of Tanganyika." In *First World Conference on National Parks,* edited by A. B. Adams. Washington, D.C.: U.S. Government Printing Office, 1962.

Peet, R., and M. Watts. "Development Theory and Environment in an Age of Market Triumphalism." *Economic Geography* 69 (1993): 227–53.

Peluso, N. L. "Coercing Conservation? The Politics of State Resource Control." *Global Environmental Change* 3 (1993): 199–217.

———. "Whose Woods Are These: Counter-Mapping Forest Territories in Kalimantan, Indonesia." *Antipode* 27, no. 4 (1995): 383–406.

Raikes, P. "The State and the Peasantry in Tanzania." In *Rural Development: Theories of Peasant Economy and Agrarian Change,* edited by J. Harriss. London: Hutchinson, 1982.

Resnick, I. N. *The Long Transition: Building Socialism in Tanzania.* New York: Monthly Review Press, 1981.

Schroeder. "Contradictions Along the Commodity Road to Environmental Stabilization: Foresting Gambian Gardens." *Antipode* 27, no. 4 (1995): 325–42.

Scott, J. C. *Weapons of the Weak: Everyday Forms of Peasant Resistance.* New Haven: Yale University Press, 1985.

———. "Resistance Without Protest and Without Organization: Peasant Opposition to the Islamic Zakat and the Christian Tithe." *Comparative Studies in Society and History* (1987): 417–52.

Shivji, I. *Tourism and Socialist Development.* Dar Es Salaam: Tanzania Publishing House, 1973.

———. *Class Struggles in Tanzania.* New York: Monthly Review Press, 1976.

———. "The Politics of Liberalization in Tanzania: The Crisis of Ideological Hegemony." In *Tanzania and the IMF: The Dynamics of Liberalization,* edited by H. Campbell and H. Stein. Boulder: Westview Press, 1992.

Timberlake, L. *Africa in Crisis: The Causes, the Cures of Environmental Bankruptcy.* Philadelphia: New Society Publishers, 1986.

United Republic of Tanzania. *Report of the Presidential Committee of Inquiry into Land Matters.* Vol. 1. Dar Es Salaam: Government Printers, 1992.

United States Department of State. *1993 Human Rights Report for Tanzania.* Washington, D.C.: U.S. Government Printing Office, 1994.

Watts, M. J. "Visions of Excess: African Development in an Age of Market Idolatry." *Transition* 51 (1991): 125–41.

———. "Development II: The Privatization of Everything?" *Progress in Human Geography* 18 (1994): 371–84.

Weaver, J. H., and A. Kronemer. "Tanzanian and African Socialism." *World Development* 9 (1981): 839–49.

Wells, M., and K. Brandon. *People and Parks: Linking Protected Area Management with Local Communities.* Washington, D.C.: World Bank, WWF, USAID, 1992.

11

IS THIS THE WAY TO SAVE AFRICA'S WILDLIFE?

Victoria Butler

Three years ago, John Tendengdende, headman of Dete village in Zimbabwe, watched helplessly as his corn seedlings, cotton plants, and chili bushes shriveled and died during the country's worst drought in living memory. Tens of thousands of other subsistence farmers in the Zambezi River valley lost their crops. At the bleakest moment, Tendengdende and his people stood on the threshold of starvation, all corn stocks gone, and the future hinging on whether enough rain would fall to nurture the next crop.

Then an innovative wildlife conservation program gave the villagers a lifesaving windfall. "For the first time we got money for our wild animals," explains Tendengdende. "I used my money to buy 50 kilograms (110 lbs.) of cornmeal. That's how we survived."

The program that helped to save the Dete villagers is the Communal Areas Management Program for Indigenous Resources (CAMPFIRE). Not only does it add a new dimension to wildlife conservation in Zimbabwe by helping to make wildlife valuable to local people, but it also offers a management model that other nations are examining closely.

Under CAMPFIRE, the government has transferred ownership of wildlife on communal lands to the communities, which sell hunting or photographic concessions to safari companies. The money goes directly to the communities, whose members decide how it will be spent. Zimbabwe's Department of National Parks and Wildlife Management sets the hunting quotas and trophy fees in each communal area, while local authorities, with support from the department, bear responsibility for wildlife protection and management.

This article originally appeared in *International Wildlife* 25 (March/April 1995): 38–43. Reprinted by permission. Copyright © 1995 National Wildlife Federation. All rights reserved.

Wildlife is perhaps the greatest treasure of the Zambezi River valley, a 900-kilometer (560-mile) swath of acacia thornbush that stretches across Zimbabwe from Victoria Falls in the northwestern corner of the nation to the border with Mozambique in the east. The valley covers some of south-central Africa's most remote and rugged landscape and produces large herds of elephants, buffalo, and antelope, as well as populations of lions, leopards, and other wildlife. Consequently, about half of the valley's 56,230 square kilometers (21,710 square miles) have been set aside in national parks, reserves, and forests to protect wildlife.

The other half is divided into tribal communal lands, where some 325,000 people subsist on crops of corn, cotton, and a few vegetables. The thin gray topsoil, rocky terrain, and inadequate rainfall make for uncertain crops, but about 20 percent of the communal lands also produce significant amounts of wildlife.

Ironically, the people of Dete and other villages in northern Zimbabwe's Hurungwe District did not value the animals until the drought ravaged their crops and threatened them with destitution. Then the influx of funds from CAMPFIRE, which they had recently joined, showed the villagers that their wildlife could be vitally important to them.

The program also is important to wildlife conservation. Many communal holdings border national parks, state forests, and reserves, which cover 14.5 percent of the country. For years, people have settled illegally on protected lands. Poaching has been a persistent problem as people killed wildlife to supplement income or to provide food for the pot.

But thanks to CAMPFIRE, a new era is dawning in the Zambezi valley. The $13 that each of the 574 heads of household in his village received in the year of the drought has changed local attitudes toward wildlife, says Arius Chipere, a member of a village wildlife committee. "Ten years ago, we liked the animals, of course," he explains. "But now we like them more because we are getting money for them."

Every CAMPFIRE village has an "animal reporter" who monitors wildlife movements and reports poachers. "Local poaching is a menace," complains Champion Machaya, chairman of Dete's wildlife committee. "We have people from other areas coming in and taking our animals. But our people have stopped poaching. They understand that a buffalo is worth much more if it is killed by a foreign hunter."

Ray Townsend, the thirty-five-year-old boss of a safari camp, sees a shift in community attitudes since CAMPFIRE started. "We're looking at a slightly reduced poaching problem, and people are starting to complain that they don't have enough animals," he says.

CAMPFIRE is rooted in the 1980s, when the limited resources available for the fight against encroachment on protected areas forced officials of Zimbabwe's Department of National Parks and Wildlife Management to consider new ways to conserve the country's animals and plants. With the support of the Worldwide Fund for Nature (WWF), the United States Agency for International Development (USAID), The Zimbabwe Trust, and the University of Zimbabwe's Center for Applied Social Sciences, they settled on CAMPFIRE.

The program started officially in 1989, when the parks department granted two districts authority over their wildlife. Since then, nearly half of Zimbabwe's fifty-five local districts have signed on. In 1993, twelve districts nationwide, with a combined human population of nearly 400,000, earned $1,516,693 in trophy fees. They received an additional $97,732 from tourism, culling, and problem animals that had to be shot.

Though the idea that hunting can help save wildlife may seem ironic, hunting and conservation have a long history of mutual support. In the United States, for example, some of the earliest efforts to protect vanishing species such as bison, elk, and deer were initiated by hunters who, at the close of the nineteenth century, feared they would lose the animals that offered them sport. Under modern management, careful monitoring helps ensure that hunted populations remain stable or increase.

The Zimbabwe government regulates sport hunting by setting local quotas based on annual wildlife surveys. All foreign sportsmen must be accompanied by a professional hunter licensed by Zimbabwe after completing a rigorous apprenticeship and passing state exams. A national parks game scout accompanies hunters to ensure that quotas are observed. Hunting is banned between dusk and dawn.

Most hunters book safaris months in advance. Safari operators meet their clients at the airport and whisk them away to bush camps that boast varying degrees of luxury in the wild. Some camps offer tents, some thatch huts, and some lodges of natural stone. In all camps, a staff of cooks and waiters attends the clients, who happily pay up to $1,000 per day to hunt in the African bush. A single hunter can spend more than $40,000, with half going to local communities.

One concern ever on the minds of CAMPFIRE advocates is the increasingly vociferous antihunting lobby in Western countries. Although Jon Hutton—an ecologist and director of Africa Resources Trust, a nongovernmental organization for conservation and human development—is encouraging the development of photo safaris and other nonconsumptive uses of wildlife, he says it will be a long time before communal lands will have the infrastructure to cater to large numbers of tourists. "Hunters have a different view than tourists," explains Hutton. "They just need a bush camp and a fire, without all the fancy facilities."

Zimbabwean ecologists argue that hunting is both good conservation and sound financial sense. The Zambezi valley, for example, supports at least twenty-two thousand elephants with an annual population growth rate of 4 to 5 percent. The valley's eight communal districts have a combined quota of only fifty-eight elephants, roughly 5 percent of the growth rate in 1993. Hurungwe was entitled to seven, but foreign hunters shot only six, yielding $54,825 to the district. The safari operator distributed all meat to local villagers. The Department of National Parks and Wildlife Management, in contrast, periodically culls elephants to maintain an ecological balance in wilderness areas. Although villagers get the meat, they do not receive any money.

WWF estimates that CAMPFIRE has increased household income in communal areas by 15 to 25 percent. At the end of 1992, the thirty-one thousand people in Hurungwe District received $119,342 through CAMPFIRE. In 1993, they received $145,519.

Each village decides at year's end how it will use the income. Some villages divide it equally among heads of household. Some put it into community projects, such as schools, grinding mills, beekeeping, or clinics. Others split it between projects and household cash dividends.

In many villages, the money goes to pay children's school fees. For financially strapped villages, this can be the difference between education and ignorance. Sign Chawabvunza and his wife, Semi, have lived in the Chundu area of Hurungwe District all their lives. They grow corn, groundnuts, and sorghum on 3.6 hectares (9 acres). Five of their eleven children live at home. "Before CAMPFIRE, we did not have enough money to buy seeds, fertilizers, and pay school fees," says Chawabvunza. In 1993, Chawabvunza received $54 from CAMPFIRE. "I used some of the money to pay school fees, and the rest I used to buy food for the family," he said. "Before CAMPFIRE, no one assisted us. We had to struggle to make ends meet."

The success of CAMPFIRE has engendered some hard feelings, primarily because not everyone shares the program's financial rewards. Only villages with wildlife resources can participate. Moreover, each village's share of the spoils depends on which animals are shot on its land. Dete, for example, received $3,534 in 1993, while the neighboring village of Chikova earned four times as much because it has more wilderness area and higher-priced animals, including elephants, buffalo, lions, leopards, and sable antelope.

Villages with no wildlife resources receive no CAMPFIRE funds. Margaret Taodzera, chairperson of the wildlife committee in Hurungwe's Chitindiva village, is an outspoken critic of the CAMPFIRE program. "We do not have any animals in this area. We are surrounded by other villages. So, we get nothing," she complains. "Every household in this district should get a share. We are all under one chief. The animals belong to God, not to this village or that village. No one is feeding those animals. The money should go to projects, not to households."

In an attempt to address such grievances, neighboring villages that received money from CAMPFIRE recently gave Chitindiva funds to buy a grinding mill. Nevertheless, people living on communal lands with wildlife argue that they deserve their compensation since they bear the sometimes heavy costs imposed by the animals. Every year crocodiles, hippopotamuses, and elephants kill people in communal areas.

The animals also destroy crops. In Chikova, where nearly one-third of the land remains a wilderness, elephants, baboons, and bushpigs regularly raid fields. Kenyas Dzokwnushure grows corn and groundnuts. "I have major problems with elephants, because when they come into the fields, they completely destroy them," he says. The elephants come at night to feed. "I start a fire and the whole family beats tin cans and buckets throughout the night to scare them off." Sometimes this doesn't work, and the lost crops mean no food and more debt.

In Hurungwe, CAMPFIRE does not reimburse farmers for losses, but it does put money into their pockets while offering some protection. As part of his deal with the Hurungwe District Council, which allows him to bring clients onto communal lands, professional hunter Ray Townsend agreed to help communities handle problem animals. Every year, at least one elephant develops an appetite for corn. Before reacting, district as well as park officials assess the extent of the damage. If they determine that an elephant has become a menace, they turn it over to Townsend. He tries to sell the animal, often at a reduced price, to a client, ensuring the greatest return for the community.

CAMPFIRE has sparked interest throughout Africa, and conservationists from many neighboring countries have visited Zimbabwe to assess the program. All are trying to devise ways for local communities in their own countries to benefit from national parks and reserves. Simon Metcalf, a development expert with the Zimbabwe Trust, says that while CAMPFIRE probably will not be duplicated, other countries may borrow some of its features.

Richard Leakey, former head of the Kenyan Wildlife Service and an outspoken conservationist, is less enthusiastic. "It works perfectly well on the communal lands in Zimbabwe, but I don't think it would work anywhere else in Africa in the same way. The communal land structure in Zimbabwe is unique."

Leakey also has doubts about sport hunting as a sustainable source of income, particularly for highly remunerative species such as elephants, which each yield at least $7,500 in trophy fees. "Sport hunting of elephants is obviously an alternative to the ivory trade at the moment, but whether it is viable in the long term, I seriously doubt. I personally believe that attitudes about elephants have and will continue to change, making it more and more likely that killing elephants will become increasingly antisocial."

For the moment, however, Dete residents are tapping the benefits of CAMPFIRE. In 1993, the villagers voted to use CAMPFIRE income to finish a desperately needed clinic. Before the clinic was completed, pregnant women, the sick, and the injured walked, went by ox cart, or caught an irregular bus to the nearest clinic 18 kilometers (12 miles) away. "A few months ago a child died on the bus on the way to the clinic," says David Mutara, a sixty-two-year-old corn farmer.

His frown melts into a smile as he points at the clinic being built in the center of the village. "I'm very happy about what CAMPFIRE did," he says.

12

WILDLIFE CONSERVATION OUTSIDE PROTECTED AREAS:
LESSONS FROM AN EXPERIMENT IN ZAMBIA

Dale Lewis, Gilson B. Kaweche, and Ackim Mwenya

INTRODUCTION

Over the past decade wildlife populations in Africa have declined drastically throughout much of their former range. This trend has occurred despite attempts to improve law enforcement and supply badly needed field equipment. While the search for more lasting, effective solutions to wildlife conservation in Africa continues, there is a growing consensus that part of the solution will require greater involvement by residents living near wildlife resources to both manage and benefit from the sustainable development of these resources. Ironically, this approach conforms to customs of traditional African societies. Prins and Marks argue that these customs were ecologically adaptive and rigidly followed under societal traditions.[1]

Modern-day Africa has introduced new rules and authorities for governing land use, and this has tended to weaken and sometimes even eliminate traditional authorities.[2] Wildlife conservation has not been exempt from such government-regulated changes, which to a large extent were inherited by postcolonial governments. One important consequence of such government intervention is the reduced level of ownership by rural residents of their wildlife resources.[3]

If any lesson can be learned from past failures of conservation in Africa, it is that conservation implemented solely by government for the assumed benefit of its people will probably have limited success, especially in countries with weakened economies. Instead, conservation for the people and by the people

---

This article originally appeared in *Conservation Biology* 4 (June 1990): 171–80. Reprinted by permission of Blackwell Science, Inc. Copyright © 1990 Blackwell Science, Inc. All rights reserved.

with a largely service and supervisory role delegated to government authorities could foster a more cooperative relationship between government and the residents living with the resource. This might reduce the costs of law enforcement and increase revenues available to other aspects of wildlife management, which could help support the needs of conservation as well as those of the immediate community. Such an approach would have the added advantage of restoring to local residents a greater sense of traditional ownership and responsibility for this resource. Convincing proof that such a partnership is possible has yet to be demonstrated and has therefore been more theoretical than pragmatic.

This paper examines how a form of partnership reduced potential conflicts between villagers and wildlife in one experimental project. Its objective was to help financially support both community improvements and wildlife conservation through sustained-yield uses of this resource with the involvement of local villages in an area outside South Luangwa National Park, Zambia.

The results of this experiment fell into four broad categories: (1) rate of illegal off-take of wildlife, (2) sustainability of economic incentives for local residents to support sustained-yield management of this resource, (3) attitude and perceptual changes by residents toward wildlife issues, and (4) level of manpower and leadership derived from the village community to properly manage wildlife resources. From these results we derived a cost-benefit analysis to compare the methods adopted in this experiment with more conventional approaches of wildlife management used elsewhere in Africa.

Description of Study Area, Village-Wildlife Conflicts, and Village Socioeconomics

The Luangwa Valley supports a world-acclaimed diversity and abundance of wildlife, the greatest concentrations occurring along the alluvial belt of the Luangwa River. Its four national parks account for approximately 20 percent of the Luangwa Valley catchment. Surrounding these parks are game-management areas more than three times as large as the parks themselves.

These game-management areas differ from parks in that they are zoned for wildlife utilization, mainly hunting, and also allow human residency. In both zones, however, wildlife is constitutionally the property of the state, and hunting in game-management areas requires licenses that are often prohibitively expensive to residents. This denial of access to protein resources undoubtedly arouses negative sentiments among local residents toward government wildlife policies. Such conflicts have been well recognized in previous work[4] and have contributed to the drastic decline of both elephants and black rhinos in the Luangwa Valley.[5]

A total of five distinct tribes with their own respective chiefdoms live in the Luangwa Valley. Located to the east of South Luangwa National Park is the Lupande Game Management Area. With a total land area of 4,849 square kilometers, Lupande is occupied by six separate chiefdoms of the Kunda tribe with a total population of approximately twenty thousand. This study was carried out principally in one of these chiefdoms, Malama, and to a lesser extent in the Chief Kakumbi and Chief Msoro areas. Chief Malama's area, totaling approximately 400 square kilometers, is the most remote of the six Kunda chiefdoms. It is also the least populated, with about seven hundred people distributed in village settlements that occupy about 5 percent of the total Malama land area. These settlements are located principally along alluvial drainages where subsistence farming and some cash farming are practiced.

Prior to this study, a socioeconomic survey was conducted in the Malama, Msoro, and Kakumbi areas to establish a baseline description of revenue sources generated from within the respective areas.[6] The results showed a rural economy based largely on agriculture, although only 3 to 7 percent of the total land area is arable.[7] The largest single revenue earner was international safari hunting, which grossed approximately $350,000 per year for the whole of Lupande. Ironically, less than 1 percent of the safari hunting revenue was returned to support local village economies[8] and a negligible amount was turned to wildlife management costs. Predictably, local attitudes toward safari hunting were negative.

During much of the 1970s and early 1980s, illegal hunting of wildlife, especially elephants and rhinos, reached unprecedented levels, with a 6 to 8 percent annual decrease for elephants.[9] The Zambian government intensified law enforcement with better-equipped wildlife scouts who were government civil servants. These scouts were not members of the local community and their presence near villages was generally unwelcome. In Malama's chiefdom, more than half of the respondents in Atkins's survey expressed negative attitudes toward Zambia's National Parks and Wildlife Service (NPWS), which in their view favored the hunting privileges of foreign hunters more than local residents.[10] As a result, poachers entering their area from more distant places were encouraged to hunt illegally as long as they shared some of the meat. The role of the NPWS government department in Lupande Game Management Area was therefore frequently contested and criticized by local residents, as is reflected in a statement in an address given by Chief Malama: "Tourists come here to enjoy the lodges and to view wildlife. Safari companies come here to kill animals and make money. We are forgotten. . . . Employment here is too low. Luamfwa Lodge employs only about four people and safari hunting employs no one. How can you ask us to cooperate with conservation when this is so?"[11]

Methods

Experimental Design

In response to these problems, an experimental design for managing wildlife was adopted by the NPWS and tested in the lower half of the Lupande Game Management Area. It was called the Lupande Development Project. The design was based on the critical premise that a share of wildlife revenues be retained by NPWS to support the management needs of this department for the area where the funds were generated. A financial institution within NPWS, the Wildlife Conservation Revolving Fund (WCRF), was created for this purpose in 1983. A second premise of this design is that NPWS be allowed to employ additional staff beyond the government-approved quota of civil servants.

The initial design of the proposed Game Manangemnet Area (GMA) structure called for a wildlife management approach in which (1) manpower requirements were drawn from the local village community, (2) manpower training and development was supervised by NPWS to ensure a high standard, (3) issues of wildlife management were dealt with in collaboration with village leaders through wildlife subcommittees for each chiefdom, (4) administration of proposed new GMA structure was charged to a NPWS officer, called a unit leader, and (5) revenue generated by the WCRF from within Lower Lupande was retained to support both wildlife management costs and local community benefits.

To develop a reliable and adequate workforce for wildlife management, men from the ages of twenty to thirty-five who resided in local villages were chosen for training under NPWS personnel in such skills as law enforcement, wildlife censusing, data collecting, and report writing. After a minimum of six months training, these recruits were designated village scouts officially employed by NPWS as non-civil servants. They were employed year-round, and they remained in their respective chiefdoms as the local "custodians" of their wildlife resources. The unit leader supervised and monitored their work. On a seasonal basis, additional workers were also recruited from the local community to assist with other management needs of the area. A crew of brickmakers, masons, and carpenters was recruited as a building maintenance team. Supervised by the unit leader, they provided the work force for constructing residences for the village scouts at remote stations in Lupande.

The management design also provided for critical input from village leaders on existing wildlife programs and on the planning of future ones. Initially, this took the form of direct talks with the local chief in the company of an unspecified number of village elders or headmen. Within the course of a year, however, local village leaders themselves recommended that a village wildlife committee, composed of all the headmen, the chief, and appropriate NPWS officials, including the unit leader, be formed to discuss issues of wildlife. Under the joint chairmanship of the unit leader and the chief, this approach was adopted.

Of critical significance to the design is the earning and handling of wildlife revenues. Two sources of revenues were identified for this project: wildlife safari hunting and sustained-yield harvesting of hippos. Revenue from the former was based on a public auction in which safari-hunting companies made private bids for the right to hunt in the area, which would be managed and policed through the joint participation of NPWS and the local community. The highest bid was accepted as a concession fee, and 40 percent was handed over by WCRF to the local Kunda chiefs for community projects, 60 percent being recycled back into the GMA for meeting wildlife management costs.

Another source of revenue was from the sustained-yield harvest of hippos and the commercial marketing of their teeth, hides, and meat. The net profit from this program was shared in the same proportions as described above between the local community and NPWS, which used its share for wildlife management costs in Lupande. The methodologies adopted for this harvesting program were labor intensive to maximize employment opportunities for local residents in Malama area.

This design assumed that awareness and understanding of wildlife management by local residents would increase as revenue benefits to the community increased and village participation in the program expanded. The design encouraged a greater administrative role in wildlife management by local authorities under the guidance and supervision of the unit leader. In this way, cost-effective solutions to wildlife management problems that might require local input of ideas and participation would be more forthcoming, and would help diversify the sources and overall amount of wildlife revenue in the game management areas.

Testing the Design

The test of the design was based on measured responses evaluated before and after implementation. The following data were assessed: illegal killing of animals in Lower Lupande, local attitudes toward wildlife conservation, revenue earnings from wildfire resources, and the allocation of these earnings toward community development and wildlife management costs. These data were collected from 1985, the year preceding the Lupande Development Project, through 1987. Since one objective of this experimental design was to reduce the poaching (or illegal killing) of elephants and black rhinos, these two species were used as indicators of overall trends in poaching in Lower Lupande.

Various measures of law enforcement were assessed to relate them to any changes in poaching rates. One was total number of field officers active in the study area and the other was man-days, the total number of twenty-four-hour days of patrolling for a given law enforcement operation times the total number of people who were on patrol. The assessment of law enforcement effort was also related to area, since the density of enforcement personnel might be more important than just total number of personnel. To determine the area patrolled, field staff indicated on a map the area traversed during their field operations.

Assessing local attitudes toward conservation involved several indirect indicators: attitudes and views expressed at public forums, frequency of information volunteered by local residents in assisting with arrests of illegal hunters, and the incidence of volunteer participation in matters affecting wildlife conservation. A more direct method was used toward the end of the study period. In November 1987, thirty adult male respondents were randomly selected in Malama chiefdom to answer a set of questions designed to evaluate local community perceptions of the various programs implemented in this management design and whether local attitudes toward NPWS had become more favorable as a result. This questionnaire survey was carried out by a team of agricultural enumerators who were nonresidents and were unrelated to the Lupande Development Project. The respondents were told the survey was part of the routine work of the Department of Agriculture; for this reason, questions of crop damage by wildlife were included. This was done so that respondents would not link the questionnaire to the Lupande Development Project and feel obliged to answer in a biased way.

Revenue earnings were made fully accountable through the WCRF for earnings from safari-hunting concessions and hippo culling. Allocation of these funds to the local community was determined by the agreed percentage of net revenues that went directly to the Kunda chiefs from the WCRF. Through their own forum for deciding how these funds should be spent, the chiefs then allocated these revenues to community projects.

Results

Wildlife Management: Effort and Results

In 1985 effective law enforcement within the designated area of this Experimental Management program was limited to only 200 square kilometers. Five civil servant wildlife scouts and six village scout trainees were allocated to this area, and they conducted law enforcement operations only when gunshots of suspected illegal hunters were heard. These personnel were deployed from a single camp. In the surrounding area effective law enforcement was nil. In 1986 the recruitment of wildlife management personnel from the local community as village scouts intensified. By the second half of 1986 a total work force of eighteen policed an area of approximately 275 square kilometers. With continued recruiting and training of village scouts, the number of field staff reached twenty-six by August 1987, and the total area of surveillance was approximately 400 square kilometers. Of this area, 30 percent overlapped inside the South Luangwa National Park.

Prior to the village scout program, a workforce comprised almost exclusively of civil servants policed the area. The increase in manpower from eleven to twenty-six between 1985 and 1987 resulted from recruitment of local residents as village scouts. From November 1986 to July 1987, total man days of field operations for civil servants and village scouts revealed a disproportionately greater effort by village scouts to patrol and conduct field operations than by civil servant scouts. This difference was 717 versus 176 man-days, respectively. Contributing to this discrepancy was the higher rate of absenteeism among civil servants, who left their stations for supplies, medical assistance, or collection of salaries. With village scouts, salaries were paid directly by the unit leader and visits away from their stations were normally for short-duration trips to their villages, which were near their camps.

Based on their relatively high man-day effort, village scouts' contribution to law enforcement accounted for much of the increase in number of arrests and firearms seized during the period of 1985 to 1987. Consistent with these trends in law enforcement was the downward trend in poaching mortality for both elephants and black rhinos. Annual mortality from the illegal hunting of these two species, expressed as a number of poached carcasses found per year per square kilometer, decreased at least tenfold from 1985 to 1987.

In addition to law enforcement, other important functions of wildlife management were carried out by village scouts in the experimental program under the supervision of unit leaders. For instance, village scouts accompanied safari hunters and their clients throughout the hunting season to fill out daily data sheets for evaluating hunting effort and trophy yields and to record information on wildlife numbers in the areas hunted. Additional duties included controlled early burning in May and animal counts along fixed transects during September. Village scouts' particular skills useful for wildlife management were also developed in this program; these skills included vehicle maintenance, typing, and driving.

Wildlife Earnings and Allocation of Revenue to Wildlife Conservation

Revenue. A key focus of this study was whether conservation costs for the program area could be annually balanced from wildlife revenue earned by WCRF from the same area. Five sources of revenue, excluding profits to the private sector, resulted from this experimental program: safari-hunting concession fees, safari-hunting licenses, collection of ivory, fines from court convictions for illegal hunting, and sales from hippo harvest. These revenues were shared among those entities with a political influence on wildlife conservation (i.e., the Central Treasury) as well as with the resident community participating in this program. Table 1 shows the complete breakdown and intended destination of the total earnings for 1987, which equaled K [Zambian Kwacha] 694,574. The total amount allocated specifically for wildlife management under the Lupande Development Project was K 212,067, as derived from safari hunting and a projected income from hippo harvesting. A reduced hippo harvest in 1987 caused by an anthrax epidemic lowered the actual revenue allocated to wildlife management for Lower Lupande in 1987 to K 146,000. This amount was earned from less than 25 percent of the total area of Lower Lupande, although the remaining area has the potential to become another safari hunting concession block if former wildlife numbers are restored.

Wildlife management costs. To determine the affordability of this management program, expenditures on all activities included in this experimental program were monitored for 1987 and balanced against revenue earned to pay for these management costs. Six categories were used to describe the recurrent costs of this program: village scouts, housing construction and camp improvements, office supplies, vehicle maintenance, public relations, and capital replacement costs. The recurrent cost for implementing the village scout program was

Table 1. A breakdown of revenue earnings from wildlife in Lower Lupande GMA excluding those from the private sector. Indicated also are the destinations of these funds.

Category	Destination	Revenue
Safari-hunting concession fee	Wildlife management costs	K 146,000
	Local community projects	97,400
	Total	243,400
Safari-hunting license fee	Central Treasury	289,040
Elephant ivory	NPWS Revolving Fund	18,200
Hippo utilization	Wildlife managements costs	65,667
	Local community projects	65,667
	Total	438,574
Fines from arrested poachers	Central Treasury (approx)	12,000
	Total revenue:	K 693,974

K 63,685, which exceeded all other management costs (see Table 2) and covered salaries, uniforms, ammunition, and so on. In addition to the village scouts, building and camp improvement projects required a major fraction of the total budgets allocated to employment. In accordance with the design of this management program, only local residents were considered for employment. The total revenue spent on subsidies for these two projects for 1987 was K 50,747, or 60 percent of their combined total budget. Total revenue directed to the local community as salaries from all wildlife activities in this management program equaled K 70,212 (see Table 3).

Overall, recurrent costs for 1987 equaled K 141,212, which was below the realized revenue for wildlife management costs for that year. The positive balance was only 4 percent of the total revenue, not enough to expand this management program to unaffected areas elsewhere in Lower Lupande. This underlined the importance of the potential revenue expected from the sustained yield harvesting of hippopotamuses as a way to diversify and increase revenue from Lupande. Funds from overseas conservation grants were used for capital equipment purchases, which equaled K 181,240. These capital purchases were special case purchases; they were not necessary to implement the wildlife management program but were required to help monitor the wildlife development of this area to relate it to other game management areas.

Table 2. Recurrent wildlife management costs from November 1986 to October 1987 in the project area.

Category	Items	Total expenditure	Summary
Village Scout Program	Salaries & bonuses	K 41,200	
	Rations for patrols	1,165	
	Operational expenses (fuel, ammunition)	14,765	
	Additional requirements (boots, uniforms)	6,555	
			K 63,685
Building construction and camp improvements	Salaries	9,547	
	Materials	8,142	
	Rations	890	
	Transport	2,150	
			20,729
Office supplies			2,993
Vehicle & other equipment maintenance	Repairs	12,820	
	Spares	13,485	
	Oils & grease	4,000	
			30,305
Public relations	Display boards	1,400	
	Wildlife subcommittee	2,100	
			3,500
Per annum depreciation on capital items			20,000
		Total recurrent costs:	K 141,212

Attitude Changes Among Local Residents

Earlier work revealed that the majority of local residents in the area to be affected by this project had negative sentiments toward NPWS.[12] Many residents felt this department served the needs of tourists and safari hunters and even the resource itself, more than it served the needs of local residents. The people interviewed in these earlier studies had little appreciation of the level of legal benefits wildlife could bring into their area.

The role of research in trying to better manage and develop wildlife resources was seen as a ploy to convert lands being used by villages into another national park. Local residents held strong traditional feelings of

Table 3. Employment records from November 1986 to October 1987 showing number of people employed for particular job categories, salaries paid out, and the source of employment.

Category	Total number employed	Total salaries paid out	Source of employment
Building projects	20	K 11,715	NPWS Rev. Fund
Safari hunting	46	10,897	Safari company
Village scouts	16	41,200	NPWS Rev. Fund
Village scout trainees	12	3,200	NPWS Rev. Fund
Hippo utilization	20	3,200	NPWS Rev. Fund
Totals:	114	70,212	

authority over their lands and challenged the role of NPWS in advising local villagers on how to manage their wildlife resources.

It appeared that socioeconomic factors contributed to these strong sentiments, since the most common means available to local residents to profit directly from wildlife was illegal hunting, and existing low levels of income made such practices a question of survival for some residents. The solution adopted by this program was to employ local residents in the wildlife management effort on a permanent and seasonal basis and to provide a significant share of the wildlife revenue to the Kunda chiefs in Lupande Game Management Area. As stated earlier, local employment in 1987 was K 70,212. Employment from safari hunting is included in this figure because such employment was a condition of the terms for the safari company to hunt in Lower Lupande. In 1986 a relatively short hunting season was concessioned to a safari-hunting company in Lower Lupande for K 159,000. Based on the revenue-sharing formula, 40 percent of these funds went directly to the local chiefs. In early 1987 a total of K 63,600 was made available to the Kunda chiefs in Lupande for sponsoring community projects.

In December 1986 the first wildlife subcommittee for Chief Malama's area was convened and a review of the wildlife management program was presented to the committee members. Unlike earlier meetings between local leaders and National Park staff in Malama area, this meeting ended with an audience convinced that benefits from wildlife were possible through sustained-yield management and that NPWS could promote such benefits as an extension service to the local community. A resolution by the committee members was that local headmen should form a security committee to prevent poachers from entering their area. In March 1987, Chief Malama convened a meeting of all headmen, instructing his village not to cooperate with poachers and to report to him the presence of any poachers who entered his area. Also in 1987, Chief Msoro, whose village is located east of Chief Malama's area, expressed a desire to have village scouts in his chiefdom, and later that year he personally arrested a resident within his village for poaching an elephant in the eastern half of Lupande outside the area of this project. In addition, the presence of poachers became known through the assistance of residents who volunteered the information. From January to August 1987, the arrest of three groups of poachers resulted from such information. Such assistance from local residents was a rare occurrence in previous years.

This positive change in local attitudes toward wildlife resources and the improved perceptions of NPWS was confirmed by the November 1987 questionnaire survey. Recognition of NPWS's unit leader as an appropriate channel to resolve conflicts over wildlife issues was suggested by the fact that 66 percent of the respondents said they seek the unit leader (or a senior NPWS officer) for such help. In addition, 63 percent approved of the village scout program. The most common reason for this approval was the reduction of poaching it has brought to their area. When asked whether the sustained-yield culling program should be continued, 86 percent replied yes and 68 percent correctly perceived the role of this program as a way of bringing employment and development to their area. When asked where residents now go to find employment in their area 64 percent answered they seek work through some form of wildlife management activity now being promoted by NPWS in Lupande.

Discussion

The Luangwa Valley, like many other wildlife areas in Africa, has a history of wildlife protection policies that are implemented to a large extent by civil servants who are not residents of the areas where they work. These policies have been both expensive and difficult to finance on a regular basis. Recent trends of elephant and black rhino poaching in the Luangwa Valley suggest that this form of law enforcement is limited in its effectiveness, is extremely expensive, and may be inappropriate for an area where the ratio of potential poachers to the number of arresting civil servant officers present is high or rapidly increasing.[13]

Lupande Game Management Area supports both village residents and wildlife populations, whereas national parks allow no residency by villagers. Integrating local residents as part of the management solution to protect wildlife resources is an option available to game management areas. Furthermore, areas surrounding national parks are conduits for illegal hunters entering parks. If the value of conservation is fully appreciated by local communities in game management areas, the residents themselves might discourage illegal hunters from entering parks as well as their own areas. In theory, effective wildlife management outside protected areas should lower the costs of wildlife protection within these protected areas.

Involving local residents through wildlife employment and sharing wildlife revenues with tribal authorities are not new ideas.[14] A cost-benefit analysis for how such a program can be designed and budgeted is far less common, especially for the case where subsistence farming is practiced and wildlife is often considered a liability to farmers. The principal aim of this paper was to conduct such an analysis to evaluate the affordability and cost-effectiveness of this experimental design of wildlife management.

Improved Wildlife Management

Deployment of village scouts yielded a net increase of per capita man-day effort in law enforcement operations. In addition to a superior knowledge of the surrounding countryside, village scouts appeared to have an advantage over non-resident civil servant scouts in carrying out operations. This was apparent from the dramatic increase in arrests and firearm confiscations after the village scouts entered the work force. In addition to improved law enforcement, village scouts provided the badly needed services of monitoring safari hunting, conducting animal counts, and improving public relations between NPWS and the local community. Salary incentives for these scouts, though less than those of civil servant scouts, were above expected income levels for the area,[15] and this undoubtedly contributed to their relatively high motivation for work. The overall cost of this program was K 63,685, which provided a manpower coverage of 8 square kilometers per scout for the program area or approximately K 80 (or $10) per square kilometer.

Based on the decline of poached elephant carcasses found during the period of this study and the existing density of this species,[16] potential revenues from future sustained-yield uses of elephants would more than meet the total costs of this program as based on current earnings in Zimbabwe.[17] In addition, approximately half of the cost for supporting the village scouts was equivalent to the total sum derived from revenue earned by ivory collection during scout patrols and by court fines from poachers arrested. Though these sources of revenue did not go back into the resource management costs of Lower Lupande, they do illustrate the amount of money the Zambian government could recover though this form of local involvement in wildlife management.

Public Attitudes and Perceptions

Public reaction in the local village community to the village scouts was initially negative but became supportive once the revenue benefits from wildlife to the local community became apparent. Providing local leaders with the information that more than 90 percent of the poachers apprehended by village scouts were people living outside their area appeared to encourage a more protective attitude for their wildlife. It also precipitated greater appreciation among residents for the role of village scouts.

Organizational and administrative leadership of this program was provided by the NPWS unit leaders. Their leadership was essential in promoting public awareness of how the involvement of local residents in management can contribute to improved economic benefits for residents from wildlife. Unit leaders also served as technical advisors to the local community on wildlife topics. During the local wildlife management committee meetings for each chiefdom, leaders were able to present the chief and committee members with information and ideas helpful to their area's wildlife development. Membership of this committee was primarily limited to the chief's council as a show of respect for traditional leadership authorities in the villages. Such conformity with local customs may have increased the level of appreciation among residents of the increased employment and the shared wildlife revenues provided by this program. The perceptional changes expected would include greater understanding of the values of conservation and of the legitimacy of NPWS as a wildlife extension service. The results of the attitude survey suggested that such changes were evolving in the community.

Affordability of Conservation Based on the Lupande Model

Attempts to define the financial inputs necessary to sustain wildlife conservation costs in Africa are based largely on estimates from countries whose economies allow as much as $200 per square kilometer per year.[18] Estimates range from three to five times higher for more developed non-African nations. In less developed countries whose economies are much weaker, conservation must be carefully budgeted since it generally receives a low priority of government spending. In Zambia, for instance, the average recurrent investment in conservation for its wildlife estate in the early 1980s was about $2 to $4 per square kilometer.

This investment has been used to support a management approach whose design is basically modeled after much wealthier nations and requires input fifty times greater to be effective. Hence, there is great justification for examining alternative approaches to conservation in developing countries.

The experimental design described and tested in this study has revealed that wildlife conservation in areas outside protected areas can be made more cost-effective by combining the efforts of local residents with those of NPWS personnel, rather than engaging only NPWS civil servants. The overall recurrent costs for this program was K 175 (or $22) per square kilometer, and in less than three years, law enforcement efforts reduced illegal hunting of both elephants and black rhinos in the program area by approximately 90 percent. Such improvements in conservation also helped sustain future revenues from wildlife for local community benefits and helped meet wildlife management costs.

In 1987 total wildlife revenue exceeded the total annual operating budget for this management program by almost four times. More than half of the revenue entered Zambia's Central Treasury in the form of payments for hunting licenses. The revenue sharing formula adopted in this design, however, provided K 146,400 for wildlife management costs, which exceeded recurrent cost by K 5,198. This marginally positive balance would have easily been a substantially positive one had the hippo utilization scheme not been canceled due to the anthrax outbreak. The anthrax outbreak that depleted the hippo population emphasized the need for a multiple-use approach to wildlife resources to minimize dependence on a single source of income. As a result, the Lupande Development Project subsequently embarked on a multispecies utilization scheme and a self-catering tourist enterprise owned and operated by the local community. This has helped to diversify and maintain wildlife earnings at levels necessary to meet the overall needs of conservation and community benefits.

Results from this study suggested that maximizing revenue earnings for wildlife management costs may not be in the best interest of wildlife conservation. For instance, the revenue-sharing formula guaranteed significant shares of wildlife profits to the local community and the central government through its appropriate institutions. This revenue-sharing formula, combined with the high priority given to local employment, increased both public and political support for the economic importance of wildlife. This approach, which maximizes local benefits, contrasts favorably with the more traditional system of using wildlife earnings to finance law enforcement activities to control illegal uses of wildlife.

The design also allowed market forces to dictate the actual earnings for conservation and community benefits since the hunting concession was based on public bids among the companies competing for this area. This approach ensured that the private sector was not overcharged for the concession and it also placed greater responsibility on local residents to help maintain this market value for their own community welfare.

Summary

An experimental program in Lupande Game Management Area, Zambia, tested the feasibility of allowing local residents to participate in managing wildlife resources through a joint partnership with the National Parks and Wildlife Service. Such a partnership, in the form of village wildlife management committees, deployment of village scouts, a sustained-yield wildlife utilization scheme managed by local villagers, and wildlife-related employment for local residents, increased traditional authorities' involvement in planning for their wildlife resources. This approach was far more acceptable to the local community than past efforts to manage wildlife by relying entirely on government authorities.

The cost-effectiveness of wildlife law enforcement increased dramatically with the establishment of village scouts, who were all local residents trained under this program to manage and police the wildlife resources in their respective chiefdoms.

Revenue earned by charging a concession fee for a self-sustained and carefully regulated off-take of wildlife trophies by safari hunting sportsmen was sufficient to meet the recurrent costs of this program.

A multiple-use approach to wildlife management was found to be desirable to minimize risks of cash shortfalls in maintaining the costs of this program on an annual basis.

A management linkage with tribal authorities as was used in this model study will reduce the costs of protecting wildlife resources in adjacent national parks where human cohabitation with wildlife is not allowed. Once monetary benefits from wildlife are made legally available to local residents living outside the national park, nonresident intruders who threaten this resource will be unwelcome and be discouraged from entering the area, even to gain access to the national park.

Notes

1. G. Prins, *The Hidden Hippopotamus* (Cambridge: Cambridge University Press, 1980); S. A. Marks, *Large Mammals and a Brave People* (Seattle: University of Washington Press, 1976).

2. A. R. Roberts, *A History of the Bemba* (Madison: University of Wisconsin Press, 1973); G. C. Bond, *The Politics of Change in a Zambian Community* (Chicago: University of Chicago Press, 1976); A. J. Willis, *An Introduction to the History of Central Africa*, 4th ed. (Oxford: University of Oxford Press, 1985).

3. Marks, *Large Mammals and a Brave People*.

4. D. M. Lewis and G. B. Kaweche, "The Luangwa Valley of Zambia: Preserving Its Future by Integrated Management," *Ambio* 14 (1984): 362–65.

5. D. M. Lewis, *A Management Study of the Luangwa Valley Elephants* (Lusaka, Zambia: National Parks and Wildlife Service, 1986); N. Leder-Williams, "Black Rhino in South Luangwa National Park," *Oryx* 19 (1985): 27–34.

6. S. Atkins, "Socio-Economic Aspects of the Lupande Game Management Area," in *Proceedings of the Lupande Development Workshop*, ed. D. B. Dalal-Clayton and D. M. Lewis (Republic of Zambia, 1984).

7. D. B. Dalal-Clayton, R. N. Magai, and W. J. Veldkamp, "Preliminary Land Classification and Land Evaluation in Lupande Game Management Area," in *Proceedings of the Lupande Development Workshop*.

8. Atkins, "Socio-Economic Aspects of the Lupande Game Management Area."

9. Lewis, "A Management Study of the Luangwa Valley Elephants."

10. Atkins, "Socio-Economic Aspects of the Lupande Game Management Area."

11. G. Malama, "Welcoming Address by the Honorable Chief Malama," in *Proceedings of the Lupande Development Workshop*.

12. Atkins, "Socio-Economic Aspects of the Lupande Game Management Area."

13. Lewis, "A Management Study of the Luangwa Valley Elephants"; Leder-Williams, "Black Rhino in South Luangwa National Park."

14. D. Western, "Amboseli National Park: Enlisting Landowners to Conserve Migratory Wildlife," *Ambio* 5 (1982): 302–308; Rowan Martin, personal communication with the authors; A. Vedder, "In the Hall of the Mountain Gorilla," *Animal Kingdom* 92 (1984): 31–43.

15. Atkins, "Socio-Economic Aspects of the Lupande Game Management Area."

16. Lewis, "A Management Study of the Luangwa Valley Elephants."
17. Martin, personal communication.
18. IUCN/WWF/NYZS Elephant/Rhino Specialist Group Meeting 1986.

References

Atkins, S. L "Socio-Economic Aspects of the Lupande Game Management Area." In *Proceedings of the Lupande Development Workshop*, edited by D. B. Dalal-Clayton and D. M. Lewis. Republic of Zambia, 1984.

Bond, G. C. *The Politics of Change in a Zambian Community*. Chicago: University of Chicago Press, 1976.

Dalal-Clayton, D. B., R. N. Magai, and W. J. Veldkamp. "Preliminary Land Classification and Land Evaluation in Lupande Game Management Area." In *Proceedings of the Lupande Development Workshop*, edited by D. B. Dalal-Clayton and D. M. Lewis. Republic of Zambia, 1984.

Leder-Williams N. "Black Rhino in South Luangwa National Park." *Oryx* 19 (1985): 27–34.

Lewis, D. M. *A Management Study of the Luangwa Valley Elephants*. Lusaka, Zambia: National Parks and Wildlife Service, 1986.

Lewis, D. M. and G. B. Kaweche. "The Luangwa Valley of Zambia: Preserving Its Future by Integrated Management." *Ambio* 14 (1984): 362–65.

Malama, G. "Welcoming Address by the Honorable Chief Malama." *Proceedings of the Lupande Development Workshop*, edited by D. B. Dalal-Clayton and D. M. Lewis. Republic of Zambia, 1984.

Marks, S. A. *Large Mammals and a Brave People*. Seattle: University of Washington Press, 1976.

Martin, Rowan. Personal communication with the authors.

Prins, G. *The Hidden Hippopotamus*. Cambridge: Cambridge University Press, 1980.

Roberts, A. R. *A History of the Bemba*. Madison: University of Wisconsin Press, 1973.

Vedder, A. "In the Hall of the Mountain Gorilla." *Animal Kingdom* 92 (1989): 31–43.

Western, D. "Amboseli National Park: Enlisting Landowners to Conserve Migratory Wildlife." *Ambio* 5 (1982): 302–308.

Willis, A. J. *An Introduction to the History of Central Africa*. 4th ed. Oxford: Oxford University Press, 1985.

13

WILDLIFE CONSERVATION IN TRIBAL SOCIETIES

Raymond Hames

INTRODUCTION

It is widely alleged that unacculturated tribal peoples lived in harmony with the environment.[1] This very general concept is most frequently used to imply that aboriginal use of the environment approached a steady state such that demands for renewable resources did not exceed environmental replenishment. This claim is most forcefully made in reference to exploitation of wild game, fish, and plant resources by hunter-gatherers or low-density shifting cultivators (e.g., native Amazonians) with high dependence on wild species.[2] Three lines of independent evidence are used to support this claim. First, there is little contemporary evidence (until recently) that relatively unacculturated native tribal populations have caused local extinctions of fish and game species upon which they depended for their subsistence.

Second, the population densities of many such tribal groups appear to be in equilibrium, i.e., are not substantially increasing or decreasing. For example, Birdsell has argued that an equilibrium density of Amazonian aboriginal hunter-gatherers is achieved by cultural means such as infanticide, abortion, and sexual regulations that prevent individual groups from placing heavy demands on the ecosystem.[3] Sahlins has similarly argued that, in addition to cultural population regulation mechanisms, culturally determined desires or wants in tribal societies are moderate and thus prevent tribal people from overtaxing their resource base.[4] These twin processes of population regulation and moderate desires led to

---

From *Biodiversity: Culture, Conservation, and Ecodevelopment*, edited by Margery L. Oldfield and Janis B. Alcorn. Reprinted by permission. Copyright © 1991 Raymond Hames. All rights reserved.

the maintenance of tribal populations at approximately 50 percent of the carrying capacity of the environment.[5]

Third, native ideological systems have been adduced as evidence for wildlife conservation. Martin argued that reverence of game animals in many North American native societies was designed specifically for conservation.[6] Reichel-Dolmatoff suggested for the Colombian Tukano and many other native Amazonian groups that religion promotes a balance between humans and wildlife by preventing humans from overexploiting game resources.[7] McDonald and to some extent Ross suggested that among many Amazonian groups, animal taboos represent an effective mechanism designed to prevent overhunting.[8]

It is easy to conclude that native peoples were or are conservationists given the paucity of contemporary evidence that native peoples caused local extinction of game species, the fact that such societies apparently exist below the carrying capacity of their environment and that the veneration of game animals is a common ideological element in their societies. Even negative evidence for conservation among native people has been used to demonstrate aboriginal conservation. There is clear evidence that the introduction of guns to Plains Indians led, in part, to local extinctions of bison.[9] However, since the gun was nonnative, native peoples did not have adequate time to adjust to this remarkable advance in hunting technology. Among subarctic Indians, the steel trap and a commercial market for furs led to the local decimation of valuable furbearers.[10] In this case, there was not only a technological introduction, but more importantly, Indians were forced into a nonnative economic system that stressed individualistic aggrandizement. In a convoluted thesis, Martin argued that North American native peoples began slaughtering the species they formerly conserved because of the introduction of European diseases.[11] In a sense, these exceptions prove the rule: Instances of nonconservation are the result of loss or disruption of aboriginal culture as a result of Western acculturation.

The aim of this chapter is to put the issue of native wildlife conservation on a firm empirical and theoretical foundation. A number of serious methodological, empirical, and theoretical problems with claims about native systems of conservation are discussed and illustrated with recent data on Amazonian groups—contemporary peoples who tend to have had the least contact with the Western world. The goal is not to generalize about the existence of native conservation, but rather to develop ideas about the conditions under which a conservation ethic will evolve, the kinds of evidence required to determine whether conservation is practiced in a particular society, and the degree to which native peoples may be able to successfully exploit game resources if their land becomes a park or reservation.[12]

Methodological and Theoretical Problems

Deducing conservation from ideology alone is dubious at best. Such analyses rest on no sure scientific foundation; this can lead to highly divergent interpretations. For example, the exchange between two of the best known anthropologists of recent times, Levi-Strauss and Harris, on their radically different interpretations of the meaning of a particular Bella Bella myth clearly indicates the lack of consensus in this area of anthropological research. Harris suggested that the myth provides ecologically adaptive information on clam exploitation, whereas Levi-Strauss claimed it defines critical social relationships.[13] Even if consensus is reached about the proper way to interpret the meaning of native ideology, one still must demonstrate how ideology provides an effective code for behavior. To bring this point home, what can one conclude about Christian North America by reading the Bible? White suggested that Christianity presents an anticonservation ideology, while Berry claimed that the Bible has a conservationist program.[14] The White-Berry and Harris–Levi-Strauss debates are clear demonstrations that diametrically opposite interpretations regarding conservation can be derived from a common ideological source. Tuan, in a similar vein, points out that Confuscian ideology stresses harmony with the environment, yet this has not prevented widespread environmental degradation in China during premodern times.[15]

Callicott recently reviewed a number of symposium papers dealing with the ideological bases of conservation among a variety of tribal peoples. He cogently shows that researchers' analyses of similar ideological systems claim a variety of motives behind conservation ranging from utilitarian (designed for optimum sustained yield), religious reverence (with conservation as a side-effect), ecological awareness (understanding environmental dynamics and wanting to preserve all regardless of economic importance), and environmental ethics (animals must be treated as if they were people, therefore one should follow the golden rule when dealing with game).[16] Martin's widely cited thesis argues that precontact North American native peoples were conservationists and only began indiscriminate slaughter of the species they formerly conserved because of the introduction of European diseases.[17]

A serious blow to Martin's 1978 thesis was provided by Brightman. Using subarctic ethnohistoric data, he directly counters three of Martin's most fundamental claims about native North American conservation. While there is limited early eyewitness evidence for conservation and management among Canadian Indians, Brightman says, "The above findings must be placed in the context of a far greater body of evidence for the lack of conservation and intentional management and for a proclivity to kill animals indiscriminately in numbers beyond what were needed for exchange or domestic use."[18] Like the North American bison hunters, the Canadian hunters stripped delicacies (tongues, livers) from large game, leaving most of the carcass to rot. Ideologically, Cree and other Algonkian Indians believed that the more animals they killed, the more that would be available, or that animals immediately reincarnated themselves after proper disposal of blood and bones. Brightman argued that conservation did emerge in the early 1800s as a result of fur trading. For similar critiques of Martin's thesis as it relates to subarctic foragers, the edited collection of Krech should be consulted.[19]

There are several lessons one can draw from attempts by researchers to use ideological data to make claims about the existence of conservation. First, there is no generally accepted way to ascertain the meaning of ideological information contained in myth and ritual. This is especially difficult when one is attempting to say something about the relationship between the Western concept of conservation and native concepts. Second, there should be some correspondence between a belief system and the behavioral system it allegedly determines, but examination of this relationship has low priority for those who analyze ideology. Most importantly, if the behavior of the people in question satisfies stipulated theoretical expectations, this is sufficient to establish the reality of conservation. If a people have a conservation ideology but do not act as conservationists, then they are not conservationists. Conservation deals with how humans treat the resources in their environment. How a hunter feels about the deer he hunts and how those feelings may have been engendered by his religion are interesting psychological questions. But the hunter's actual treatment of deer is the only thing that matters.

Brightman and Berkes suggested that ideological and behavioral evidence must be combined to determine whether conservation or management of wild-derived resources occurs.[20] They argued that for conservation to occur, hunters must understand the relationship between hunting activities and game depletion and must be motivated to regulate their activity in order to maintain ecological balance.[21] That is, there must be some sort of conscious intent. Although conscious intent based on some understanding of game population dynamics may enhance the probability of successful conservation or management, at the same time, it is neither sufficient nor necessary for conservation. Conservation is a matter of performance—not intent.

According to Brightman, modern Cree who engage in conservation believe "that ritual procedures for disposing of animal bones and blood prefigure and influence animal regeneration and reincarnation."[22] But such rituals need not have any effect on game populations. Suppose, for example, that the sparing of a mated pair of beavers was done by Cree to avoid supernatural punishment.[23] This sort of belief is presumably adaptive because it will serve to effectively maintain the breeding stock, and thus the beaver population. The relationship between supernatural retribution and killing of a mated pair of beavers is false. The ideological rationale for engaging in a practice need not correspond to ecological reality. It is merely sufficient that the belief leads to behavior that has the effect of conservation or management. Brightman and Berkes appeared to be troubled by a lack of agreement between scientific and local native explanations of native behavior.[24] Brightman, for example, noted that "the Cree are not practitioners of scientific reductionism."[25] The evolutionary-ecological explanations pursued here are not models of how natives think any more than optimal foraging theory is a model how a praying mantis perceives its prey.[26]

Ideological approaches to the problem of conservation in tribal society have suffered from a number of theoretical shortcomings because they have not adequately dealt with conditions, maintenance, design, and alternative hypotheses. Almost invariably, the conditions under which conservation should evolve have not been specified. It is never made clear why conservation is practiced by one society, but not another. The cause of conservation is often seen as a group's ethos or ideology, but the conditions that favor the development of such belief systems are never made specific. How conservation is maintained by insuring that people do not cheat is often not examined. Many studies simply have assumed that people are slaves to ideology. Other studies pointed out, for example, that the consumption or killing of a taboo animal will lead to sickness. However, these same studies indicate that not everyone follows the rules, and in most tribes (as opposed to chiefdoms and states[27] and below), legal enforcement of such rules is impossible.

Finally, most of the ideological studies assumed implicitly or explicitly that conservation is adaptive for the group. If a behavior is adaptive, then one must show that it is well designed for the task and it is not the result of some other process.[28] This failure has already been noted by Callicott.[29] For the last twenty years, modern evolutionary biology has generally recognized that behavioral or other traits do not evolve in a population because they enhance the well-being of the group (i.e., as a consequence of group selection). Instead they are hypothesized to persist primarily because they are advantageous for individuals within populations. Anthropologists Lees and Bates have convincingly shown that population regulation models such as those proposed by Birdsell and Sahlins are based on a group selection model.[30] The following sections present my own recommendations about proper method, data evaluation, and theory in dealing with the complex problem of conservation.

Comparative Case Studies of Tribal Foraging Strategies in Amazonia

The best way to determine whether aboriginal humans practiced deliberate conservation, as opposed to conservation as a side effect of poor technology or low population density, is to examine archaeological, paleontological, and ethnological behavioral data on the relationship between human predation and demography of game populations. The archaeological and paleontological data are sparse and controversial, while the ethnographic data are extensive. As a result the prehistoric data will be quickly reviewed, and then I will focus on recent Amazonian tribal data. The Amazonian research data were chosen for several reasons. First, most groups studied have had weak to no contact with external markets such that their hunting and fishing activities are largely aimed at domestic consumption. Second, low-density Amazonian populations have achieved little in the way of landscape modification that, in itself, has serious consequences for game populations. And third, Amazonian researchers have collected some of the finest behavioral data available on human predation.[31]

The idea that humans were in large part responsible for the extinction of Pleistocene megafauna in North America was first put forward in a series of publications by Martin in what has become known as the "Pleistocene Overkill" hypothesis.[32] This hypothesis was based on a correlation between human entry and expansion into the New World and the timing of megafaunal extinctions. Martin forcefully argued that North American megafauna succumbed because they had not evolved adaptations to cope with the sudden appearance of this new "superpredator." Initially, this hypothesis raised a great deal of controversy, but more recently, a large number of paleontologists appeared to agree that humans may have played at least a partial role in these Pleistocene extinctions.[33] However, recent archaeological evidence on human antiquity in the New World suggests that Martin will have to modify the "blitzkrieg" portion of his hypothesis.[34]

Perhaps the best archaeological and paleontological data on human-caused extinctions of animal species comes from islands. Data from New Zealand on the extinction of moas,[35] Madagascar—on prosimians,[36] and insular Mediterranean—on pygmoid relatives of mainland animals[37] suggest that human hunting (as opposed to landscape modification from agriculture) has led directly to extinction. The balance of the island studies indicate that hunting, coupled with landscape modification through agriculture plus the introduction of foreign or exotic species (e.g., pigs and rats), caused the extinction of numerous Hawaiian bird species,[38] birds and marsupials in Melanesia,[39] and a variety of vertebrates in the West Indies.[40] That insular extinctions seem to be more common than their mainland counterparts may result from several factors, some of which may interact. First, islands have lower species diversity than mainland areas of comparable latitudes, forcing hunters to take a more restricted range of game.[41] Second, since islands are small, delimited ecosystems, local extinction is tantamount to total extinction because recruitment from neighboring islands is difficult over the short term. And third, human population density is often high in island ecosystems, leading to extreme pressure on local resources. Because of their small size and limited biological diversity, island ecosystems intensify human ecological and evolutionary constraints and enhance probabilities of extinction.

The best available way to determine whether modern-day tribal populations practice conservation is to observe their behavior in the face of declining resource yields.[42] Unfortunately, today there are no tribal populations that have not been affected by nation-states. Thus, it is impossible to claim that such populations operate under the same conditions that have characterized humanity for most of its history as a species. Nevertheless, tribal populations that engage in little or no cash cropping and manage their own political affairs provide the clearest view of how social and economic life were conducted prior to the advent of the nation-state. One could argue that technological introductions, e.g., use of shotguns instead of bows[43] or the introduction of Christianity, radically alter tribal environmental relations. But the most dramatic kind of environmental change occurs when resource procurement ceases to be aimed at subsistence and is instead market driven, especially when accompanied by loss of a traditional authority that regulated resource exploitation. That is, demand for resources is not based on local nutritional or technological needs, but rather on the opportunities presented by an external market.

The data presented below on hunting and fishing efficiency and time allocation come from Amazonian tribal populations that are fundamentally subsistence-based economies, i.e., they directly appropriate resources to satisfy local needs rather than the needs of an external market. All things being equal, resource depletion is governed by time allocated to resource acquisition. If human foragers are interested in decreasing resource depletion and thus conserving resources, then time allocated to acquisition should decline at some point as resources are depleted, to allow fish and game resources to rebound. If, on the other hand, foragers have no conservation strategy and instead are interested in maintaining a relatively stable level of fish and game consumption, then time allocated to fishing and hunting should increase as depletion increases. While there are many studies that have documented time allocated to hunting and fishing, there are none that have coupled such observations with direct measures of fish and game densities. However, if hunting and fishing efficiency (kilograms of game or fish harvested per hour of foraging effort) is a fair index of protein density, then comparative data are available. A positive correlation between hunting/fishing time and hunting/fishing efficiency should provide evidence for conservation, while a negative correlation between the two should indicate lack of a conservation strategy. In a previous study, I demonstrated a significant negative correlation between foraging time and foraging efficiency in a sample of fifteen native Amazonian groups.[44] Thus, generally, Amazonians react to depletion by increasing time allocated to hunting or fishing. While this correlation provides evidence against a conservation ethic, it is by no means conclusive. One could argue that levels of game depletion have not yet reached sufficiently critical levels for native hunters to reduce their predatory impact on fish and game populations. Such crosscultural comparisons do not have the inherent controls found in studies of neighboring village populations of the same culture, or in longitudinal comparisons of foraging activities of a single village over time. While local tests are less general than crosscultural tests, they are likely to be more reliable. Below, a number of local level studies are analyzed to evaluate the relationship between time allocation and foraging efficiency.

A number of Yanomamo villages in Brazil experienced fish and game depletion resulting from the building of a spur of the TransAmazon highway. Fish and game were taken by road crews and were driven out by road building. Streams were blocked, diverted, and muddied by runoff.[45] Although the crews have departed, the damage remains. Saffirio collected data on time allocation and hunting and fishing returns of Yanomamo hunters and fishers who lived adjacent to the road after departure of the road crews. At the same time, he collected the same kind of data on hunters and fishers from a village unaffected by the road building. For the highway villagers, hunting and fishing time was greater than for forest villagers who had greater hunting/fishing efficiency. Therefore, the Yanomamo responded to resource depletion (albeit not of their own making) by increasing their exploitation pressure. A similar comparison can be made between different Yanomamo villages studied by Lizot and Colchester.[46] Those studied by Lizot gained game at a rate of 2.8:1 (kcal output per kcal input) and hunted 74 minutes per day, while those studied by Colchester gained game at about half Lizot's rate (1.8:1) but hunted about twice as intensively—at 139 minutes per day. Again, there is clear evidence of a negative correlation between hunting time and efficiency which indicates that the Yanomamo respond to game depletion by increasing their hunting efforts.

Baksh's longitudinal study of Machiguenga fishing clearly showed an increase in fishing time as yields declined.[47] After the Machiuenga set up a new village, fishing returns were high because they exploited undepleted streams near the village. Over time, fish became depleted, and the villagers were forced to fish further from the village. In addition, villagers were forced to turn to labor-intensive poison (*barbasco*) fishing as yields declined. But even labor-intensive techniques produced diminishing yields. For example, one stream yielded 768 kilograms when it was first poisoned in the wet season (when poison is minimally effective). It was poisoned again during the dry season (when poison is maximally effective), but yielded less than 100 kilograms. Baksh concluded that the Machiguenga coped with declining yields by working longer hours. Other Amazonian groups such as the Ye'kwana also engage in frequent barbasco fishing. However, they will not poison the same stream twice in one year—not because they fear depleting resources, according to informants, but because the low yield will not be worth the effort.[48]

Vickers's ten years of longitudinal research on the Siona-Secoya of Ecuador provide the best available data on the process of game depletion over time. His data revealed a consistent negative correlation between hunting efficiency and time allocated to hunting.[49] Furthermore, hunting led to near local extinction of woolly monkeys, curassows, and trumpeters in heavily exploited areas.[50] Most importantly, he showed that game populations have reached an equilibrium (with the exception of the three species mentioned above) such that foraging efficiency and effort appear to have leveled. This important study suggests that the Siona-Secoya are attempting to maintain a stable game intake, and have not caused the extinction of basic game resources even though they have no apparent environmental ethic or conservation policy. One might conclude this is because of the low demand the Siona- Secoya make on game resources (which may be a function of Siona-Secoya population density).

Qualitative and quantitative data on the Ye'kwana and Yanomamo of the Venezuelan Amazon may provide some insight into the attitudes and behaviors these Indians manifest toward game species that supply them with 80 percent of their high-quality protein intake.[51] The Yanomamo, like many Amazonian groups, have taboos against the consumption and/or killing of certain wild animals. For example, in most areas of their tribal distribution capybara, deer, and otter are taboo. However, this does not prevent the killing of these animals should the situation present itself. Once, while traveling in a canoe with two Yanomamo companions, we spotted a herd of capybara on the shore. I was entreated to shoot one, which I did successfully. Upon disembarking, both of my companions began poking the dead animal with sticks. I asked who was going to butcher the animal, and was told that no Yanomamo would touch such a "filthy" animal, much less prepare it for consumption. I asked why they had asked me to kill it. One said he wanted to see a big one up close, while the other said he wanted to see how it died.

On another occasion, I was hunting in the forest with a Ye'kwana. His dogs cornered a giant anteater—an animal tabooed as a source of food. After failing in his attempt to divert the dogs from their quarry in order to move on to proper game, he killed the anteater. He even prevented the dogs from eating the animal, claiming this would only encourage them to hunt it again and that it would cause them to become very ill just like a human. Many similar examples of treatment of taboo animals could be cited. Kracke, in reference to the Tupian Kagwahiv, stated "but their exclusion from the diet does not prevent Kagwahiv from killing them for their skins whenever they are encountered."[52] The point is that at least among the Ye'kwana and Yanomamo, tabooing is not associated with conservation attempts or any general reverence for animal life.[53]

Finally, Rambo's monographic account of the Malaysian Semai, shifting cultivators with a high reliance on foraging of wild animals, directed attention to the environmental impact a tribal population can have on its environment.[54] Although the Semai are not Amazonian, they live in a similar environment and have similar ecological adaptations. Rambo documented widespread habitat destruction and befoulment, and noted that the switch from blowpipe hunting to the blunderbuss led to the local decimation of many game species.[55]

A preponderance of the evidence indicates that Amazonian tribal populations make no active or concerted effort to conserve fish and game resources. At the same time, it is clear in most cases there may be no need for a conservation policy, because current local subsistence demands on resources have not led to severe resource shortages. At this point, it would be useful to formulate empirical hypotheses about how foragers should behave if their goal were one of conservation rather than maximization of foraging success, and specify the conditions under which a conservation ethic is likely to evolve.

Difficulties in the Development of Empirical Hypotheses

An important goal in ecological anthropology should be the formulation of hypotheses that will allow determination of the foraging practices of tribal peoples designed[56] to accomplish conservation as opposed to those that will maximize foraging efficiency or some other goal.[57] A potential problem is that maximization of hunting efficiency may lead to conservation. The importance of determining whether tribal practices are designed for conservation can be illustrated by the World Conservation Strategy definition endorsed by Berkes: "Conservation is the maintenance of essential ecological processes and life support systems, the preservation of genetic diversity, and the sustainable utilization of species and ecosystems."[58] Should environmental conditions change or resource demand increase, hunters whose practices are designed to maximize efficiency may cause environmental destruction while hunters whose practices are designed for sustainable yields will not upset the ecological balance. Below, a series of paired hypotheses are described that may help to determine whether foraging practices are aimed at conservation or the opportunistic maximization of efficiency. As will become clear, the task is not simple because many of the paired predictions are identical. This leaves us with the possibility that maximization of foraging efficiency and conservation are not mutually exclusive goals in some environments.[59]

In central place foraging theory, differential allocation of time in hunting patches leads to patterns of radial game depletion. After a site is settled, hunters hunt close to the village since game is equally dense in all areas. Hunting near the village brings the highest rate of return because travel time is minimized. Through time, areas close to the village become depleted of game, and hunters extend their range because the added travel to distant zones is repaid by higher rates of return in those zones. This process leads to a depletion gradient, with game populations densest in the most distant hunting zones and most sparse in the closest zones. Game densities tend to increase with increasing distance from the village. Given these conditions, conservationists should allocate more time to distant areas with the greatest density of game, thereby allowing the closest zones to recover by releasing hunting pressures. However, hunters who wish to maximize their rate of return in hunting would follow the same strategy since this would yield the highest rate of return. The Yanomamo and Ye'kwana appear to be following this pattern since both groups of hunters allocate more time to rich than poor hunting zones.

If a hunter wishes to hunt in a game-rich distant area, he must first travel through depleted hunting areas adjacent to the village. This leads to another pair of predictions. Hunters interested in conservation should not take game in depleted areas while they are in transit to more distant, undepleted areas. On the other hand, hunters interested in maximizing their efficiency should take game wherever encountered, even in depleted areas. Unfortunately, there are no quantitative data to evaluate this hypothesis. However, in numerous hunts with both Ye'kwana and Yanomamo hunters, I always observed them to pursue game in depleted areas while they were en route to more distant hunting areas.

Tabooing by its very nature should aid in conserving species. Ross suggested that variability in game taboos among Amazonian tribal populations presents an important insight into the dynamics of tropical forest adaptations.[60] He seemed to simultaneously maintain that tabooing is designed to allow hunters to maintain the demographic integrity of game animal populations (particularly those that reproduce slowly) and to also hunt at maximal efficiency.[61] Following Slobodkin's "prudent predator" model, tabooing should be designed to facilitate long-term maximization.[62] According to Ross, tabooing would be most likely to occur among those groups with stable settlement patterns (e.g., the Achuara Jivaro of the Peruvian Montana), but least likely among expanding populations, e.g., the Yanomamo.

If hunters instituted a taboo on a species to conserve its population size(s), then one would expect the taboo to work in a density-dependent manner. Over time, tabooing would be switched on or off—depending on a species' level of depletion. But hunters out to maximize their hunting efficiency might also institute a taboo on depleted species, not because they are depleted but because their rate of return upon encounter was declined. In this case, tabooing is designed to maximize efficiency, and conservation is a side effect of efficient hunting. Species that have been subjected to heavy predation frequently exhibit behavioral depression.[63] That is, they become more wary, nocturnal, or refuge themselves in habitats (e.g., thickets) that make hunting more difficult. As a result their net rate of return knocks them out of a hunter's optimal diet breadth.[64] With a conservation strategy, such animals should lose their tabooed status when they lose those attributes that made their cost of acquisition high. Maximizers differ in response from conservationists, because for the former it is the cost of acquisition that is important, while in the latter it is game density.

The only evidence for taboos being switched on and off comes from Kensinger's work on the Cashinahua. Inedible animals become edible "when a village is old and the preferred game supply has been depleted by years of hunting to the point that a dependable supply of the *kuin* [edible] animals is unavailable."[65] No data other than the above statement was supplied. Unfortunately, Kensinger's observation is not detailed enough to understand why certain animals were tabooed initially and whether taboos are lifted because of changing rates of return or depletion.

Kent Redford in a personal communication argues that tabooing could not lead to the maximization of rates of return in hunting. He points out that if one, for example, stumbled onto a sleeping tapir (Ye'kwana and Yanomamo hunters have recounted instances to me) it should be taken since the likely rate of return upon encounter is extremely high. There are two ways of dealing with this criticism. First, calculations of the optimal diet breadth (the game species one should always go after upon encounter) are based on average rates of return, which means that an animal is either in or out of the diet breadth independent of circumstances of encounter. However, this response exposes a weakness in the classic simple optimal diet breadth model: It does not deal with variance in rates of return upon encounter that are reliably correlated with environmental cues.[66] Nevertheless, optimal diet breadth models can be easily modified to take this condition into account. Second and most important, the claim that one strategy is more adaptive than another is not the same as saying that it is the best strategy conceivable.[67] For the taboo to be adaptive (never take the animal upon encounter), it must merely be superior to the alternative (always take the animal upon encounter).

While Ross and others are concerned with general or blanket taboos, or those that extend to all members of a population,[68] many Amazonian groups have taboos on consumption of game species limited to specific population segments.[69] For example, pregnant women, children, and men undergoing certain long-term initiation rites may be prohibited from consuming certain kinds of fish or game that others may consume freely. Furthermore, whether the taboo is general or restricted, tabooed animals may be killed, although they may not be consumed by certain people. The obvious effect of restricted taboos may be to depress (as opposed to reducing to zero under blanket prohibitions) the rate at which certain game species are taken. Thus, they may function like a quota system or "bag limit" in Western hunting systems. These sumptuary regulations are extremely common throughout lowland Amazonia, and there is no simple reason why a hunter out to maximize his net rate of return would want to be subjected to such regulations. However, a hunter prohibited from consuming a particular game species could be motivated to hunt that species, even if his wife and family were not able to consume it, if he could gain prestige by giving it to another family. In most Amazonian groups, hunters are expected to share large game, e.g., among the Yanomamo, approximately 60 percent of all game consumed by a family was acquired by a hunter who is not a family member.[70] This figure is consistent with other quantitative examinations of exchange in Amazonian societies.[71]

A satisfactory materialistic explanation of restricted versus generalized hunting and fishing taboos may prove extremely difficult. Ross's hypothesis, as mentioned previously, is the only rigorous ecologically based model available;[72] but it does not deal with restricted taboos. It may be that fish and game taboos, whether general or restricted, have more to do with psychology than anything else. Both the Ye'kwana and Yanomamo appear to be extremely reluctant to eat foods with which they have no familiarity. Once a Yanomamo explained to me that he was reluctant to eat a mata-mata turtle even though many in his village consumed it because he had not "learned to eat it yet"; he then implied it might cause him to become sick. In addition, the Yanomamo sometimes associate illnesses with things they have recently consumed, especially nongarden foods.[73] Conservatism in eating habits may be adaptive in an environment where food spoils easily and where plants are loaded with toxic or unpalatable secondary compounds that probably serve as a defense against herbivory, i.e., it may be that many food taboos have nothing to do with conservation or hunting efficiency.

A final way that conservation might be obtained is by sparing females. As is well known, the demographic viability of a population is dependent on the number of reproductive females. Game laws in most states regulate the taking of females in species as diverse as pheasants and deer. Therefore, tribal hunters interested in maintaining game populations at a sustained level of harvest should spare mature females. Hunters out to maximize their rates of return, on the other hand, should take game independent of the sex or reproductive status of the species. Ye'kwana and Yanomamo hunters have no compunction in taking females. In fact, a common deer-hunting tactic comes into play when a hunter encounters a fawn frozen in undergrowth. A hunter strangles the fawn, and then waits for the mother to return so she may be killed. Perusal of the literature on Amazonian hunting tactics fails to supply any evidence that females are spared.

Determining whether conservation occurs is not as theoretically important as specifying the conditions under which conservation will evolve and be maintained. This has obvious implications for the development of effective conservation policies, and it would immensely aid in the interpretation of the design of tribal foraging patterns. Two mechanisms that are dependent on an ecological variable must be in place to allow conservation to develop. From an ecological-evolutionary perspective, humans, like all other organisms, should be genetically selfish. That is, they should maximize their inclusive fitness.[74] As a result, they should forego attempts to maximize their net rate of return while foraging only if it is in their net reproductive interest to do so. This assumes there is a positive correlation between foraging efficiency and reproductive success.[75] They can be either "prudent predators,"[76] interested in maximizing hunting returns over a long period of time, or opportunistic predators interested in maximizing their instantaneous rates of return.

The problem with prudence is that a coresident may cheat and take selfish advantage of another hunter's restraint.[77] As a result, the first requirement for conservation is a social mechanism that punishes or threatens punishment for those who fail to practice prudent predation. In states, and to some extent chiefdom-level societies, there are legal bureaucracies that can enforce such regulations through supernatural means or physical coercion.[78] However, all the Amazonian societies reviewed above are tribes that lack such institutions at the present time. Leaders in tribes are headmen who must rely primarily on verbal persuasion to induce people to follow rules. Among the Yanomamo and other Amazonian groups some people adhere to taboos while others do not.[79] The only consistent punishment appears to be supernatural and personal. For example, for a northern Yanomamo dialect group, Taylor noted, "For the Sanuma, as is the case for so many different societies, breaking a food taboo can result in a state of illness or bodily disorder."[80] In other cases the consumer of a tabooed animal is considered odd or disgusting. On the other hand, according to my Ye'kwana informants, the killing of an anaconda becomes a social threat since it is believed that this will lead to disastrous flooding of the entire village. This last sort of mechanism, threat to the social group, may be more effective than the threat of individual supernatural punishment because the one who did not follow the rule endangers everyone.

The second requirement for effective conservation is some sort of social mechanism to prevent outsiders from poaching. Hardin was one of the first to stress that some means of community control of resources is necessary for effective conservation.[81] Further, it is obvious that mechanisms of control must not only extend legally to community members through some system of tenure, but also to neighbors through political means. Although members of a village may behave in ways that prevent overexploitation, their efforts can be nullified if a neighboring group takes endangered game in their customary hunting areas. To succeed, the social group would have to be territorial.

According to ecological-evolutionary theory, territories tend to develop in areas where resources are relatively dense and predictable.[82] Brown's economic model of territorial defendability posits that the costs of defending a resource (patrolling, fighting, and advertisement) must be less than the benefits of having unrestricted access to resources in a given area. This model has been evaluated on human populations with promising results.[83] Areas in which resources are relatively dense and predictable are less costly to defend because less travel time is required to maintain a perimeter by detecting poachers than attempting to defend a sparse resource distributed over a larger area. Brown's model is a cost-benefit model, and costs and benefits may vary in different ecological circumstances. Thus, density and predictability are indices—not direct measures—of costs and benefits.

In sum, it seems unlikely that many Amazonian tribal societies meet the specific conditions that should lead to conservation of wild game and fish populations. Amazonian game resources tend to be sparsely rather than densely distributed and, with the exception of the white-lipped peccary,[84] game tends to be spatially and temporally predictable, with most species as year-round residents. The issue of territoriality among Amazonian groups has not been systematically addressed although some have related it to warfare and protection of hunting areas.[85] Others have strongly denied that war is a matter of competition over scarce game resources through territorial defense.[86] In addition, modern-day native Amazonians do not possess the social institutions to enforce conservation among covillagers, although there are indications that chiefdoms may have once existed in Amazonia prior to European conquest.[87] The model developed here suggests a more careful look at chiefdom-level societies that depend heavily on resources that are economically worth defending.

CONSERVATION IN OCEANIC CHIEFDOMS

Conditions necessary for the development of native conservation systems; i.e., institutionalized enforcement of rules and territoriality seem to be met among chiefdom-level societies in Micronesia and Polynesia. Chiefdoms differ from tribes, such as the egalitarian Amazonian tribal societies described above, in at least two fundamental ways. First, social status is based on heredity; only those having the proper status can rise to become leaders. And second, the most powerful position of leadership is the chief, instead of the headman in tribal egalitarian society. Chiefs, unlike headmen, have the ability to enforce their decisions and punish transgressors through a variety of social mechanisms unavailable to headmen.[88] Territoriality, as manifested socially by a variety of resource tenure practices, extends not only to agricultural land but also to fishing areas.

According to Chapman, chiefs and their councils in Micronesia and Polynesia habitually regulated the exploitation of communally owned fishing areas through a variety of mechanisms such as rotation of closed and open fishing areas, restriction of effective fishing technology, and limitations on the number of fishermen in particular coastal fisheries.[89] In a comparative study of Micronesian tenure, Sudo observed "it is common to all societies that some persons (chiefs) or organizations (councils) exist to control and conserve marine resources. Those [sic] have rights to regulate the use of particular sea sections, and have responsibilities to protect against the exhaustion of food resources."[90] Importantly, Chapman pointed out that when open access to fisheries occurs, depletion follows because chiefs can no longer manage resources for the benefit of the local community. The importance of existence of adequate social mechanisms to enforce conservation was made more compelling when Chapman compared the absence of conservation and management in Melanesian fisheries.[91] She pointed out that Melanesian societies are less politically complex than chiefdoms. Instead of chiefs, Melanesian groups are led by "big men," a kind of glorified headman. Like headmen, big men lack the coercive abilities of a chief. As a result, they do not have the ability to conserve or manage resources, even if they so desired.

It is highly possible that conservation of resources occurred among the hunting, fishing, and gathering chiefdoms of the Northwest Coast of North America—a coastal area stretching from northern California to southern Alaska. In nearly all of the chiefdoms, economically important fishing, hunting, and gathering areas were owned. Chiefs and other high-status individuals determined who, when, and how intensively individuals could exploit resource areas.[92] Unfortunately, it has proven difficult to uncover precise information on the relationship between resource depletion and harvesting restraint, although it is frequently implied.[93] Of interest is Forde's observation on the Nootka, "If a run in a particular stream began to fail they actually restocked it, obtaining spawn from another river at breeding season and carrying it back in moss-lined boxes to start a new generation in the depleted stream."[94] Clearly, the Northwest Coast is worthy of further investigation to determine the extent of native systems of conservation.

Can Native Peoples Coexist with Parks?

Throughout much of human history, and especially in the last five hundred years of European expansion, tribal peoples have had their land and lives taken from them by state peoples. Contact has led to reduction of their areal base, introduction of new technology and market exchange, and settlement concentration. The strength of these variables and the ways in which they interact are the main determinants of a tribal people's ability to maintain an equilibrium with their environment. The balance of this chapter is devoted to a description of these factors and how they effect tribal adaptation. The aim here is to inform those responsible for setting up parks or reservations who seek to maintain cultural and/or biological diversity.

Native bows are much less efficient in taking game than shotguns or rifles.[95]

The common effect of general-purpose weapons, such as the shotgun, is reduction of hunting time,[96] which suggests that demand for game is relatively constant. This parallels the introduction of steel axes in New Guinea: The amount of forest cleared for gardens remained the same, but the time required to do the work decreased.[97] Although game depletion occurs sometimes,[98] it may be that it is conditioned by the interaction of other factors.

Specific-purpose weapons, especially those designed to exploit a narrow range of species, can lead to severe game depletion. The Ye'kwana of the Padamo[99] and Yanomamo who live around the missions of the upper Orinoco have heavily depleted riverine game such as capybara, herons, egrets, otters, and caiman by using outboards on their dugouts and headlamps. Headlamps and outboards, coupled with shotguns, make hunting of riverine game so efficient that hunters specialize. It may be that technology that reduces hunting costs over a wide range of game does not lead to game depletion, whereas technology that permits specialized cost reductions for a small range of game does lead to depletion.

Entrance into an external market economy changes the value of game animals. Aboriginally, game value is based on local subsistence demand. External markets increase game value, depending on their utility for nonaboriginals as biological specimens, pets, and their products, e.g., skins, feathers, and furs. The increase in value leads to increased hunting and depletion,[100] not infrequently among species that had no previous local value. Entrance into a market economy is undoubtedly the most potentially devastating change in aboriginal environmental relations.

Agents of contact such as missionaries and government officials routinely install trading posts and other sites such as medical clinics and schools that are by their very nature designed to attract and concentrate large numbers of natives. For example, among the Yanomamo, a mission was established in 1968 at the mouth of the Mavaca River. This mission initially attracted a village of approximately 100 Yanomamo,[101] but today there are seven villages, with a population of approximately 550, in the same immediate area. As a result, hunting and fishing pressures probably have increased more than fivefold (missionaries, their staff, and visitors are frequently supplied with local fish and game). Having worked at the mouth of the Mavaca for about three months during 1985–87, it is my impression from observations of food exchange and casual discussion with longtime hunters of the area that hunting there is less rewarding than in nonmission areas. Just as importantly, populations such as the Yanomamo tend to have a mobile settlement patterns; villages are commonly relocated every five years or so.[102] Mobile settlement patterns reduce game depletion (although they are probably not specifically designed to do so). But Western presence leads to a less mobile settlement pattern,[103] thus intensifying the problem of game and resource depletion. These twin processes result in unrelieved high hunting and fishing pressure.

Nevertheless, simultaneous conservation of ethnic and biological diversity is possible. Native peoples can exist in equilibrium with game populations if the technology they use is regulated, if hunting and fishing are aimed solely at meeting local nutritional demands, and if settlement patterns remain dispersed, mobile, and at low population densities. The simplest way to achieve such a goal is to regulate the activities of nonnatives. Recently, it has been suggested that natives be barred from biological reserves toward the end of maintaining the natural functioning of neotropical ecosystems. This policy is erroneous for two reasons. First, native peoples have been a part of neotropical ecosystems for thousands of years in their role as top predators. They are therefore an integral part of it as much as natural predators such as the endangered jaguar and puma. Second, Indians have been systematically deprived of their land for the sake of "economic development" ever since the entrance of whites in the neotropics. Now other nonnatives are asking them to give up their land for the sake of "biological conservation."

Given strong competing interests among missionaries, anthropologists, governments, conservationists, and commercial interests for native peoples, nonintervention is unlikely. Added to this are the desires of native peoples for contact, change, and the material accoutrements the Western world can give them. Technically, the maintenance of cultural and biological diversity is a simple problem. Politically, it is, and will probably continue to be, enormously difficult.

NOTES

1. See for example J. Bodley, *Anthropology and Contemporary Human Problems* (Menlo Park, Calif.: Benjamin Cummings Publishing, 1976); C. Martin, *Keepers of the Game* (Berkeley: University of California Press, 1978); and J. Clad, "Conservation and Indigenous Peoples: A Study of Convergent Interests," *Cultural Survival Quarterly* 8, no. 4 (1984): 68–73.

2. M. Sahlins, "Notes on the Original Affluent Society," in *Man the Hunter*, ed. R. B. Lee and I. DeVore (Chicago: Aldine, 1968), pp. 85–89.

3. J. Birdsell, "On Population Structure in Generalized Hunting and Collecting Populations," *Evolution* 12 (1958): 189–205; and Birdsell, "Some Predictions for the Pleistocene Based on Equilibrium Systems Among Recent Hunter-Gatherers," in *Man the Hunter*, pp. 229–40.

4 Sahlins, "Notes on the Original Affluent Society"; and Sahlins, *Stone Age Economics* (Chicago: Aldine, 1972).

5. Ibid.

6. Martin, *Keepers of the Game*.

7. G. Reichel-Dolmatoff, "Cosmology as Ecological Analysis: A View From the Rain Forest," *Man* 11 (1976): 307–18.

8. E. Ross, "Food Taboos, Diet, and Hunting Strategy: The Adaptation to Animals in Amazonian Cultural Ecology," *Current Anthropology* 19 (1978): 1–36.

9. J. Ewers, *The Horse in Blackfoot Indian Culture*. Bureau of American Ethnology Publ. No. 23 (Washington, D.C.: Smithsonian Institution, 1955).

10. E. Leacock, *The Montagais "Hunting Territory" and the Fur Trade*, American Anthropological Association Memoir No. 18 (Washington, D.C.: American Anthropologist, 1954).

11. Martin, *Keepers of the Game*.

12. K. Redford and J. Robinson, "Hunting by Indigenous Peoples and Conservation of Game Species," *Cultural Survival Quarterly* 9 (1985): 41–44.

13. See C. Levi-Strauss, "Structuralism and Ecology," *Social Science Information* 12, no. 1 (1973): 7–24; M. Harris, "Levi-Strauss et La Pariourde," *L'Homme* 26 (1976): 5–22.

14. L. White, "The Historic Roots of Our Ecologic Crisis," *Science* 155 (1967): 1203–1207; W. Berry, "The Gift of Good Land," *Sierra Club Newsletter* (Nov./Dec. 1979): 20–26.

15. Y. F. Tuan, "Discrepancies Between Environmental Attitude and Behaviour: Examples From Europe and China," *Canadian Geographer* 12, no. 3 (1968): 176–91.

16. J. B. Callicott, "American Indian Land Wisdom? Sorting Out the Issues," in *The Struggle for the Land*, ed. P. Olson (Lincoln: University of Nebraska Press, 1990), p. 44–69.

17. Martin, *Keepers of the Game*.

18. R. Brightman, "Conservation and Resource Depletion," in B. McCay and J. Acheson, eds., *The Question of the Commons*, ed. B. McCay and J. Acheson (Tucson: University of Arizona Press, 1987), p. 123.

19. S. Krech, ed., *Indians, Animals, and the Fur Trade* (Athens: University of Georgia Press, 1981).

20. Brightman, "Conservation and Resource Depletion"; F. Berkes, "Common-Property Resource Management and Cree Indian Fisheries in Subarctic Canada," in *The Question of the Commons*, pp. 66–91.

21. See also R. Nelson, "Athapaskan Subsistence Adaptation in Alaska," in Y. Kotani and W. Workman, eds., *Alaska National Museum of Ethnology*, ed. Y. Kotani and W. Workman (Osaka, Alaska: 1979), pp. 205–32.

22. Brightman, "Conservation and Resource Depletion," p. 131.

23. Ibid., p. 123.

24. Ibid.; and Berkes, "Common-Property Resource Management."

25. Brightman, "Conservation and Resource Depletion," p. 89.

26. E. Charnov and G. Orians, "Optimal Foraging: Some Theoretical Explanations," (University of Utah, Salt Lake City, 1973, mimeographed).

27. See E. Service, *Origins of the State and Civilization* (New York: W. W. Norton and Company, 1975).

38. G. Williams, *Adaptation and Natural Selection* (Princeton: Princeton University Press, 1966).

29. Callicott, "American Indian Land Wisdom? Sorting Out the Issues."

30. S. Lees and D. Bates, "The Myth of Population Regulation," in N. Chagnon and W. Irons, eds., *Evolutionary Biology and Human Social Behavior*, ed. N. Chagnon and W. Irons (North Scituate, Md.: Duxbury, 1979), pp. 273–89.

31. R. Hames and W. Vickers, *Adaptive Responses of Native Amazonians* (New York: Academic Press, 1983).

32. P. S. Martin, "Africa and Pleistocene Overkill," *Nature* 212 (1967): 339–42; Martin, "Prehistoric Overkill," in *Pleistocene Extinctions*, ed. P. S. Martin and H. Wright (New Haven: Yale University Press, 1967), pp. 75–120; Martin, "The Discovery of America," *Science* 179 (1973): 969–74; and Martin, "Prehistoric Overkill: The Global Model," in *Quaternary Extinctions—A Prehistoric Revolution*, ed. P. Martin and R. Klein (Tucson: University of Arizona Press, 1984), pp. 354–403.

33. See J. Diamond, "News and Views," *Nature* 298 (1982): 787; Diamond, "Historic Extinctions: A Rosetta Stone for Understanding Prehistoric Extinctions," in *Qua-*

ternary Extinctions, pp 824–62; and Diamond, "The Environmentalist Myth," *Nature* 324 (1986): 19–20.

34. N. Guidon and G. Delibrias, "Carbon-14 Dates Point to Man in the Americas 32,000 Years Ago," *Nature* 321 (1986): 769–71.

35. A. Anderson, "The Extinction of Moas in Southern New Zealand," in *Quaternary Extinctions*, pp. 728–40; M. Trotter and B. McCulloch, "Moas, Men, and Middens," in *Quaternary Extinctions*, pp. 221–39.

36. A. Richard and R. Sussman, "Future of the Malagasy Lemurs: Conservation or Extinction?" in *Lemur Biology*, ed. I. Tattersall and R. Sussman (New York: Plenum Press, 1975) p. 86–102; R. Dewar, "Extinctions in Madagascar," in *Quaternary Extinctions*, pp. 574–91.

37. S. Davis, *The Archaeology of Animals* (New Haven: Yale University Press, 1987).

38. M. McGleone, *Archaeology of Oceania* 18 (1983): 11; S. L. Olson and H. James, "The Role of Polynesians in the Extinctions of the Avifauna of the Hawaiian Islands," in *Quaternary Extinctions*, pp. 768–80.

39. R. Cassels, "The Role of Prehistoric Man in the Faunal Extinctions of New Zealand and Other Pacific Islands," in *Quaternary Extinctions*, pp. 741–67.

40. D. W. Steadman, G. K. Pregill, and S. L. Olson, "Fossil Vertebrates From Antigua, Lesser Antilles: Evidence for Late Holocene Human-Caused Extinctions in the West Indies," *Proceedings of the National Academy of Sciences* 81 (1984): 4448–51.

41. R. MacArthur and E. O. Wilson, *The Theory of Island Biogeography* (Princeton: Princeton University Press, 1967).

42. Redford and Robinson, "Hunting by Indigenous Peoples and Conservation of Game Species."

43. R. Hames, "A Comparison of the Shotgun and the Bow in Neotropical Forest Hunting," *Human Ecology* 7 (1979): 219–52.

44. R. Hames, "Game Depletion and Hunting Zone Rotation Among the Ye'kwana and Yanomamo of Amazonas, Venezuela," in *Studies in Hunting and Fishing in the Neotropics* (Bennington, Vt.: Bennington College, 1980), p. 66.

45. G. Saffirio and R. Hames, "The Forest and the Highway," Working Papers on South American Indians No. 6 and Cultural Survival Occasional Paper No. 11. Cultural Survival, Inc., Cambridge, Mass., 1983.

46. See J. Lizot, "L'economie Primitive," *Libre* 4 (1978): 69–113; M. Colchester, "Rethinking Stone Age Economics: Some Speculations Concerning Pre-Columbian Yanoama Economy," *Human Ecology* 12 (1984): 291–314.

47. M. Baksh, "Faunal Food as a 'Limiting Factor' on Amazonian Cultural Behavior: A Machiguenga Example," *Research in Economic Anthropology* 7 (1985): 145–77.

48. R. Hames, *A Behavioral Account of the Division of Labor Among the Ye'kwana*, Ph.D. thesis, University of California at Santa Barbara, 1978.

49. W. Vickers, "An Analysis of Amazonian Hunting Yields as a Function of Settlement Age," in *Studies in Hunting and Fishing in the Neotropics*, pp. 7–29.

50. W. Vickers, "Game Depletion Hypothesis of Amazonian Adaptation: Data From a Native Community," *Science* 239 (1988): 1521–22.

51. N. Chagnon and R. Hames, "Protein Deficiency and Tribal Warfare in Amazonia: New Data," *Science* 203 (1979): 910–13.

52. K. Kracke, "Don't Let the Piranha Bite Your Liver: A Psychoanalytic Approach to Kagwahiv (Tupi) Food Taboos," in K. Kensinger and W. Kracke, eds., *Food Taboos in Lowland Amazonia: Working Papers on South American Indians*, ed. K. Kensinger and W. Kracke (Bennington, Vt.: Bennington College, 1981), p. 109.

53. See D. McDonald, "Food Taboos: A Primitive Environmental Protection Agency," *Anthropos* 72 (1977): 734–48; E. Ross, "Food Taboos, Diet, and Hunting Strategy: The Adaptation to Animals in Amazonian Cultural Ecology," *Current Anthropology* 19 (1978): 1–36.

54. A. Rambo, *Primitive Polluters: Semang Impact on the Malaysian Tropical Rain Forest Ecosystem*, Museum of Anthropology, Anthropological Papers No. 76 (Ann Arbor: University of Michigan, 1985).

55. A. Rambo, "Bows, Blowpipes, and Blunderbusses: Ecological Implications of Weapons Change Among the Malaysian Negritos," *Malayan Nature Journal* 32, no. 2 (1978): 209–16.

56. G. Williams, *Adaptation and Natural Selection* (Princeton: Princeton University Press, 1966).

57. See E. Hunn, "Mobility as a Factor Limiting Resource Use in the Columbian Plateau of North America," in *Resource Managers: North American and Australian Hunter-Gatherers*, ed. N. Williams and E. Hunn (Boulder: Westview Press, 1982), pp. 17–43; R. Hames, "Game Conservation or Efficient Hunting?" in *The Question of the Commons*, pp. 192–207.

58. F. Berkes, "Common-Property Resource Management," p. 84.

59. Ibid., p. 89.

60. Ross, "Food Taboos, Diet, and Hunting Strategy."

61. Hames, "Game Conservation or Efficient Hunting?"

62. L. Slobodkin, "Prudent Predation Does Not Require Group Selection," *American Naturalist* 108 (1974): 665–78.

63. E. Charnov, G. Orians, and K. Hyatt, "Ecological Implications of Resource Depression," *American Naturalist* 110 (1976): 247–59.

64. Charnov and Orians, "Optimal Foraging: Some Theoretical Explanations."

65. K. Kensinger, "Food Taboos as Markers of Age Categories in Cashinahua," in *Food Taboos in Lowland Amazonia: Working Papers on South American Indians*, p. 161.

66. D. Stephens and E. Charnov, "Optimal Foraging: Some Simple Stochastic Models," *Behavioral Ecology and Sociobiology* 10 (1982): 251–63.

67. G. Williams, *Adaptation and Natural Selection* (Princeton: Princeton University Press, 1966).

68. See Ross, "Food Taboos, Diet, and Hunting Strategy"; and McDonald, "Food Taboos."

69. For example, see the contributors in Kensinger and Kracke, *Food Taboos in Lowland Amazonia*.

70. Author's unpublished data.

71. See L. Aspelin, "Food Distribution and Social Bonding Among the Mamainde of Mato Gross, Brazil," *Journal of Anthropological Research* 35 (1979): 309–27; and H. Kaplan and K. Hill, "Hunting Ability and Reproductive Success Among Male Ache Foragers: Preliminary Results," *Current Anthropology* 26 (1985): 131–33.

72. Ross, "Food Taboos, Diet, and Hunting Strategy: The Adaptation to Animals in Amazonian Cultural Ecology."

73. See Kracke, "Don't Let the Piranha Bite Your Liver," p. 102, for a similar observation on the Brazilian Kagwahiv.

74. W. Hamilton, "The Genetical Evolution of Social Behavior," *Journal of Theoretical Biology* 7 (1964): 1–52.

75. See Kaplan and Hill, "Hunting Ability and Reproductive Success Among Male Ache Foragers."

76. Slobodkin, "Prudent Predation Does Not Require Group Selection."

77. Hunn, "Mobility as a Factor Limiting Resource Use in the Columbian Plateau of North America."

78. Service, *Origins of the State and Civilization*.

79. See P. Menget, "From Forest to Mouth: Reflections on the Txicao Theory of Substance," in *Food Taboos in Lowland Amazonia*, pp. 1–17; J. Abelove and R. Campos, "Infancy Related Food Taboos Among the Shipibo," in *Food Taboos in Lowland Amazonia*, pp. 174–77; and Kracke, "Don't Let the Piranha Bite Your Liver."

80. K. Taylor, "Knowledge and Praxis in Sanuma Food Prohibitions," in *Food Taboos in Lowland Amazonia*, p. 43; see also T. Langdon, "Food Taboos and the Balance of Opposition Among the Barasana and Taiwana," in *Food Taboos in Lowland Amazonia*, p. 58.

81. G. Hardin, "The Tragedy of the Commons," *Science* 162 (1968): 1243–48; see also M. Chapman, "Environmental Influences on the Development of Traditional Conservation in the South Pacific Region," *Environmental Conservation* 12 (1985): 217–30; and R. Hames, "Game Conservation or Efficient Hunting?" in *The Question of the Commons*, pp. 192–207.

82. J. Brown, "The Evolution of Diversity in Avian Territorial Systems," *Wilson Bulletin* 6 (1964): 160–69; N. B. Davies and A. I. Huston, "Territory Economics," in *Behavioral Ecology: An Evolutionary Approach*, ed. J. R. Krebs and N. B. Davies (Sunderland, Mass.: Sinauer, 1984), pp. 148–69.

83. R. Dyson-Hudson and E. Smith, "Human Territoriality: An Ecological Reassessment," *American Anthropologist* 80 (1978): 21–41.

84. R. Kiltie and J. Terborgh, "Observations on the Behavior of Rain Forest Peccaries in Peru: Why Do Some White-Lipped Peccaries Form Herds?" *Zietschrift fur Tierpsychologie* 62 (1983): 241–55.

85. See for example Ross, "Food Taboos, Diet, and Hunting Strategy."

86. Chagnon and Hames, "Protein Deficiency and Tribal Warfare in Amazonia: New Data"; R. Hames, "The Settlement Pattern of a Yanomam Population Bloc," in *Adaptive Responses of Native Amazonians*, ed. R. Hames and W. Vickers (New York: Academic Press, 1983), pp. 192–229.

87. A. C. Roosevelt, "Resource Management in Amazonia Before the Conquest: Beyond Ethnographic Projection," in *Resource Management in Amazonia: Indigenous and Folk Strategies*, ed. D. Posey and W. Balee (New York: New York Botanical Garden Society, 1989), pp. 30–62.

88. See Service, *Origins of the State and Civilization*.

89. Chapman, "Environmental Influences on the Development of Traditional Conservation in the South Pacific Region."

90. K-I. Sudo, "Social Organization and Types of Sea Tenure in Micronesia," in *Maritime Institutions in the Western Pacific*, ed. K. Ruddle and R. Akimichi (Osaka: National Museum of Ethnology, 1984), p. 226.

91. Chapman, "Environmental Influences on the Development of Traditional Conservation in the South Pacific Region."

92. See P. Drucker, *Cultures of the Pacific Northwest* (San Francisco: Chandler Publishing Company, 1965); K. Oberg, *The Social Economy of the Tlingit Indians*, AES Monograph No. 22 (Seattle: University of Washington Press, 1973), p. 88; A. Richardson, "The Control of Productive Resources on the Northwest Coast of North America," in *Resource Managers*, pp. 93–112.

93. See for example Richardson, "The Control of Productive Resources on the Northwest Coast of North America."

94. D. Forde, *Habitat, Economy, and Society* (New York: Dutton, 1963), p. 78.

95. Rambo, "Bows, Blowpipes, and Blunderbusses"; R. Hames, "A Comparison of the Shotgun and the Bow in Neotropical Forest Hunting," *Human Ecology* 7 (1979): 219–52.

96. Hames, "A Comparison of the Shotgun and the Bow in Neotropical Forest Hunting."

97. R. Salisbury, *From Stone to Steel* (Cambridge: Cambridge University Press, 1962).

98. See L. Binford and W. Chasko, "Nunamiut Demographic History," in E. Zubrow ed. *Demographic Anthropology* (Albuquerque: University of New Mexico Press, 1976), p. 107; Rambo, "Bows, Blowpipes, and Blunderbusses."

99. Hames, "A Comparison of the Shotgun and the Bow in Neotropical Forest Hunting"; Hames, "Game Depletion and Hunting Zone Rotation Among the Ye'kwana and Yanomamo of Amazonas, Venezuela."

100. B. Neitschmann, "The Hunting and Fishing Focus Among the Miskito Indians of Eastern Nicaragua," *Human Ecology* 1 (1972): 41–67.

101. N. Chagnon, *Yanomamo: The Fierce People*, 3d ed. (New York: Holt, Rinehart & Winston, 1985).

102. Hames, "The Settlement Pattern of a Yanomam Population Bloc."

103. Chagnon, *Yanomamo: The Fierce People*.

References

Abelove, J., and R. Campos. "Infancy Related Food Taboos Among the Shipibo." In *Food Taboos in Lowland Amazonia*, edited by K. Kensinger and W. Kracke. Working Papers on South American Indians. Bennington, Vt.: Bennington College, 1981.

Anderson, A. "The Extinction of Moas in Southern New Zealand." In *Quaternary Extinctions—A Prehistoric Revolution*, edited by P. Martin and R. Klein. Tucson: University of Arizona Press, 1984.

Aspelin, L. "Food Distribution and Social Bonding Among the Mamainde of Mato Gross, Brazil." *Journal of Anthropological Research* 35 (1979): 309–27.

Baksh, M. "Faunal Food as a 'Limiting Factor' on Amazonian Cultural Behavior: A Machiguenga Example." *Research in Economic Anthropology* 7 (1985): 145–77.

Berkes, F. "Common-Property Resource Management and Cree Indian Fisheries in Subarctic Canada." In *The Question of the Commons*, edited by B. McCay and J. Acheson. Tucson: University of Arizona Press, 1987.

Berry, W. "The Gift of Good Land." *Sierra Club Newsletter* (November/December 1979): 20–26.

Binford, L., and W. Chasko. "Nunamiut Demographic History." In *Demographic Anthropology*, edited by E. Zubrow. Albuquerque: University of New Mexico Press, 1976.

Birdsell, J. "On Population Structure in Generalized Hunting and Collecting Populations." *Evolution* 12 (1958): 189–205.

———. "Some Predictions for the Pleistocene Based on Equilibrium Systems Among Recent Hunter-Gatherers." In *Man the Hunter*, edited by R. Lee and I. DeVore. Chicago: Aldine, 1968.

Bodley, J. *Anthropology and Contemporary Human Problems*. Menlo Park, Calif.: Benjamin Cummings Publishing, 1976.

Brightman, R. "Conservation and Resource Depletion." In *The Question of the Commons*, edited by B. McCay and J. Acheson. Tucson: University of Arizona Press, 1987.

Brown, J. "The Evolution of Diversity in Avian Territorial Systems." *Wilson Bulletin* 6 (1964): 160–69.

Callicott, J. "American Indian Land Wisdom? Sorting Out the Issues." In *The Struggle for the Land*, edited by P. Olson. Lincoln: University of Nebraska Press, 1990.

Cassels, R. "The Role of Prehistoric Man in the Faunal Extinctions of New Zealand and Other Pacific Islands." In *Quaternary Extinctions: A Prehistoric Revolution*, edited by P. Martin and R. Klein. Tucson: University of Arizona Press, 1984.

Chagnon, N. *Yanomamo: The Fierce People*. 3d ed. New York: Holt, Rinehart & Winston, 1985.

Chagnon, N., and R. Hames. "Protein Deficiency and Tribal Warfare in Amazonia: New Data." *Science* 203 (1979): 910–13.

Chapman, M. "Environmental Influences on the Development of Traditional Conservation in the South Pacific Region." *Environmental Conservation* 12 (1985): 217–30.

Charnov, E., and G. Orians. "Optimal Foraging: Some Theoretical Explanations." Mimeograph. Salt Lake City: University of Utah, 1973.

Charnov, E., G. Orians, and K. Hyatt. "Ecological Implications of Resource Depression." *American Naturalist* 110 (1976): 247–59.

Clad, J. "Conservation and Indigenous Peoples: A Study of Convergent Interests." *Cultural Survival Quarterly* 8, no. 4 (1984): 68–73.

Colchester, M. "Rethinking Stone Age Economics: Some Speculations Concerning Pre-Columbian Yanomam Economy." *Human Ecology* 12 (1984): 291–314.

Davies, N. B., and A. I. Huston. "Territory Economics." In *Behavioral Ecology: An Evolutionary Approach.*, edited by J. R. Krebs and N. B. Davies. Sunderland, Mass.: Sinauer, 1984.

Davis, S. *The Archaeology of Animals*. New Haven: Yale University Press, 1987.

Dewar, R. "Extinctions in Madagascar." In *Quaternary Extinctions: A Prehistoric Revolution*, edited by P. Martin and R. Klein. Tucson: University of Arizona Press, 1984.

Diamond, J. "News and Views." *Nature* 298 (1982): 787.

———. "Historic Extinctions: A Rosetta Stone for Understanding Prehistoric Extinctions." In *Quaternary Extinctions: A Prehistoric Revolution*, edited by P. Martin and R. Klein. Tucson: University of Arizona Press, 1984.

———. "The Environmentalist Myth." *Nature* 324 (1986): 19–20.

Drucker, P. *Cultures of the Pacific Northwest*. San Francisco: Chandler Publishing Company, 1965.

Dyson-Hudson, R., and E. Smith. "Human Territoriality: An Ecological Reassessment." *American Anthropologist* 80 (1978): 21–41.

Ewers, J. *The Horse in Blackfoot Indian Culture*. Bureau of American Ethnology Publication No. 23. Washington, D.C.: Smithsonian Institution, 1955.

Forde, D. *Habitat, Economy, and Society*. New York: Dutton, 1963.

Guidon, N., and G. Delibrias. "Carbon-14 Dates Point to Man in the Americas 32,000 Years Ago." *Nature* 321 (1986): 769–71.

Hames, R. "A Behavioral Account of the Division of Labor Among the Ye'kwana." Ph.D thesis, University of California at Santa Barbara, 1978.

———. "A Comparison of the Shotgun and the Bow in Neotropical Forest Hunting." *Human Ecology* 7 (1979): 219–52.

———. "Game Depletion and Hunting Zone Rotation Among the Ye'kwana and Yanomamo of Amazonas, Venezuela." In *Studies in Hunting and Fishing in the Neotropics*, edited by R. Hames. Bennington, Vt.: Bennington College, 1980.

———. "The Settlement Pattern of a Yanomam Population Bloc." In *Adaptive Responses of Native Amazonians*, edited by R. Hames and W. Vickers. New York: Academic Press, 1983.

———. "Game Conservation or Efficient Hunting?" In *The Question of the Commons*, edited by B. McCay and J. Acheson. Tucson: University of Arizona Press, 1987.

Hames, R., and W. Vickers. *Adaptive Responses of Native Amazonians*. New York: Academic Press, 1983.

———. "Optimal Foraging Theory as a Model to Explain Variability in Amazonian Hunting." *American Ethnologist* 9 (1983): 358–78.

Hamilton, W. "The Genetical Evolution of Social Behavior." *Journal of Theoretical Biology* 7 (1964): 1–52.

Hardin, G. "The Tragedy of the Commons." *Science* 162 (1968): 1243–48.

Harris, M. "Levi-Strauss et La Pariourde." *L'Homme* 16 (1976): 5–22.

Hunn, E. "Mobility as a Factor Limiting Resource Use in the Columbian Plateau of North America." In *Resource Managers: North American and Australian Hunter-Gatherers*, edited by N. Williams and E. Hunn. Boulder: Westview Press, 1982.

Kaplan, H., and K. Hill. "Hunting Ability and Reproductive Success Among Male Ache Foragers: Preliminary Results." *Current Anthropology* 26 (1985): 131–33.

Kensinger, K. "Food Taboos as Markers of Age Categories in Cashinahua." In *Food Taboos in Lowland Amazonia*, edited by K. Kensinger and W. Kracke. Bennington, Vt.: Bennington College, 1981.

Kensinger, K., and W. Kracke. *Food Taboos in Lowland South America*. Bennington, Vt.: Bennington College, 1981.

Kiltie, R., and J. Terborgh. "Observations on the Behavior of Rain Forest Peccaries in Peru: Why Do Some White-Lipped Peccaries Form Herds?" *Zietschrift fur Tierpsychologie* 62 (1983): 241–55.

Kracke, K. "Don't Let the Piranha Bite Your Liver. A Psychoanalytic Approach to Kagwahiv (Tupi) Food Taboos." In *Food Taboos in Lowland South America*, edited by K. Kensinger and K. Kracke. Bennington, Vt.: Bennington College, 1981.

Krech, S., ed. *Indians, Animals, and the Fur Trade*. Athens: University of Georgia Press, 1981.

Langdon, T. "Food Taboos and the Balance of Opposition Among the Barasana and Taiwana." In *Food Taboos in Lowland Amazonia*, edited by K. Kensinger and W. Dracke. Working Papers on South American Indians. Bennington, Vt.: Bennington College, 1983.

Leacock, E. *The Montagais "Hunting Territory" and the Fur Trade*. American Anthropological Association Memoir No. 18. Washington: American Anthropologist, 1954.

Levi-Strauss, C. "Structuralism and Ecology." *Social Science Information* 12, no. 1 (1973): 7–24.

Lees, S., and D. Bates. "The Myth of Population Regulation." In *Evolutionary Biology and Human Social Behavior*, edited by N. Chagnon and W. Irons. North Scituate, Md.: Duxbury, 1979.

Lizot, J. "L'economie primitive." *Libre* 4 (1978): 69–113.

MacArthur, R., and E. O. Wilson. *The Theory of Island Biogeography*. Princeton: Princeton University Press, 1967.

Martin, C. *Keepers of the Game*. Berkeley: University of California Press, 1978.

Martin, P. S. "Africa and Pleistocene Overkill." *Nature* 212 (1967): 339–42.

———. "Prehistoric Overkill." In *Pleistocene Extinctions*, edited by P. S. Martin and H. Wright. New Haven: Yale University Press, 1967.

———. "The Discovery of America." *Science* 179 (1973): 969–74.

———. "Prehistoric Overkill: The Global Model." In *Quaternary Extinctions: A Prehistoric Revolution*, edited by P. S. Martin and R. G. Klein. Tucson: University of Arizona Press, 1984.

Martin, P. S., and R. G. Klein, eds. *Quaternary Extinctions: A Prehistoric Revolution*. Tucson: University of Arizona Press, 1984.

McDonald, D. "Food Taboos: A Primitive Environmental Protection Agency." *Anthropos* 72 (1977): 734–48.

McGleone, M. *Archaeology of Oceania* 18 (1983): 11.

Menget, P. "From Forest to Mouth: Reflections on the Txicao-theory of Substance." In *Food Taboos in Lowland South America*, edited by K. Kensinger and W. Kracke. Bennington, Vt.: Bennington College, 1981.

Neitschmann, B. "The Hunting and Fishing Focus Among the Miskito Indians of Eastern Nicaragua." *Human Ecology* 1 (1972): 41–67.

Nelson, R. "Athapaskan Subsistence Adaptation in Alaska." In *Alaska National Museum of Ethnology*, edited by Y. Kotani and W. Workman. Osaka, Alaska, 1979.

Oberg, K. *The Social Economy of the Tlingit Indians*. AES Monograph No. 22. Seattle: University of Washington Press, 1973.

Olson, S. L., and H. James. "The Role of Polynesians in the Extinctions of the Avifauna of the Hawaiian Islands." In *Quaternary Extinctions: A Prehistoric Revolution*, edited by P. S. Martin and R. G. Klein. Tucson: University of Arizona Press, 1984.

Rambo, A. "Bows, Blowpipes, and Blunderbusses: Ecological Implications of Weapons Change Among the Malaysian Negritos." *Malayan Nature Journal* 32, no. 2 (1978): 209–16.

———. *Primitive Polluters: Semang Impact on the Malaysian Tropical Rain Forest Ecosystem*. Museum of Anthropology. Anthropological Papers No. 76. Ann Arbor: University of Michigan, 1985.

Redford, K., and J. Robinson. "Hunting by Indigenous Peoples and Conservation of Game Species." *Cultural Survival Quarterly* 9 (1985): 41–44.

Reichel-Dolmatoff, G. "Cosmology as Ecological Analysis: A View from the Rain Forest." *Man* 11 (1976): 307–18.

Richard, A., and R. Sussman. "Future of the Malagasy Lemurs: Conservation or Extinction?" In *Lemur Biology*, edited by I. Tattersall and R. Sussman. New York: Plenum Press, 1975.

Richardson, A. "The Control of Productive Resources on the Northwest Coast of North America." In *Resource Managers: North American and Australian Hunter-Gatherers*, edited by N. Williams and E. Hunn. Boulder: Westview Press, 1982.

Roosevelt, A. C. "Resource Management in Amazonia Before the Conquest: Beyond Ethnographic Projection." In *Resource Management in Amazonia: Indigenous and Folk Strategies*, edited by D. Posey and W. Balee. New York: New York Botanical Garden, 1989.

Ross, E. "Food Taboos, Diet, and Hunting Strategy: The Adaptation to Animals in Amazonian Cultural Ecology." *Current Anthropology* 19 (1978): 1–36.

Saffirio, G., and R. Hames. *The Forest and the Highway*. Working Papers on South American Indians No. 6 and Cultural Survival Occasional Paper No. 11. Cambridge, Mass.: Cultural Survival, Inc., 1983.

Sahlins, M. "Notes on the Original Affluent Society." In *Man the Hunter*, edited by R. B. Lee and I. Devore. Chicago: Aldine, 1968.

———. *Stone Age Economics*. Chicago: Aldine, 1972.

Salisbury, R. *From Stone to Steel*. Cambridge: Cambridge University Press, 1962.

Service, E. *Origins of the State and Civilization*. New York: W. W. Norton and Company, 1975.

Slobodkin, L. "Prudent Predation Does Not Require Group Selection." *American Naturalist* 108 (1974): 665–78.

Steadman, D. W., G. K. Pregill, and S. L. Olson. "Fossil Vertebrates from Antigua, Lesser Antilles: Evidence for Lake Holocene Human-Caused Extinctions in the West Indies." *Proceedings of the National Academy of Sciences* 81 (1984): 4448–51.

Stephens, D., and E. Charnov. "Optimal Foraging: Some Simple Stochastic Models." *Behavioral Ecology and Sociobiology* 10 (1982): 251–63.

Sudo, K-I. "Social Organization and Types of Sea Tenure in Micronesia." In *Maritime Institutions in the Western Pacific*, edited by K. Ruddle and T. Akimichi. Osaka: National Museum of Ethnology, 1984.

Taylor, K. "Knowledge and Praxis in Sanuma Food Prohibitions." In *Food Taboos in Lowland South America*, edited by K. Kensinger and K. Kracke. Bennington, Vt.: Bennington College, 1981.

Trotter, M., and B. McCulloch. "Moas, Men, and Middens." In *Quaternary Extinctions: A Prehistoric Revolution*, edited by P. S. Martin and R. G. Klein. Tucson: University of Arizona Press, 1984.

Tuan, Y.-F. "Discrepancies Between Environmental Attitude and Behaviour: Examples from Europe and China." *Canadian Geographer* 12, no. 3 (1968): 176–91.

Vickers, W. "An Analysis of Amazonian Hunting Yields as a Function of Settlement Age." In *Studies on Hunting and Fishing in Amazonia*, edited by R. Hames. Bennington, Vt.: Bennington College, 1980.

———. "Game Depletion Hypothesis of Amazonian Adaptation: Data from a Native Community." *Science* 239 (1988): 1521–22.

White, L. "The Historic Roots of Our Ecologic Crisis." *Science* 155 (1967): 1203–1207.

Williams, G. *Adaptation and Natural Selection*. Princeton: Princeton University Press, 1966.

Part IV
Utilizing Wildlife

14

Economics: Theory versus Practice in Wildlife Management

Raymond Rasker, Michael V. Martin, and Rebecca L. Johnson

Introduction

There is an ongoing debate concerning the use of neoclassical economics or, more accurately, the application of market solutions to the management of wildlife. In North America this issue is complicated by the fact that we live in a mixed economy. The blend of public and private ownership makes it difficult and at times inappropriate to recommend commercialization as a means of resolving conflicts in natural resource management. This is particularly so for wild species. Wildlife is a public resource, yet the habitat on which wild animals depend is often privately owned.

The fundamental problem and the major source of disagreement is a weakness many of us share: It is tempting to look for easy solutions to complex problems. Biologists and economists alike, having become enamored with the potential solutions offered by the free market, tend to oversimplify issues related to wildlife management. The purpose of this paper, therefore, is to address the application of market solutions to wildlife management in North America and to add some clarity to this all too polarized debate. The discussion will be about the commercialization of wildlife, about "free market" theory, and about potential for market failures. Remedies to market failure are then offered as tools to be used in conservation. The goal is simply to stimulate thought and discussion.

This paper is not intended as a definitive work on all topics related to the commercialization of wildlife, since such an effort would require several vol-

---

This article originally appeared in *Conservation Biology* 6 (September 1992): 338–49. Reprinted by permission of Blackwell Science, Inc. Copyright © 1992 Blackwell Science Inc. All rights reserved.

umes. Certainly there are complex issues to be addressed in this discussion. This paper focuses exclusively on economic incentives in wildlife management and habitat development. Other issues such as the commercialization of dead wildlife and the trade of wildlife parts are of immediate concern but are beyond the scope of this effort. Readers interested in these topics should begin with Geist and Lanka et al.[1]

For the last ten years we have followed, with a great deal of fascination, the debate over alternative approaches to wildlife management. It appears that one of the most controversial issues in wildlife management in North America is the trend toward the commercialization of wildlife. Among the most extreme is the idea to privatize wildlife, that is, to transfer property rights from the public to private interests for the purpose of facilitating market transactions.

A Case for Privatization: Africa's Elephants

It may appear at first that the commercialization of wildlife holds great promise. Examples from African nations offer an excellent opportunity to compare the effectiveness of private versus public ownership of wildlife in promoting conservation. The success Zimbabwe and South Africa have had in expanding their elephant populations is well known. While throughout Africa elephant numbers have fallen from 1.3 million in 1979 to 650,000 in 1989, Zimbabwe has since 1981 increased its population from 47,000 to 49,000. South Africa has been able to build its elephant population from about 500 in the nineteenth century to its present stable population of about 8,200.[2] Biologists at Kruger National Park erected an elephant-proof fence along the border with Mozambique in 1983 and have since been able to significantly reduce poaching. In 1989, the park earned $2.5 million, 10 percent of its annul budget, by selling ivory and hides from 350 elephants culled to prevent overpopulation. These funds are then used to train and arm game wardens and rangers.[3]

Much of the success of Zimbabwe and South Africa can be attributed to policies that have allowed commercialization to take place. In both countries wildlife is privately owned, and existing markets for hunting and ivory have given landowners the incentive to manage land for the benefit of wildlife. For example, the price of an average hunt in Zimbabwe, where elephants are the main trophy, is about $25,000.[4] The Department of National Parks and Wildlife Management spends over $600 per square mile to protect wildlife, and poachers are shot on sight.[5] In addition, local tribes are given incentives to participate in conservation efforts. They can sell hunting permits on their land and have the right to hunt a certain number of elephants.[6] Over the past seven years Zimbabwe has paid out some $4.5 million to local communities as part of the "private game reserves" program, wherein people living on communal lands share revenues and meat from hunting.[7]

Kenya stands in sharp contrast to the success of Zimbabwe and South Africa. There is no market for elephant hunting, and ivory trade is illegal. Also—and this is an important point—elephants compete with poor farmers and herders for scarce land. They are viewed as either competition for land resources or as a source of lucrative profits for ivory smuggling. Apparently lacking, however, is a financial incentive to enhance and protect elephant populations.

As in Zimbabwe, poachers in Kenya are shot on sight. Unfortunately, efforts to save the elephants based on disincentives alone appear to be failing. From 1979–89 the population has declined from 65,000 to 19,000. While poaching for the international ivory market is blamed, others find Kenya at fault for not commercializing their wildlife and for not creating monetary incentives for elephant protection.[8] Thus, the African experience suggests that private control and commercial incentives can be used effectively to manage endangered wildlife species. We will return to this example later in this paper.

Commercialization of Wildlife in North America

It is tempting to conclude that the system of Zimbabwe and South Africa can also be effectively transferred to North America. In Oregon, for example, we have seen fee hunting (an access fee for hunting rights on private land) serve as a powerful incentive for wildlife habitat conservation on farmland.[9] There are obvious drawbacks to fee hunting, however, mainly because the system relies on hunters to pay the bill for habitat development and assumes that landowners invest proceeds from fee leases in habitat improvements. Restrictions in the hunting season, often necessary for biological reasons (such as poor recruitment and high mortality) may erode the demand for hunting on private land and, in turn, the landowner's source of revenue for habitat development.

Many examples that link financial returns from fee hunting and game ranching to conservation can be found in the literature.[10] For example, Stier and Bishop assessed government incentive programs aimed at compensating farmers in the Horicon area of Wisconsin for crop damage from migratory geese. They concluded that existing policies were ineffective and recommended that restoration of a market approach for hunting rights would "improve farmers' attitudes towards wildlife and revive the traditional partnership that existed between farmers, hunters, and wildlife managers."[11] Not all would concur with this conclusion in light of perceived distribution effects resulting from overreliance on pure market solutions.

California's Ranch for Wildlife Program, which focuses largely on game animals, is an extreme example of how markets for hunting rights can be developed. Under this program agricultural landowners, after approval by the Fish and Game Commission, have authority to increase the bag limits on their land beyond those imposed on the rest of the state and to sell tags or permits directly to hunters.[12] The assumption is that the profit motive will lead landowners to increase wildlife populations at their own expense in exchange for flexibility from standard game regulations.[13] At the time of this writing the success or failure of this program cannot be clearly assessed.

Interest in wildlife management for profit is not limited to farms and ranchers. The southeastern division of International Paper (IP) provides an example of corporate involvement. International Paper reportedly earns 25 percent of its net profits from marketing recreation, including fee hunting.[14] The company employs wildlife biologists and publishes manuals on wildlife management for their foresters.[15] In addition to managing for game species, IP has also created habitat for bald eagles and rare woodpeckers.[16] It should be noted however, that the passage of time can lead to changes in ownership and management philosophy. This is always a risk in an environment of private property and private management.

Although interest in wildlife commercialization is growing, there is also a cadre of wildlife biologists, hunters, and other wildlife enthusiasts committed to blocking further development of fee hunting and game ranching. A lively and very serious debate is underway. We hope that open communication between the various factions will lead to a better understanding of these issues and to the design of more effective public policies. Clearly, the overriding objective should be the design of policies that lead to the protection and conservation of wildlife. The central question is: What is best possible set of mechanisms by which to accomplish this goal?

Fueling the controversy is the fact that many of us do not use the same terminology, which can lead to considerable confusion. Mention "fee hunting" or "game ranching" as an approach to wildlife management and you are likely to be met with a variety of responses. For some, fee hunting suggests the privatization or the de facto privatization of wildlife.[17] Others oppose game ranching and fee hunting because of the prospect of introducing exotic game and, as a consequence, a multitude of potential negative side effects. These include introduced diseases, competition with native species, genetic pollution, and the trade in animal parts.[18] Objections also relate to issues of land access: farmers controlling rights of way to public land[19] and ranchers excluding recreationists from access to public lands on which they have a grazing privilege.[20]

To others, fee hunting, particularly in the United States, is nothing more than a landowner exercising his or her property rights.[21] In some instances, it is argued, returns from selling hunting access can lead to a powerful incentive to manage land in a way that protects and enhances wildlife.[22]

In this paper the term *fee hunting* refers to an agreement whereby a landowner charges hunters a fee for access to private land for the purpose of pursuing publicly owned game. In *game ranching*, the private landowner owns the animals, which are kept on the property by a barrier of some sort, most often a fence.

Background Example: Fee Hunting in Oregon

The experience in Oregon typifies many elements of the ongoing debate surrounding private commercialization versus public ownership and control of wildlife.

In 1987 the Cooperative Extension Service of Oregon State University, responding to a request from farmers and ranchers, organized a workshop entitled "Developing Profitable Resource-Based Recreation on Private Land."[23] The purpose was to help farmers diversify their economic base by marketing recreational opportunities ranging from dude ranches to land-access fees to guided big-game trophy hunts. Organizers believed this could result in farmers taking greater interest in protecting and promoting wildlife. A rehash of the statistics on habitat loss due to modern agricultural practices is not required here.[24] The important point is that there seemed to be interest on the part of some farmers and ranchers in Oregon to reverse this trend. The incentive, of course, was the potential for financial return, in some cases exceeding those from farming.[25] As will be noted later, there are several mechanisms for creating these financial incentives.

One Oregon rancher provided an example of how fee hunting had influenced the way he managed his ranch. He was charging between $5,000 and $6,000 for a guided trophy bull elk hunt. He had fenced off his riparian areas to keep his cows out, reduced the size of his cattle herd, and made genuine progress in improving the habitat conditions, not only for elk but for the fishery as well. In this case the fee hunting arrangement proved significantly superior to the alternative of using this land for commercial agriculture. The topography, soil type, rainfall, and remote location suggest that rational landowners will choose any reasonable alternative to low-output agriculture.

This particular conference added fuel to a controversy that had been developing in Oregon for quite some time. Many private citizens, conservation groups, hunting groups, and individuals in state and federal wildlife management agencies were skeptical about a state university becoming involved in teaching landowners how to put together fee-hunting operations. One of the more extreme reactions came from Chuck Griffith, executive director for the Rocky Mountain region of the National Wildlife Federation. In a hunting magazine, Griffith wrote that "Certain Extension Service range management specialists in some western states were active in organizing training sessions aimed at *privatizing wildlife resources on public lands*" and questioned whether it was appropriate for the Extension Service to "become an active advocate of promoting the public ownership of wildlife on public lands into private ownership."[26] Those were strong words, typical of the language heard in the debate over fee hunting, game ranching, and the commercialization of wildlife.

There is no evidence that workshop organizers or participants intended to suggest that the ownership of wildlife would be transferred from the public sector to private landowners. It is understandable, nonetheless, that it was perceived that way by some.

In 1987 Oregon's governor appointed an eight-member task force to "determine if statewide fee hunting, regulated fee hunting, and commercial game farming on privately owned lands can be implemented to the satisfaction of the people of Oregon and for the benefit of Oregon's wildlife and habitat."[27] The public was invited to participate in meetings held throughout the state. They were asked to indicate whether they favored or opposed fee hunting. This illustrates how the complexities of the issue were reduced to a simple yes/no vote. It appears that the same oversimplification is occurring in other forums throughout the United States and Canada, and discussions of wildlife commercialization are becoming increasingly polarized. In an effort to find common ground, a review of economics and its application to wildlife management may be timely and helpful.

Economics and Wildlife Management

There is growing recognition on the part of wildlife managers of the importance of understanding market economics and the potential alternatives that can be brought to management. We often take markets for granted, particularly when they operate to our benefit. Producers and sellers compete with each other for a share of consumer expenditure. Consumers have the freedom and good fortune to choose among a wide array of products and services. The end result of market competition can be efficient and effective: Firms specialize, filling specific niches and thereby serving the diverse needs of consumers; economic incentives stimulate research and development that leads to new and better products; and, in effort to sell more, deals are offered.

Is it possible that commercialization of wildlife will lead to such desirable outcomes? Will more wildlife be "produced" at cheaper prices? An answer requires brief review of the neoclassical theory of microeconomics.

The Theory of Competitive Markets

Economics is a study of choices by individuals or groups that result from an interplay between two forces: values and opportunities. The individual is believed to pursue satisfaction (utility) through the consumption of goods, including services and amenities. While the desire to consume may be insatiable, however, the means for consumption are limited. That is, there is a restricted set of opportunities from which the individual must choose. Goods (or resources)are limited—some are scarce and others less so—and the individual must choose among these. Moreover, an individual's ability to command scarce goods or resources is limited by income or wealth.

In simple terms, the market encourages efficiency through specialization in the supply side of the market. An exchange economy develops such that efficient producers generate income which, in turn, allows them to become consumers. Contemporary market economies are certainly more complex but they operate fundamentally the same as more primitive barter economies.

The market allocates resources and determines the value of both productive input and output. It is a system that in its purest form is thought to promote growth and efficiency without the need for central planning. The market generates information on the relative value of goods and services—expressed in terms of prices—

and on the amount of goods to be produced and consumed; all of this is driven by the principle of individual choice. It is a system by which each good ends up in its highest-valued use—the hands of those willing to pay the most.

Efficiency is said to exist when the given allocation of resources is "better than another if every individual feels it is better according to his own individual values."[28] A trade is "efficient" if each individual is able to choose according to his or her own values in a way that makes him or her better off without making someone else worse off. Or, according to Mansfield, "society should note any change that harms no one and improves the lot of some people. If all such changes are carried out, this situation is termed Pareto-optimal or Pareto-efficient."[29] Under a very specific set of assumptions this efficiency can be achieved through the price system.

Economic Efficiency

Efficiency, as defined in neoclassical theory, is referred to as Pareto-efficiency (named for the economist and philosopher Wilfredo Pareto). While it is possible for efficient solutions to exist in a perfectly competitive marketplace, efficiency should not and is not meant to be used as the criterion for all social well-being. However, many economists apply it regularly in natural-resource policy, such as in cost-benefit analysis, with a devotion that implies economic efficiency is the only matter on which economists can pass scientific judgment.[30] The Pareto-efficiency criteria does not describe the "best" situation in terms of distributive justice. It is a criteria that takes as given the existing distribution of income and property rights.

One complaint against fee hunting is that it could lead to hunters being denied access to land based on socioeconomic status.[31] While the trade between a landowner and a pay-to-hunt customer may make both sides better off, it says nothing about the costs and benefits to those individuals not party to the transaction. Although technically the trade may be efficient, this does not imply that the full social value is reflected in the transaction. Is society better off as a result of trade, such as fee hunting? We don't know, at least based on the Pareto-efficiency criteria.[32] Regarding equity considerations, particularly income distribution, economists have yet to develop a viable theory within the neoclassical paradigm.[33]

Economists with a normative bent tend to argue that the market, allowed to operate as closely as possible to the dictates of theory, will result in the best possible allocation of resources. They argue that restrictions placed against the market operating free and unperturbed lead to inefficiency. Demsetz, Cheung, and Furubota and Pejovich argue, for example, that the existence of common property will lead to a tragedy of the commons as described by Hardin; failure to internalize the full social cost of grazing cattle on the "commons" will lead to overexploitation.[34] Following this theory, the extreme case argues that if only the "commons" were privately owned—if property rights were assigned and clearly defined—then the market would determine the distribution of resources.

If all property rights were correctly specified and all goods and services were traded in competitive markets, then the relative value of all goods would be known. Prices would be fully determined by supply and demand. Inputs and resources would be allocated based on market-determined values of final outputs. Those resources not commercially traded, and therefore with no price, would not be efficiently allocated. Wildlife frequently falls into this latter category as it is typically not commercially traded.

As a practical matter, the application of market solutions to wildlife is at best difficult. What do we do in those frequent instances in which the price is not known? How much is a grizzly bear or a spotted owl worth? What would the price be for nongame animals such as song bids and rodents, animals without any conceivable market? If the market is the sole determinant of how resources are distributed, then we will value resources only on the basis of their market-derived prices. In doing so, we run the danger of severely undervaluing nonmarket resources. As an alternative, we can admit that other forms of demand also play a role in society and that those are not expressed in dollar terms.[35] The preferences of society can also be expressed via the political process. As in economics, politics is about choices.

Separating faith in the market from reality is a continuing dilemma for economists which although intellectually stimulating, has led in some instances to a dangerous oversimplification of the challenge of wildlife management. Harvard economist Thomas Schelling phrased the conflict well in an article entitled "Economic Reasoning and the Ethics of Policy":

> Nothing distinguishes economists from other people as much as a belief in the market system, or what some people call the free market system. A perennial difficulty in dealing with economics and policy is the inability of people who are not economists, and some who are, to ascertain how much of an economist's work is faith and how much is analysis and observation. How much is due to. . . observing the way the markets work and judging actual outcomes, and how much is belief that the process is right an just?[36]

Most economists recognize that for various reasons, including distributive justice and potential for monopoly power, certain functions could not be left entirely to the market. National defense, maintenance of justice among individuals, and public works such as education were seen as proper and necessary functions of the state.[37]

So even in the early history of "free-market" thinking, philosophers and economists left room for the realization that society can perhaps best succeed as a mix of free markets and state control. The market operates well in some circumstances, poorly in others. Just as the neoclassical theory teaches us the potential benefits of competitive markets, economists continue to alert us to the possibility of market failures.

Market Failure

By definition, market failure occurs when incentives created in the market system fail to adequately reflect the present and future economic interests of consumers or society as a whole.[38] Imagine a situation where a landowner in Montana charges a fee for hunting access to his land, which happens to be a prime elk-hunting area. He has the good fortune to have elk on his property during the hunting season but on other ranches during the remainder of the year. Here fee hunting may lead to market failure: The cost of feeding the animals throughout most of the year may be borne by someone other than the one deriving the benefits. While the elk may be a liability to one rancher, competing for pasture with cattle, they are an unearned profit (rent) for another.

In such a case fee hunting may not necessarily lead to habitat improvements on the part of the individual charging the fee. Jordan and Workman found that in Utah less than 25 percent of landowners who charged a fee for hunting improved wildlife habitat or demonstrated an active interest in wildlife on their property.[39] Marion warns that "the wildlife resource and lower-income hunters do not necessarily benefit from [fee hunting]."[40] Further, Burger and Teer conclude that "despite the growing commercial hunting systems and increasing values of wildlife in Texas, few landowners invest much capital and other resources into management of wildlife."[41]

Markets are more likely to operate efficiently if the goods traded are highly divisible, mobile, and lacking external effects when produced or consumed. Also, markets work best when entitlement to property is clearly defined and enforceable.[42] But can migratory waterfowl, or wild deer and elk with travel patterns that cross different property boundaries, be privately owned? The fugitive nature of wildlife as a resource makes it difficult if not impossible to create a true market.

The market also cannot fully capture nonuse values of a resource, two of which are *existence values* and *option values*. Existence values are the psychic returns associated with knowing that wildlife exists. Option values are returns the resource holder receives from maintaining the option of reallocating the resource to another use in the future.

One way around wildlife mobility, of course, is to put a fence around the property to contain privately-owned animals, as with game ranching. Although it eliminates the fugitive problem, it creates others, not the least of which falls more into the realm of philosophy than of economics. That is, when we fence in animals do we take the "wild" out of wildlife?

In some parts of the world, where conservation of wildlife has reached crisis proportions, game ranching may be the only option. As a strategy for preserving elephants, the choice made by Zimbabwe and South Africa may offer the only viable solution: Put up a fence to keep the animals from wandering off, sell hunting permits to wealthy hunters, sell the ivory and hides, and use the proceeds to arm paramilitary-style park rangers. In North America, however, we have not reached this level of desperation.

Sources of Market Failure

Understanding the sources of market failure will bring us closer to an outline of policies that may be effective in influencing conservation efforts in North America. As noted earlier, one source of market failure is the characteristics of some resources that prevent private ownership. Other reasons why markets fail to function in society's best interest have to do with problems of information, imperfect competitive markets, and externalities.[43]

While conservation efforts are ideally aimed at improving future as well as current conditions, economic interests usually discount future benefits and costs in favor of present consumption. Because information about the future is limited, a premium is put on the present; therefore, short-term profits may be favored over long-term, and hence uncertain, profits of the future.[44]

Clark put this dichotomy of private versus public discount rates into perspective in an analysis of the economics of killing blue whales. The policy question was whether it would be advisable, from a strictly economic point of view, to halt the harvest of these whales in order to give the population a chance to recover to levels where a sustainable harvest would be possible. The results of his analysis indicated that the rational economic decision was to harvest every blue whale as fast as possible, even to the point of extinction.[45]

Clark concluded that in situations where the discount rates are high and growth rates are low (as with whale populations) the rational economic choice is to harvest and invest the money in areas with a higher rate of return. He emphasized that his model was not intended as a welfare model and did not imply that extinction is socially optimal. The conclusion was that "extermination of the entire population may appear as the most attractive policy, even to an individual resource owner" when harvesters "prefer present over future revenues."[46]

Another problem that may lead to market failure is the potential existence of associated externalities. Those who benefit by exploiting wildlife may not be the ones bearing the full cost. Conversely, individuals may act as "free riders" by deriving benefits for which they have not paid. This problem has its roots, again, in a lack of exclusive ownership of the resource. Migratory waterfowl may be a cost to the farmer in Canada and a source of profit for the duck-club operator in Louisiana. There is no mechanism by which the farmer can charge for his or her share of the costs of "producing" the resource. Even if the free-rider in Louisiana was willing to pay, there is no market mechanism to facilitate the transaction and, if there were, the cost of carrying out the transaction could outweigh any efficiency gains made in the trade. So not only does the possibility exist for external effects, but the cost of internalizing them may be prohibitive.

Remedies Relative to Private Land

A particularly challenging task for the wildlife manager in North America is to design incentives for private landowners to manage their land for the benefit of wild animals. Unless properly compensated for the costs associated with raising wildlife, a profit-motivated land manager may be better off reducing or even eliminating all habitat. This is especially true when hunters trespass and liability becomes a significant problem,[47] when wildlife causes damage to pasture and crops,[48] and when the returns from farming outweigh any benefits from preserving land resources for wildlife. In the absence of incentive or disincentive programs—such as subsidies or taxation, education, cost-share programs, damage compensation, farm programs, and liability relief—the only option available to the landowner may be the commercialization of wildlife. Charging hunters an access fee may be the only method for some to recover the costs of having wildlife on their land.

When profit incentives offered by fee hunting and game ranching lead to market failure, however, alternatives must be explored. Concern over the expansion of fee hunting may be a symptom of a different problem: There are insufficient incentives (other than charging an access fee) for the landowner to take wildlife into consideration when land-use decisions are made. Farmers and ranchers respond to many different incentives, of which profit may be only one. Other incentives for wildlife management may come from aesthetic appreciation, a sense of pride and public recognition for public services provided when wildlife interests are integrated into the farming or ranching operation, and the landowners' desire to hunt.[49] Some landowners may have altruistic motives. To be practical, however, we should not expect them to produce wildlife for free. Cost recovery may be sufficient for some; others may operate from a more profit-oriented position.

Several avenues can be followed in the search for ways to manage wild animals on private land, and many of these have been explored by the Leopold Committee.[50] One of the most immediate forms of action is land acquisition. Although this approach requires sizable state or federal expenditures, public ownership facilitates stability in management over time, since it is less vulnerable than private ownership to short-term economic trends.[51]

The discrete nature of private land holdings may not promote optimal habitat development in light of the migratory patterns and habitat needs of wildlife. Therefore, some form of cooperation and coordination between individual contiguous landowners may be required to fully meet the needs of wildlife. Again, this may be induced by incentive arrangements or by government regulation.

While it is beyond the scope of this article to review all ways to stimulate participation in wildlife management, two other approaches that have shown some promise should be mentioned: government farm programs and tax incentives.

Farm Programs

In the last decade, farm policies in the United States have been refocused in many ways. Beginning in the 1930s, Congress established a complex set of farm programs with the principal purpose of stabilizing the agricultural economy. Incentives built into these programs resulted in the expansion of farm production. As a result, many of the nation's prime wildlife-producing areas, including marshes, swamps, woodlands, and prairie grasslands, were eliminated to make way for the "highest and best" use of land. This often meant large-scale monoculture farming operations.

In the early 1980s changing economic conditions gave rise to efforts to further regulate agricultural production. Rising surpluses, declining commodity prices, and increased international competition have led to the development of new policies which intend to (1) reduce agricultural surpluses, (2) promote agricultural exports, and (3) encourage conservation practices.[52] Certain programs under both the 1985 and 1990 farm bills have been hailed by conservationists as a boon to wildlife and the first realistic opportunity for cooperation between wildlife interests and farmers.[53] It appears that these provisions will remain central to future farm legislation as well.

The Conservation Reserve Program (CRP) allows for direct payments to farmers who retire eligible cropland on a ten-year basis and who plant this land with cover crops. Congress targeted forty-five to fifty million highly erodible acres for enrollment into the CRP, and by February 1989 over twenty-eight million acres had been enrolled. The CRP provides cover and food for upland game, ground-nesting birds, and other grassland-dwelling species, and for some waterfowl.[54]

The CRP program has not been without flaws. Hays et al. monitored the effects of the CRP on farmland and warned that the program "could have adverse impacts for some wildlife species if it eliminates needed winter food formerly available on cropland, or if disturbances occur such as mowing during nesting."[55] Miller and Bromley conducted a survey in Iowa and Virginia and found that farmers there did not have a high interest in wildlife.[56] They suggest that close contact between wildlife professionals and farmers may help motivate participation in farm programs that benefit wildlife.

Taxation

Another way to encourage conservation practices is through tax programs. Economists often favor tax incentives because they can be designed to work in much the same way as market prices.[57] They are also less subject to the fickle nature and frequent shifts in direction common in farm programs. Cook, for example, warns that the advances made in the 1985 bill may evaporate once farmers' ten-year CRP contracts expire.[58] (The 1990 farm bill acknowledged this concern and extended the CRP program. The U.S. Department of Agriculture is currently working on transitional programs for CRP when it finally draws to a conclusion.)

To illustrate how taxes can be implemented to alleviate market failures, imagine a situation where the actions of an individual result in a cost to others, such as a farmer draining a wetland valuable for waterfowl production. In order to equate social and private costs, a tax could be levied on the farmer equal to the amount of damage done to the public good, the waterfowl. Theoretically, the landowner will respond to avoid or minimize this additional cost.[59] If the tax is large enough, the price of draining the wetland may be too high and it may be more profitable to leave it intact.

From a practical standpoint, taxation may be difficult to apply. Not only are politicians leery of taxation, it may also be difficult to determine the extent of the "social cost" tax equivalent. As a result, tax regulations may require a trial-and-error approach. Also, taxes run the danger of being perceived as a penalty rather than an incentive, not always an effective means to enlist the cooperation of private landowners.

An interesting and promising way to encourage wildlife-habitat preservation is through positive tax incentives. Conservation easements are one type of tax incentive. A landowner can donate a conservation easement to a charitable organization and, in return, qualify for a tax deduction.[60] The potential income-tax savings can be substantial, while complete ownership and management authority is maintained by the owner. The only limitations to land use are those pertaining to the easement, and these are specifically designed for each landowner. The one strict requirement is that the land covered under the easement be used for conservation purposes.[61]

Recently in Montana, Ted Turner (of cable-television fortune) donated a substantial conservation easement to The Nature Conservancy. The easement will ban mining and timber harvesting and will prohibit subdivision of 107,000 acres of land within the Greater Yellowstone ecosystem. Currently The Nature Conservancy has about 150,000 acres of land in Montana under conservation easements from thirty-five to forty different landowners.[62] The Montana Land Reliance holds an additional 72,646 acres in easements, which are estimated to protect about 19,000 acres of elk habitat and offer 145 miles of river and streambank protection.[63] Easements donated to this organization in the Yellowstone ecosystem alone amount to 30,000 acres, protecting valuable wintering habitat and blue-ribbon trout streams and preventing the common western phenomena of ranch subdivision.[64]

Besides tax incentives and government farm programs, a myriad of other potential solutions exist that do not rely on the commercialization or privatization of wildlife.

Discussion and Conclusion

Two questions posed at the beginning of this paper were: Which mechanisms work best for the protection and conservation of wildlife? and: Will market solutions to wildlife management lead to the desirable results seen in other markets? A simple response to either is not possible. In some instances market solutions may be the best alternative. For Zimbabwe and South Africa, commercialization of elephant populations via the granting of private management control may be the most effective solution given the severity of the poaching problem and competing uses for the land. It may also be that budget priorities of these countries necessitates a system by which wildlife management pays for itself. In reality, the merits of market incentives, government intervention, and regulation must be tailored to each national situation.

The apparent success some countries have had in wildlife conservation via market incentives does not imply, however, that wildlife commercialization is the best solution for North America. In some instances markets may indeed work. In others, public ownership may work best. There is no single mechanism that will work in all instances.

In this paper some of the basics of neoclassical theory have been reviewed and the potential for market failure has been explored. The cynic might be tempted to discount the use of economics solely on the basis of the limiting assumptions inherent in the theory. This would be a mistake. Economic theory is, most importantly, a powerful device for deductive reasoning and as such is useful as a tool for analysis and description. It can be a useful instrument in the wildlife manager's toolbox.

The field of resource economics is largely concerned with the study of when and why markets fail. At issue is not whether or not to recommend market solutions as a tool for wildlife management but rather when and where to recommend application of the free market, and identifying when markets fail. Without perfect information—when all costs to society are not accounted for by the market—and when resources are mobile and difficult to privatize, the task of the advising economist is much more complicated than dispensing market solutions as a panacea. The existence of market failure necessitates a more complex and sophisticated approach. It is more difficult yet more satisfying to devise policies that, besides being economically efficient, are also socially equitable and based on sound biological judgment.

Some promoters of the market system are quite persuasive in extolling the virtues of private enterprise. Nevertheless, few economists ever recommend a "pure" private enterprise solution. Even the most ardent supporters of private enterprise support some form of regulation, justify certain kinds of subsidy, and even admit the need for direct governmental intervention in certain activities.[65]

It is advisable to know the limitations of markets as a social institution, particularly regarding the management of wildlife on private land in North

America. For example, to expect that individuals behave as theory assumes would be unrealistic and ill-advised. Kellert suggests that "history is replete with examples of humans foregoing substantial material rewards because their acquisition was somehow perceived as ethically wrong or incompatible with a more desirable lifestyle and society."[66] Arrow asserts that, in addition to the market, society is also governed by "invisible institutions": the principles of ethics and morality.[67] In fact, Arrow argues that the price system can be attacked on the grounds that it "harnesses motives which our ethical systems frequently condemn. It makes a virtue of selfishness." Further, Bromley reminds us that "market processes are derivative of a larger social system; they do not supersede that system."[68]

Much of this paper has focused on market failure. The possibility of "government failure" is certainly as real. Government officials operate on different sets of incentives and bureaucracies can be slow and ineffectual. Fortunately, we are not faced with an absolute choice. Wildlife management in North America will most likely always be a blend of private and public ownership of resources and a mix of market- and public-oriented management strategies. Between the extremes lie potentially successful solutions to some of our most difficult resource-management dilemmas.

To explore further this middle ground and to bring the lessons from this discussion into perspective, let's revisit the dilemma of Kenya's elephants. Specifically, should a market be developed for elephant hunting and should ivory trade be legalized in Kenya? Is that the only option available? Certainly not. To begin, it is doubtful that current elephant populations in Kenya could sustain a harvest. In Zimbabwe and South Africa elephants are harvested as a means to control the population, bringing the herd size in line with the carrying capacity of limited land resources. This biological surplus then enters the market in the form of hides, meat, ivory, and the marketing of hunting permits. The surplus allows hunting without damage to the population. Without a healthy population in Kenya, however, alternatives need to be explored. A few observations from Amboseli National Park offer an example.

Tourism is Kenya's largest export—its most important source of foreign currency. Amboseli National Park earns about $100 per acre, most of it from tourism. Using the travel cost approach, it has been estimated that each lion is worth $27,000 per year in visitor attraction, and each elephant is worth $610,000 per year unrelated to hunting or trade in hides or ivory.[69] Brown and Henry estimated that elephant viewing in Kenya is worth about $25 million annually.[70] While these estimates are open to debate, the fundamental point is that wildlife is a significant generator of income in countries such as Kenya. An important question is, are native peoples sharing in this income?

Since conflicting uses of land for wildlife and cattle threaten the integrity of the Amboseli ecosystem, a package of incentives has been developed. First, cattle-watering areas have been established outside the park, thereby alleviating grazing pressure within Amboseli. Second, the local Masai landowners are compensated for wildlife that graze on their land. Finally, the tribe receives a subsidy for the development of tourist campsites and lodges.[71]

Therefore, landowners are compensated for the cost of raising wildlife on their land, and financial incentives are offered to help the tribe reap part of the profits of Kenya's tourist trade. It is estimated that the annual monetary gain to the park from the use of Masai land is $500,000. Benefits to the Masai from the park result in an income "85 percent greater than what they could obtain from livestock alone after full commercial development."[72] These earnings totaled over $750,000 in 1990, derived primarily from concession and guide fees and from a proportion of the entrance fees. The Masai are also employed by the Park Service and by private lodges.[73] The compromise solution for Amboseli National Park relies on developing financial incentives, yet it does not call for the privatization of wildlife or trade in wildlife parts.

It is important that policy makers remain open-minded on issues related to wildlife management. A number of solutions do not rely on privatization or on commercialization of wildlife, and some of the more innovative ones can be found between the opposite extremes of public ownership and the profit-motivated incentives offered by the market.

Notes

1. V. Geist, "Game Ranching: Threat to Wildlife Conservation in North America," *Wildlife Society Bulletin* 13 (1985): 594–98; V. Geist, "How Markets in Wildlife Meat and Parts, and the Sale of Hunting Privileges, Jeopardize Wildlife Conservation," *Conservation Biology* 2, no. 1 (1988): 15–26; V. Geist, "Legal Trafficking and Paid Hunting Threaten Conservation," *Transactions of the North American Conference on Wildlife and Natural Resources* 54 (1989): 171–78; and B. Lanka et al., *Analysis and Recommendations on the Applications by Mr. John T. Dorrance III to Import and Possess Native and Exotic Species* (Cheyenne: Wyoming Game and Fish Department, 1990).

2. G. Coetzee, "Conspiracy of Silence?" *South African Panorama*, October 10–14, 1989.

3. Coetzee, "Conspiracy of Silence?"; and T. R. Simmons and U. P Kreuter, "Herd Mentality: Banning Ivory Sales Is No Way to Save the Elephant," *Policy Review* (fall 1989): 46–49.

4. Simmons and Kreuter, "Herd Mentality."

5. Ibid.; M. Knox, "Horns of a Dilemma," *Sierra* 74, no. 6 (1989): 58–67.

6. Simmons and Kreuter, "Herd Mentality."

7. G. Child, "Economic Incentives and Improved Wildlife Conservation in Zimbabwe," paper presented at Workshop on Economics, International Union for the Conservation of Nature General Assembly, Costa Rica, February 4–5, 1988; J. A. McNeely, *Economics and Biological Diversity: Developing and Using Economic Incentives to Conserve Biological Resources* (Gland, Switzerland: International Union for the Conservation of Nature, 1988).

8. Simmons and Kreuter, "Herd Mentality."

9. R. Rasker, "Agriculture and Wildlife: An Economic Analysis of Waterfowl

Habitat Management on Farms in Western Oregon," Ph.D. diss., College of Forestry, Oregon State University, Corvallis, 1989; R. Rasker, R. L. Johnson, and D. Cleaves, *The Market for Waterfowl Hunting on Private Agricultural Land in Western Oregon*, Research Bulletin 70 (Corvallis: Forest Research Laboratory, Oregon State University, 1991).

10. See G. V. Burger and J. G. Teer, "Economic and Socioeconomic Issues Influencing Wildlife Management on Private Land," in *Wildlife Management on Private Lands*, ed. R. T. Dumke, G. V. Burger, and J. R. March (Madison: Wildlife Society, Wisconsin Chapter, 1981); J. G. Teer, G. V. Burger, and C. Y. Deknatel, "State Supported Habitat Management and Commercial Hunting on Private Lands in the United States," *Transactions of the North American Conference on Wildlife and Natural Resources* 48 (1983): 445–56; R. J. White, *Big Game Ranching in the United States* (Mesilla, N.M.: Wild Sheep and Goat International, 1986); L. L. Langner, "Hunter Participation in Access Hunting," *Transactions of the North American Conference on Wildlife and Natural Resources* 52 (1987): 475–82; E. W. Schenck et al., "Commercial Hunting and Fishing in Missouri: Management Implications of Fish and Wildlife 'Markets,'" *Transactions of the North American Conference on Wildlife and Natural Resources* 52 (1987): 516–24; and D. E. Wesley, "Socio-Duckonomics," in *Valuing Wildlife: Economic and Social Perspectives*, edited by J. D. Decker and G. R. Goff (Boulder: Westview Press, 1987).

11. J. C. Stier and R. C. Bishop, "Crop Depredation By Waterfowl: Is Compensation the Solution?" *Canadian Journal of Agricultural Economics* 29 (July 1981): 159–70.

12. W. Long, "Wildlife Management Programs on Private Land in California," in *Developing Profitable Resource-Based Recreation on Private Land*, ed. R. Basker and T. E. Bedell, proceedings of the 1987 Pacific Range Management Short Course (Corvallis: Oregon State University Cooperative Extension Service, 1987); J. D. Massie, California Department of Fish and Game, Wildlife Management Division, Sacramento, personal communication with the authors, 1988.

13. J. B. Loomis and L. Fitzhugh, "Financial Returns to California Landownwers for Providing Hunting Access: Analysis and Determinants of Returns and Implications to Wildlife Management," *Transactions of the North American Conference on Wildlife and Natural Resources* 54 (1989): 196–201.

14. T. Anderson, "Panel Discussion on Fee-Based Hunting and the Public Trust in Montana," in *Fee-Based Hunting and the Public Trust*, proceedings of the Cinnabar Symposium (Bozeman: Montana State University, 1989).

15. J. L. Buckner and J. L. Landers, *A Forester's Guide to Wildlife Management in Southern Industrial Pine Forests*, Technical Bulletin No. 10 (Bainbridge, Ga.: Southlands Experiment Forest, International Paper Co., 1980).

16. Anderson, "Panel Discussion on Fee-Based Hunting and the Public Trust in Montana."

17. J. P. Ernst, "Privatization of Wildlife Resources: A Question of Public Access," paper presented at the 42d Annual Meeting of the South Dakota Wildlife Federation, Pierre, S.D., 1988, unpublished; Geist, "Legal Trafficking and Paid Hunting Threaten Conservation."

18. Geist, "How Markets in Wildlife Meat and Parts, and the Sale of Hunting Privileges, Jeopardize Wildlife Conservation."

19. M. Frentzel, "Who's Keeping You Off Public Lands?" *Western Outdoors* 33, no. 8 (1986): 42–44, 74–76; T. Roederer, "Access to Public Land: The Keystone Dialogue Project," *Transactions of the North American Conference on Wildlife and Natural Resources* 54 (1989): 162–63.

20. Ernst, "Privatization of Wildlife Resources."

21. E. D. Benson, "Holistic Ranch Management and the Ecosystem Approach to Wildlife Conservation," *Proceedings of the Privatization of Widlife and Public Lands Access Symposium* (Casper: Wyoming Game and Fish Department, 1987); D. Pineo, "Characteristics and General Requirements for Successful Recreation Enterprises," in *Developing Profitable Resource-Based Recreation on Private Land.*

22. A. Leopold, "Report to the American Game Conference on an American Game Policy," *Transactions of the Conference on American Game* 17 (1930): 284–309; Schenck et al., "Commercial Hunting and Fishing in Missouri."

23. See Rasker and Bedell, *Developing Profitable Resource-Based Recreation on Private Land.*

24. For example, see L. D. Allen, "Private Lands as Wildlife Habitat: A Synthesis," in *Wildlife Management on Private Lands*; R. W. Tiner, *Wetlands of the United States: Current Status and Recent Trends* (Washington, D.C.: U.S. Fish and Wildlife Service, 1984); N. Sampson, "The Availability of Private Lands for Recreation," in *Recreation on Private Lands—Issues and Opportunities*, workshop convened by the Presidential Task Force on Recreation on Private Lands, Dirksen Senate Office Building, Washington, D.C., March 10, 1986; J. H. Goldstein, "The Impact of Federal Programs and Subsidies on Wetlands," *Transactions of the North American Conference on Wildlife and Natural Resources* 53 (1988): 436–43.

25. See D. C. Belt and G. Vaughn, *Managing Your Farm for Lease Hunting and a Guide to Developing a Hunting Lease*, Extension Bulletin No. 147 (Newark: Delaware Cooperative Extension, University of Delaware, 1988).

26. C. Griffith, "Turning Wildlife Private," *Oregon Bowhunter* (July 1987): 4–5.

27. Fee Hunting Task Force, Report presented to the Fish and Wildlife Commission of the State of Oregon, Portland, Ore., September, 1988, unpublished.

28. K. J. Arrow, *The Limits of Organization* (New York: W. W. Norton, 1974), p. 19.

29. E. Mansfield, *Micro-Economics*, 3d ed. (New York: W. W. Norton, 1979), p. 444.

30. W. D. Bromley, "Land and Water Problems: An Institutional Perspective," *American Journal of Agricultural Economics* 64, no. 12 (1982): 834–44.

31. R. C. Bishop, "Economic Considerations Affecting Landowner Behavior," in *Wildlife Management on Private Lands*; Burger and Teer, "Economic and Socioeconomic Issues Influencing Wildlife Management on Private Land"; and R. W. Marion, "Hunting Leases for Additional Profits from the Land," in *Proceedings of the Florida-Georgia Forest Landowner Seminar: Alternative Enterprises for Your Land* (Gainesville: Cooperative Extension Service, University of Florida, 1988).

32. Arrow, *The Limits of Organization*; Bromley, "Land and Water Problems"; Bishop, "Economic Considerations Affecting Landowner Behavior."

33. T. D. Savage et al., *The Economics of Environmental Improvement* (Boston: Houghton Mifflin, 1974).

34. H. Demsetz, "Toward a Theory of Property Rights," *American Economic Review* 57 (1967): 409–13; S. N. S. Cheung, "The Structure of a Contract and the Theory of a Non-exclusive Resource," *Journal of Law and Economics* 13 (1970): 49–70; E. Furubota and S. Pejovich, "Property Rights and Economic Theory: A Survey of Recent Literature," *Journal of Economic Literature* 10 (1972): 1137–62; and G. Hardin, "The Tragedy of the Commons," *Science* 162 (1968): 1243–48.

35. W. D. Bromley, "Public and Private Interests in the Federal Lands: Toward Conciliation," in *Public Lands and the U.S. Economy: Balancing Conservation and Development*, edited by G. M. Johnston and P. M. Emerson (Boulder: Westview Press, 1984).

36. T. Schelling, "Economic Reasoning the the Ethics of Policy," *Public Interest* 63 (1981): 59.
37. J. Viner, "The Intellectual History of Laissez-Faire," *Journal of Law and Economics* 3 (October 1960): 45–69.
38. Bishop, "Economic Considerations Affecting Landowner Behavior."
39. L. A. Jordan and J. P. Workman, "Economics and Management of Fee Hunting for Deer and Elk in Utah," *Wildlife Society Bulletin* 17 (1989): 482–87.
40. Marion, "Hunting Leases for Additional Profits from the Land," p. 60.
41. Burger and Teer, "Economic and Socioeconomic Issues Influencing Wildlife Management on Private Land," p. 261.
42. A. Randall, *Resource Economics: An Economic Approach to Natural Resource and Environmental Policy* (New York: John Wiley and Sons, 1981); Bromley, "Public and Private Interests in the Federal Lands."
43. C. F. Runge, "An Economist's Critique of Privatization," in *Public Lands and the U.S. Economy*.
44. J. E. Swenson, "Free Public Hunting and the Conservation of Public Wildlife Resources," *Wildlife Society Bulletin* 11, no. 3 (1983): 300–303; Runge, "An Economist's Critique of Privatization"; D. Ehrenfeld, "Why Put a Value on Biodiversity?" in *Biodiversity*, ed. E. O. Wilson (Washington, D.C.: Academy Press, 1988); McNeely, *Economics and Biological Diversity*.
45. C. Clark, "Profit Maximization and the Extinction of Species," *Journal of Political Economy* 81 (1973): 950–61; C. Clark, "The Economics of Overexploitation." *Science* 181 (1973): 630–34.
46. Clark, "Profit Maximization and the Extinction of Species," p. 950.
47. M. Higbee, "Farmers and Wildlife: Why Is There a Rift and How Can We Bridge It?" in *Wildlife Management on Private Lands*; O. D. Hyde, "Public Lands, Private Lands and Oranges," *Ambit* (March 1986): 22–24; B. A. Wright and D. R. Fesenmaier, "Modeling Rural Landowners' Hunter Access Policies in East Texas, U.S.A.," *Environmental Management* 12, no. 2 (1988): 229–36.
48. Stier and Bishop, "Crop Depredation By Waterfowl"; D. B. Nielsen and D. D. Lytle, "Who Wins (or Loses) When Big-Game Uses Private Lands? *Utah Science* (summer 1985): 48–50; D. A. Wade, "Economics of Wildlife Production and Damage Control on Private Lands," in *Valuing Wildlife*.
49. R. Shelton, "Motivating the Landowner/manager to Manage for Wildlife," in *Wildlife Management on Private Lands*; H. Rolston, "Beauty and the Beast: Aesthetic Experience of Wildlife," in *Valuing Wildlife*; Rasker, "Agriculture and Wildlife."
50. Leopold, "Report to the American Game Conference on an American Game Policy."
51. Bishop, "Economic Considerations Affecting Landowner Behavior."
52. U.S. Department of Agriculture, Food Security Act of 1985, Public Law 99-198, December 23, 1985; M. Martin et al., "The Impacts of the Conservation Reserve Program on Rural Communities: The Case of Three Oregon Counties," *Western Journal of Agricultural Economics* 13, no. 2 (1988): 225–32.
53. B. Isaacs and D. Howell, "Opportunities for Enhancing Wildlife Benefits Through the Conservation Reserve Program," *Transactions of the North American Conference on Wildlife and Natural Resources* 53 (1988): 222–31.
54. L. L. Langner, "Land-Use Changes and Hunter Participation: The Case of the Conservation Reserve Program," *Transactions of the North American Conference on Wildlife and Natural Resources* 54 (1989): 382–90.

55. R. L. Hays, R. P. Webb, and A. H. Farmer, "Effects of the Conservation Reserve Program on Wildlife Habitat: Results of 1988 Monitoring," *Transactions of the North American Conference on Wildlife and Natural Resources* 54 (1989): 162–63.

56. E. J. Miller and P. T. Bromley, "Wildlife Management on Conservation Reserve Program Land: The Farmer's View," *Transactions of the North American Conference on Wildlife and Natural Resources* 54 (1989): 377–81.

57. Bishop, "Economic Considerations Affecting Landowner Behavior."

58. K. A. Cook, "The 1985 Farm Act and Wildlife Conservation: Outlook for 1990," *Transactions of the North American Conference on Wildlife and Natural Resources* 54 (1989): 409–13.

59. Bishop, "Economic Considerations Affecting Landowner Behavior"; Savage et al., *The Economics of Environmental Improvement*.

60. P. F. Noonan and M. D. Zagata, "Wildlife in the Marketplace: Using the Profit Motive to Maintain Wildlife Habitat," *Wildlife Society Bulletin* 10, no. 1 (1982): 46–49.

61. S. J. Small, *Preserving Family Lands: A Landowner's Introduction to Tax Issues and Other Considerations* (Boston: Powers and Hall Professional Corporation, 1990).

62. H. Zackheim, The Nature Conservancy, Helena, Mont., personal communication with the authors, April 4, 1990.

63. J. Wilson, Montana Land Reliance, Helena, Mont., personal communication with the authors, April 4, 1990.

64. Ibid.

65. Arrow, *The Limits of Organization*; R. R. Nelson, "Assessing Private Enterprises: An Exegesis of Tangled Doctrine," *Bell Journal of Economics* 12 (1981): 93–110.

66. S. R. Kellert, "Wildlife and the Private Landowner," in *Wildlife Management on Private Lands*.

67. Arrow, *The Limits of Organization*.

68. Bromley, "Public and Private Interests in the Federal Lands."

69. D. Western, "Amboseli National Park: Human Values and the Conservation of a Savanna Ecosystem," in *National Parks, Conservation and Development: The Role of Protected Areas in Sustaining Society*, ed. J. A. McNeely and K. R. Miller (Washington, D.C.: Smithsonian Institution Press, 1984).

70. G. Brown Jr. and W. Henry, "The Economic Value of Elephants," Paper 89-2 (London: London Environmental Economics Centre, 1989).

71. Western, "Amboseli National Park."

72. McNeely, *Economics and Biological Diversity*.

73. Western, "Amboseli National Park."

References

Allen, L. D. "Private Lands as Wildlife Habitat: A Synthesis." In *Wildlife Management on Private Lands*, edited by P. T. Dumke, G. V. Burger, and J. R. March. Madison: Wildlife Society, Wisconsin Chapter, 1981.

Anderson, T. "Panel Discussion on Fee-Based Hunting and the Public Trust in Montana." In *Fee-Based Hunting and the Public Trust*. Proceedings of the Cinnabar Symposium. Bozeman: Montana State University, 1989.

Arrow, K. J. *The Limits of Organization*. New York: W. W. Norton, 1974.

Belt, D. C., and G. Vaughn. *Managing Your Farm for Lease Hunting and a Guide to Developing a Hunting Lease*. Extension Bulletin No. 147. Newark: Delaware Cooperative Extension, University of Delaware, 1988.

Benson, E. D. "Holistic Ranch Management and the Ecosystem Approach to Wildlife Conservation." In *Proceedings of the Privatization of Wildlife and Public Lands Access Symposium*. Casper: Wyoming Game and Fish Department, 1987.

Bishop, R. C. "Economic Considerations Affecting Landowner Behavior." In *Wildlife Management on Private Lands*, edited by R. T. Dumke, G. V. Burger, and J. R. March. Madison: Wildlife Society, Wisconsin Chapter, 1981.

———. "Economic Values Defined." In *Valuing Wildlife: Economic and Social Perspectives.*, edited by J. D. Decker and G. R. Goff. Boulder: Westview Press, 1987.

Bromley, W. D. "Land and Water Problems: An Institutional Prspective." *American Journal of Agricultural Economics* 64, no. 12 (1982): 834–44.

———. "Public and Private Interests in the Federal Lands: Toward Conciliation." In *Public Lands and the U.S. Economy: Balancing Conservation and Development*, edited by G. M. Johnston and P. M. Emerson. Boulder: Westview Press, 1984.

Brown, G. Jr., and W. Henry. *The Economic Value of Elephants*. Paper 89-12. London: London Environmental Economics Centre, 1989.

Buckner, J. L, and J. L. Landers. *A Forester's Guide to Wildlife Management in Southern Industrial Pine Forests*. Technical Bulletin No. 10. Bainbridge, Ga.: Southlands Experiment Forest, International Paper Company, 1980.

Burger, G. V., and J. G. Teer. "Economic and Socioeconomic Issues Influencing Wildlife Management on Private Land." In *Wildlife Management on Private Lands*, edited by R. T. Dumke, G. V. Burger, and J. R. March. Madison: Wildlife Society, Wisconsin Chapter, 1981.

Cheung, S. N. S. "The Structure of a Contract and the Theory of a Non-exclusive Resource." *Journal of Law and Economics* 13 (1970): 49–70.

Child, G. "Economic Incentives and Improved Wildlife Conservation in Zimbabwe." Paper presented at Workshop on Economics, International Union for the Conservation of Nature General Assembly, Costa Rica, February 4–5, 1988.

Clark, C. "Profit Maximization and the Extinction of Species." *Journal of Political Economy* 81 (1983): 950–61.

———. "The Economics of Overexploitation." *Science* 181 (1973): 630–34.

Coetzee, G. "Conspiracy of Silence?" *South African Panorama*, October 10–14, 1989.

Cook, K. A. "The 1985 Farm Act and Wildlife Conservation: Outlook for 1990." *Transactions of the North American Conference on Wildlife and Natural Resources* 54 (1989): 409–13.

Demsetz, H. "Toward a Theory of Property Rights." *American Economic Review* 57 (1967): 409–13.

Ehrenfeld, D. "Why Put a Value on Biodiversity?" In *Biodiversity*, edited by E. O. Wilson. Washington: National Academy Press, 1988.

Ernst, J. P. "Privatization of Wildlife Resources: A Question of Public Access." Paper presented at the 42d Annual Meeting of the South Dakota Wildlife Federation, Pierre, S.D., August 22, 1987.

Fee Hunting Task Force. Report presented to the Fish and Wildlife Commission of the State of Oregon, Portland, Ore., September, 1988. Unpublished.

Frentzel M. "Who's Keeping You Off Public Lands?" *Western Outdoors* 33, no. 8 (1986): 42–44, 74–76.

Furubota, E., and S. Pejovich. "Property Rights and Economic Theory: A Survey of Recent Literature." *Journal of Economic Literature* 10 (1972): 1137–62.

Geist, V. "Game Ranching: Threat to Wildlife Conservation in North America." *Wildlife Society Bulletin* 13 (1985): 594–98.

———. "How Markets in Wildlife Meat and Parts, and the Sale of Hunting Privileges, Jeopardize Wildlife Conservation." *Conservation Biology* 2, no. 1 (1988): 15–26.

———. "Legal Trafficking and Paid Hunting Threaten Conservation." *Transactions of the North American Conference on Wildlife and Natural Resources* 54 (1989): 171–78.

Goldstein, J. H. "The Impact of Federal Programs and Subsidies on Wetlands." *Transactions of the North American Conference on Wildlife Natural Resources* 53 (1988): 436–43.

Griffith, C. "Turning Wildlife Private." *Oregon Bowhunter* (July 1987): 4–5.

Hardin, G. "The Tragedy of the Commons." *Science* 162 (1968): 1243–48.

Hays, R. L., R. P. Webb, and A. H. Farmer "Effects of the Conservation Reserve Program on Wildlife Habitat: Results of 1988 Monitoring." *Transactions of the North American Conference on Wildlife and Natural Resources* 54 (1989): 162–63.

Higbee, M. "Farmers and Wildlife: Why is There a Rift and How Can We Bridge It?" In *Wildlife Management on Private Lands*, edited by R. T. Dumke, G. V. Burger, and J. R. March. Madison: Wildlife Society, Wisconsin Chapter, 1981.

Hyde, O. D. "Public Lands, Private Lands and Oranges." *Ambit* (March 1986): 22–24.

Isaacs, B., and D. Howell. "Opportunities for Enhancing Wildlife Benefits Through the Conservation Reserve Program." *Transactions of the North American Conference on Wildlife and Natural Resources* 53 (1988): 222–31.

Jordan, L. A., and J. P. Workman. "Economics and Management of Fee Hunting for Deer and Elk in Utah." *Wildlife Society Bulletin* 17 (1989): 482–87.

Kellert, S. R. "Wildlife and the Private Landowner." In *Wildlife Management on Private Lands*, edited by R. T. Dumke, G. V. Burger, and J. R. March. Madison: Wildlife Society, Wisconsin Chapter, 1981.

Knox, M. "Horns of a Dilemma." *Sierra* 74, no. 6 (1989): 58–67.

Langner, L. L. "Hunter Participation in Access Hunting." *Transactions of the North American Conference on Wildlife and Natural Resources* 52 (1987): 475–82.

———. "Land-Use Changes and Hunter Participation: The Case of the Conservation Reserve Program." *Transactions of the North American Conference on Wildlife and Natural Resources* 54 (1989): 382–90.

Lanka, B., R. Guenzel, G. Fralick, and D. Thiele. *Analysis and Recommendations on the Applications by Mr. John T. Dorrance III to Import and Possess Native and Exotic Species*. Cheyenne: Wyoming Game and Fish Department, 1990.

Leopold, A. "Report to the American Game Conference on an American Game Policy." *Transactions of the Conference on American Game* 17 (1930): 284–309.

Long, W. "Wildlife Management Programs on Private Land in California." In *Developing Profitable Resource-Based Recreation on Private Land*, edited by R. Rasker and T. E. Bedell. Proceedings of the 1987 Pacific Range Management Short Course. Corvallis: Oregon State University Cooperative Extension Service, 1987.

Loomis, J. B., and L. Fitzhugh. "Financial Returns to California Landowners for Providing Hunting Access: Analysis and Determinants of Returns and Implications to Wildlife Management." *Transactions of the North American Conference on Wildlife and Natural Resources* 54 (1989): 196–201.

Mansfield, E. *Micro-Economics*. 3d ed. New York: W. W. Norton, 1979.

Marion, R. W. "Hunting Leases for Additional Profits from the Land." In *Proceedings of the Florida-Georgia Forest Landowner Seminar: Alternative Enterprises for Your Land*. Gainesville: Cooperative Extension Service, University of Florida, 1988.

Martin, M., H. Radtke, B. Eleveld, and S. D. Nofziger. "The Impacts of the Conservation Reserve Program on Rural Communities: The Case of Three Oregon Counties." *Western Journal of Agricultural Economics* 13, no. 2 (1988): 225–32.

Massie, J. D. California Department of Fish and Game, Wildlife Management Division, Sacramento. Personal communication with the authors. 1988.

McNeely, J. A. *Economics and Biological Diversity: Developing and Using Economic Incentives to Conserve Biological Resources*. Gland, Switzerland: International Union for the Conservation of Nature, 1988.

Miller, E. J., and P. T. Bromley. "Wildlife Management on Conservation Reserve Program Land: The Farmer's View." *Transactions of the North American Conference on Wildlife and Natural Resources* 54 (1989): 377–81.

Nelson, R. R. "Assessing Private Enterprises: An Exegesis of Tangled Doctrine." *Bell Journal of Economics* 12 (1981): 93–110.

Nielsen, D. B., and D. D. Lytle. "Who Wins (or Loses) When Big-Game Uses Private Lands?" *Utah Science* (summer 1985): 48–50.

Noonan, P. F., and M. D. Zagata. "Wildlife in the Marketplace: Using the Profit Motive to Maintain Wildlife Habitat." *Wildlife Society Bulletin* 10, no. 1 (1982): 46–49.

Pineo, D. "Characteristics and General Requirements for Successful Recreation Enterprises." In *Developing Profitable Resource-Based Recreation on Private Land*, edited by R. Rasker and T. E. Bedell. Proceedings of the 1987 Pacific Range Management Short Course. Corvallis: Oregon State University Cooperative Extension Service, 1987.

Randall, A. *Resource Economics: An Economic Approach to Natural Resource and Environmental Policy*. New York: John Wiley and Sons, 1981.

Rasker, R. "Agriculture and Wildlife: An Economic Analysis of Waterfowl Habitat Management on Farms in Western Oregon." Ph.D. diss., College of Forestry, Oregon State University, Corvallis, 1989.

Rasker, R., and T. E. Bedell, eds. *Developing Profitable Resource-Based Recreation on Private Land*. Proceedings of the 1987 Pacific Northwest Range Management Short Course. Corvallis: Oregon State University Cooperative Extension Service, 1987.

Rasker, R., R. L. Johnson, and D. Cleaves. *The Market for Waterfowl Hunting on Private Agricultural Land in Western Oregon*. Research Bulletin 70. Corvallis: Forest Research Laboratory, Oregon State University, 1991.

Roederer, T. "Access to Public Land: The Keystone Dialogue Project." *Transactions of the North American Conference on Wildlife and Natural Resources* 54 (1989): 162–63.

Rolston, H. "Beauty and the Beast: Aesthetic Experience of Wildlife." In *Valuing Wildlife: Economic and Social Perspectives*, edited by J. D. Decker and G. R. Goff. Boulder: Westview Press, 1987.

Runge, C. F. "An Economists' Critique of Privatization." In *Public Lands and the U.S. Economy: Balancing Conservation and Development*, edited by G. M. Johnston and P. M. Emerson. Boulder: Westview Press, 1984.

Sampson, N. "The Availability of Private Lands for Recreation." In *Recreation on Private Lands—Issues and Opportunities*. Workshop convened by the Presidential Task Force on Recreation on Private Lands. Dirksen Senate Office Building, Washington, D.C., 1986.

Savage, T. D., M. Burke, J. D. Coupe, T. D. Duchesneau, D. F. Wihry, and J. A. Wilson. *The Economics of Environmental Improvement*. Boston: Houghton Mifflin, 1974.

Schelling, T. "Economic Reasoning and the Ethics of Policy." *Public Interest* 63 (1981): 37–61.
Schenck, E. W., W. Arnold, E. K Brown, and D. J. Daniel. "Commercial Hunting and Fishing in Missouri: Management Implications of Fish and Wildlife 'Markets.'" *Transactions of the North American Conference on Wildlife and Natural Resources* 52 (1987): 516–24.
Shelton, R. "Motivating the Landowner/Manager to Manage for Wildlife." In *Wildlife Management on Private Lands*, edited by R. T. Dumke, G. V. Burger, and J. R. March. Madison: Wildlife Society, Wisconsin Chapter, 1981.
Simmons, T. R., and U. P. Kreuter. "Herd Mentality: Banning Ivory Sales Is No Way to Save the Elephant." *Policy Review* (fall 1989): 46–49.
Small, S. J. *Preserving Family Lands: A Landowner's Introduction to Tax Issues and Other Considerations*. Boston: Powers and Hall Professional Corporation, 1990.
Stier, J. C., and R. C. Bishop. "Crop Depredation By Waterfowl: Is Compensation the Solution?" *Canadian Journal of Agricultural Economics* 29 (July 1981): 159–70.
Swenson, J. E. "Free Public Hunting and the Conservation of Public Wildlife Resources." *Wildlife Society Bulletin* 11, no. 3 (1983): 300–303.
Teer, J. G., G. V. Burger, and C. Y. Deknatel. "State Supported Habitat Management and Commercial Hunting on Private Lands in the United States." *Transactions of the North American Conference on Wildlife and Natural Resources* 48 (1983): 445–56.
Tiner, R. W. *Wetlands of the United States: Current Status and Recent Trends*. Washington: U.S. Fish and Wildlife Service, 1984.
U.S. Department of Agriculture. Food Security Act of 1985. Public Law 99-198, December 23, 1985.
Viner, J. "The Intellectual History of Laissez-Faire." *Journal of Law and Economics* 3 (October 1960): 45–69.
Wade, D. A. "Economics of Wildlife Production and Damage Control on Private Lands." In *Valuing Wildlife: Economic and Social Perspectives*, edited by J. D. Decker and G. R. Goff. Boulder: Westview Press, 1987.
Wesley, D. E. "Socio-Duckonomics." In *Valuing Wildlife: Economic and Social Perspectives*, edited by J. D. Decker and G. R. Goff. Boulder: Westview Press, 1987.
Western, D. "Amboseli National Park: Human Values and the Conservation of a Savanna Ecosystem." In *National Parks, Conservation and Development: The Role of Protected Areas in Sustaining Society*, edited by J. A. McNeely and K. R. Miller. Washington, D.C.: Smithsonian Institution Press, 1984.
———. Personal communication with the authors. June 13, 1990.
White, R. J. *Big Game Ranching in the United States*. Mesilla, N.M.: Wild Sheep and Goat International, 1986.
Wilson, J. Montana Land Reliance, Helena. Personal communication with the authors April 4., 1990.
Wright, B. A., and D. R. Fesenmaier. "Modeling Rural Landowners' Hunter Access Policies in East Texas, U.S.A." *Environmental Management* 12, no. 2 (1988): 229–36.
Zackheim, H. The Nature Conservancy, Helena, Montana. Personal communication with the authors. April 4, 1990.

15

Elephants and Economics

Graeme Caughley

Rasker, Martin, and Johnson are to be congratulated on their instructive paper describing the theory and practice of economics as applied to wildlife management.[1] I have no argument with most of it but suggest that one of their examples, introduced by the heading "A Case for Privatization: Africa's Elephants," is an inappropriate illustration of their arguments.

They suggest that elephants are reasonably secure in Zimbabwe and South Africa, in contrast to their parlous state in Kenya, because these animals are privately owned in the first area and publicly owned in the second. I argue that the present difference between east Africa and southern Africa in the conservation status of elephants has little or nothing to do with systems of ownership, but that if absolute private ownership were instituted for wild elephants it would very likely trigger a further decline.

Clark's book on the economics of harvesting natural resources shows unambiguously that the best biological strategy and the best economic strategy coincide only when a population's maximum rate of increase is relatively high.[2] May suggested this as the reason why whales were consistently overharvested.[3] The argument goes like this. Suppose your uncle bequeaths you $10,000 to be paid now, or $20,000 to be paid in five years time if you refrain from smoking tobacco. The condition is irrelevant because you have never previously smoked and have no plans to start. The decision therefore comes down to estimating whether $10,000 now is better than $20,000 deferred. The decision is reached by estimating how much $20,000 in five years time is worth now. Capital tends to increase at a rate (real growth plus inflation) of about 10 percent per year, and so the present value of

---

This article originally appeared in *Conservation Biology* 7 (December 1993): 943–45. Reprinted by permission of Blackwell Science, Inc. Copyright © 1993 Blackwell Science, Inc. All rights reserved.

$20,000 delivered in five years time is about $12,420 (20,000/1.1^5). That calculation is an application of the "discount rate," the rate by which future earnings must be discounted to estimate their present value. You would be marginally better in this example, other things being equal, if you waited patiently for the future reward. But if you are a mountaineer or a mercenary, or if you are desperately short of ready cash, the intelligent decision might well be to take the lesser sum now. That decision represents the adoption of a discount rate higher than would be contemplated by a well-capitalized company. (It should be noted that this analogy, close though it may be, is not exact when applied to natural resources because money depreciates via inflation whereas goods do not.)

The decision on the uncle's gift might have required thought. Not so this one: You are bequeathed shares displaying no prospect of capital gain and yielding a derisory 4 percent per year on capital invested. What do you do? You sell the shares and reinvest the money in a company with a better record of fiscal enterprise. Likewise, if you are bequeathed a population of elephants, you can convert it to a capital sum that is reinvested elsewhere, or you can harvest it continually at maximum sustained yield to provide a dividend. That yield is, for big animals that produce offspring infrequently, about half the population's maximum rate of increase multiplied by half the size of the population when unharvested. The intrinsic rate of increase of elephants is 6–7 percent per year,[4] and so the maximum sustained yield is about 3–4 percent per year. The highest estimate of maximum sustained yield is Craig's 4.8 percent per year, but he preferred a lower estimate, made by a different method, of 3.9 percent.[5] Thus, elephants are terrestrial whales. For such animals, when the discount rate substantially exceeds their sustainable harvesting rate, the highest return in both the short and long term comes from converting the population to cash and reinvesting that capital.

If I have got that right, it follows that Rasker et al. have got it backwards. A fully informed and free market, underpinned by absolute ownership, will drive to extinction, or at least to very low density, a resource yielding 3–4 percent per year. So why are elephants not being reduced in Zimbabwe? I offer two possibilities: The form of ownership over elephants exercised by the local people has been in force for only a short time, and the "ownership" is hedged with conditions having more to do with conservation than with the imperatives of a free market. The essence of complete ownership is that the owners can do with the goods whatever they wish, including disposing of them entirely by gift or sale. That does not come close to describing the relationship between people and elephants in Zimbabwe.

It is important to differentiate what the law says from what the law does in the matter of ownership of wild animals. A landowner in England, although barred from owning game under the legal doctrine of *res nullius* is granted all perquisites of ownership of that game by that country's convoluted laws of trespass.[6] Similarly, New Zealand's statutory law states explicitly that a holder of land possesses no title to the wild deer on that land but then, in the same section, it bestows upon the landholder all the perquisites of ownership over them.[7] Alternatively, the law may assign "ownership" of wildlife to a landholder but either withhold the perquisites of private ownership or grant them subject to case-by-case administrative approval. The phrase identifying that artifice is usually something like "in consultation with the relevant government department."

Legislation was passed in 1975 to make wildlife private property on private land in Zimbabwe. It had little bearing on the management of elephants because elephants seldom live on farms. Most of Zimbabwe's elephants are in national parks and game reserves, with fewer on unfarmed areas of communally owned land. This legislation did not grant ownership to the occupier of the wildlife on communal land. That discrepancy was righted by legislation in 1989[8] as part of the CAMPFIRE project. "CAMPFIRE is a philosophy of sustainable rural development that enables rural communities to manage, and benefit directly from, indigenous wildlife. It is essentially an entrepreneurial approach to development, based on wildlife management, that uses market forces to achieve economic, ecological and social sustainability."[9]

The granting of limited and conditional ownership of wild animals to the people of Zimbabwe under CAMPFIRE is restricted as of December 1992 to twenty-two communal areas, only about half of which contain more than a few elephants. The villagers may sell elephants to safari operators. About a hundred are taken in this way each year. The quota is proposed by the villagers but must be approved by the Department of National Parks and Wildlife Management (DNPWM). The law continues to be enforced by the DNPWM in its well-known competent and methodical manner. Zimbabwe has always had a game department whose efficiency has been the envy of other countries in southern and central Africa, even though, in absolute terms, it is severely underfunded. Very early, the paramilitary DNPWM adopted a "shoot without challenge" policy in its attempt to control hunting of elephants and rhino. In its favor was a dearth of those ethnic groups most often identified further north with large-scale ivory hunting and trading. Even so, it failed to stop the hunting of elephants and rhino in the Zambesi Valley on Zimbabwe's northern border, and few rhino remain there today; further, unauthorized killing of elephants in other areas of Zimbabwe has accelerated over the last eighteen months.[10] Nonetheless, elephant numbers in Zimbabwe have been, within the limited accuracy of aerial survey methodology, much the same or increasing slowly from about 1980 onward.[11] The change in the legal ownership of elephants in some districts of Zimbabwe since 1989 has not coincided, so far as I can ascertain, with any marked change in their conservation status.

The comparison of Zimbabwe with Kenya can be augmented by a comparison of Zimbabwe with South Africa. There is no true private ownership of elephants in South Africa and nothing akin to the CAMPFIRE scheme of Zimbabwe. Yet, as Rasker et al. rightly pointed out, its elephant populations are not endangered. The secure conservation status of elephants in South Africa clearly has little to do with private ownership but a lot to do with government law enforcement.

In summary, the differences in conservation status of elephants among South

Africa, Zimbabwe, and Kenya are not associated strongly with forms of ownership. Further, if elephants were owned absolutely (and they are not in Zimbabwe), the expected free-market outcome is conversion of most of them to cash as soon as possible, the money being reinvested in an enterprise that pays a higher rate of return. Perhaps that is what happened in Kenya where, if one ignores the economically irrelevant niceties of law, the level of actual ownership held by the people over their elephants is considerably higher than it is in Zimbabwe.

NOTES

1. In this volume.
2. C. W. Clark, *Mathematical Bioeconomics: The Optimum Management of Renewable Resources* (New York: Wiley Interscience, 1976).
3. R. M. May, "Harvesting Whale and Fish Populations," *Nature* 263 (1976): 91–92.
4. See A. Hall-Martin, "Elephant Survivors," *Oryx* 15 (1980): 355–62 ; G. W. Calef, "Maximum Rate of Increase in the African Elephant," *African Journal Of Ecology* 26 (1988): 323–27.
5. G. C. Craig, "Population Dynamics of Elephants," in *Elephant Management in Zimbabwe*, ed. R. B. Martin, G. C. Craig, and V. R. Booth (Harare, Zimbabwe: Department of National Parks and Wildlife Management, 1989), pp. 67–72.
6. G. Caughley, *The Deer Wars: The Story of Deer in New Zealand* (Auckland, New Zealand: Heinemann, 1983), chap. 10.
7. Ibid., chap. 11.
8. S. J. Thomas, *The Legacy of Dualism and Decision-making: The Prospects for Local Institutional Development in "CAMPFIRE"* (Harare, Zimbabwe: University of Zimbabwe, 1991).
9. The Zimbabwe Trust, *People, Wildlife and Natural Resources- the CAMPFIRE Approach to Rural Development in Zimbabwe* (Harare, Zimbabwe: The Wildlife Trust, 1990).
10. Rowan Martin, personal communication.
11. V. R. Booth "The Number of Elephants Killed in Zimbabwe:1960–1988," in *Elephant Management in Zimbabwe*.

REFERENCES

Booth, V. R. "The Number of Elephants Killed in Zimbabwe: 1960–1988." In *Elephant Management in Zimbabwe*, edited by R. B. Martin, G. C. Craig, and V. R. Booth. Harare, Zimbabwe: Department of National Parks and Wildlife Management, 1989.
Calef, G. W. "Maximum Rate of Increase in the African Elephant." *African Journal of Ecology* 26 (1988): 323–27.
Caughley, G. *The Deer Wars: The Story of Deer in New Zealand*. Auckland: Heinemann, 1983.
Clark, C. W. *Mathematical Bioeconomics: The Optimum Management of Renewable Resources*. New York: Wiley Interscience, 1976.
Craig, G. C. "Population Dynamics of Elephants." In *Elephant Management in Zimbabwe*,

edited by R. B. Martin, G. C. Craig, and V. R. Booth. Harare, Zimbabwe: Department of National Parks and Wildlife Management, 1989.

Hall-Martin, A. "Elephant Survivors." *Oryx* 15 (1980): 355–62.

May, R. M. "Harvesting Whale and Fish Populations." *Nature* 263 (1976): 91–92.

Thomas, S. J. *The Legacy of Dualism and Decision-making: The Prospects for Local Institutional Development in "CAMPFIRE."* Harare: University of Zimbabwe, 1991.

The Zimbabwe Trust. *People, Wildlife and Natural Resoures—The CAMPFIRE Approach to Rural Development in Zimbabwe.* Harare, Zimbabwe: The Wildlife Trust, 1990.

16

Kangaroo Harvesting and the Conservation of Arid and Semiarid Rangelands

Gordon Grigg

It may sound unlikely from a conservation-minded biologist, but since June 1987 I have been advocating a marketing drive to increase the selling price of meat from Australia's three large species of kangaroos. My argument is that, if this were to happen, there would be a revitalization of Australia's overgrazed semiarid lands, less illegal killing of kangaroos by inexpert shooters, and, in the long term, improved conservation of kangaroos by conserving their prime habitat.

Australia needs new ideas to solve two major rural problems, land degradation and the kangaroo dilemma. The kangaroo harvesting idea offers some hope of solving both these problems, and the idea has been attracting a lot of favorable attention within Australia. It was recently the subject of a major conference sponsored by the Royal Zoological Society of New South Wales.[1]

Before expanding on the harvesting proposal let me explain the two problems, land degradation and the kangaroo dilemma. Land degradation is regarded as Australia's worst rural problem. According to a recent survey by the Commonwealth Scientific and Industrial Research Organization, Division of Soils, more than 25 percent (1.85×10^6 square kilometers) of the continent is degraded from overgrazing by domestic stock, particularly sheep. One-quarter of that area (432,000 square kilometers) is at risk of becoming permanent desert. As 30 percent of Australia has an annual rainfall of less than 200 millimeters and can be recognized already as natural desert, more than one-third of the remainder is now threatened. Significantly, most of our large kangaroos live in these degraded grazing lands.

The kangaroo dilemma is complex. There is a problem because of the public

---

This article originally appeared in *Conservation Biology* 3 (June 1989): 194–97. Reprinted by permission of Blackwell Science, Inc. Copyright © 1989 Blackwell Science, Inc. All rights reserved.

conflicts that arise from the large kangaroos being simultaneously perceived as both sacred cult objects and serious pests to agriculture. Kangaroos are the symbol of Australia and therefore are of important totemic significance to Australians. Because of their beauty and elegance and, particularly, their spectacular hopping gait, they are among the world's most loved and admired animals. Yet they are regarded as pests whenever and wherever they occur in vast numbers, competing with domestic stock for food and water and damaging fences. A recent report estimates that, in some years, each kangaroo costs the landholder about $20 in lost revenue. Hence, relevant Australian governments license the annual killing of about three million red, eastern gray, and western gray kangaroos, for pest control. This is undertaken by a restricted kangaroo industry that markets the hides and the meat, mainly for pet food, up to an annual quota, which is based on annual population surveys. Apart from kangaroos killed and taken by the industry, kangaroos outside the commercial zones are killed on "destruction permits" given to landholders to reduce their pest problem. These animals may not enter trade and are usually left to rot in paddocks. Many more are killed without permits, and it is claimed that much of the illegal killing is carried out by inexpert shooters, often for "sport," posing an animal welfare problem.

Opponents of the kangaroo industry are vocal both nationally and internationally, mainly on ethical grounds, claiming that kangaroos may not be as serious a pest as graziers claim, that the industry is rife with cruelty to kangaroos, that it is morally wrong to have an industry based on native wildlife, and that the industry is not effective at pest control because it focuses on the larger, male animals. Some claim, despite all the evidence to the contrary, that the kangaroo industry is threatening kangaroos with extinction. The kangaroo debate is long running, always emotional, and often heated. The present status quo leaves a lot to be desired and we should be striving for an improved system of kangaroo management that enjoys consensus support and gives kangaroos a better deal.

Let us turn to the harvesting proposal. Since June 1987 I have been promoting this for public discussion because I think it addresses both land degradation and the kangaroo problem. In a nutshell, the idea is that we should undertake a marketing drive for kangaroo products, raising the price to such an extent that graziers will find it worthwhile to reduce their traditional hard-footed stock in favor of free-range kangaroos. I argue that this point of view deserves the support of conservation-minded people—not only because it offers the hope of rangeland revegetation and recovery, but also because it addresses all of the major problems perceived by opponents of the present kangaroo management practices, including the animal welfare concerns.

My interest stems from experience since 1975 in aerial surveys of the three species of large kangaroos. I began to get very restless in 1982 about the misinformation being spread by opponents of the kangaroo industry who claim that the big kangaroos are being threatened with extinction. We have plenty of information showing how abundant all three species of large kangaroos are, and it was being (conveniently) ignored. So in 1984 I wrote an article in the Australian Museum's quarterly, *Australian Natural History*, in which I presented the results of aerial surveys and reviewed the numerical and ethical aspects of the kangaroo debate.[2] I also made then the rather tentative suggestion that we should be considering the idea of kangaroos as a resource rather than a pest. The article stirred up a bit of a hornet's nest in some quarters, but was well received for the most part. The aerial survey data got publicity in other ways too, and gradually it has become accepted in Australia that there really are plenty of kangaroos and that the populations are not threatened by harvesting. Overseas, however, organizations such as Greenpeace still trumpet the myth that kangaroos are threatened with extinction, choosing to ignore the facts because it suits them. I think Greenpeace did a lot of good for whales and does a good job in many areas, but is quite wrong about kangaroos.

Apart from knowledge about the size of kangaroo populations, the aerial surveys also gave me a close-up view of the extent of habitat degradation in our semiarid lands. These so-called sheep rangelands are chenopod shrublands, which occupy about 20 percent of the continent. Here are found about 15 percent of our sheep and *most* of our kangaroos—about twenty million sheep and at least ten million kangaroos. After thousands of hours flying over it at low level on aerial surveys, the strongest impression I am left with is the huge impact of hard-footed, hard-feeding sheep on the landscape. It is difficult to find a scene on which the imprint of hard hooves is not clearly visible. The vegetation has been ground underfoot, exposing the fragile ancient soil to wind and water erosion. Soil has filled in the creeks, which now flow under the sand. Compaction of the soil (the sheep's-foot-roller effect) has changed the drainage properties of the soil, modifying its suitability for plant growth. Where trees remain, the understory has been mostly replaced by bare earth. The ground is crisscrossed by a maze of anastomosing tracks, and hills have become stepped horizontally with winding footpads. Whatever your perspective, habitat degradation is a disaster. We must do something about it.

And now there is a new threat. Sheep grazing in many parts of the rangelands is said to be only marginally profitable and there is increasing interest in intensified farming of goats, which now run feral throughout the area. If intensive goat farming becomes established, we can expect much of the sheep rangeland, already barren and overgrazed, to take the final step into desert.

What can be done? Clearly the salvation for this habitat would be the removal, or at least the significant reduction, of sheep in much of the rangelands. This will not happen unless there is some alternative (or significant supplementary) economic base, and I believe that kangaroos can provide that base.

Kangaroos, unlike sheep and goats, have evolved in the Australian habitat, are soft-footed, and do not take the grass as low down when they feed. Further, most seeds pass undamaged through kangaroos, whereas sheep grind seeds and so destroy the seed "bank."

In my view, all that is needed to bring about a significant reduction in sheep numbers is a solid, well-organized marketing drive for kangaroo products. Agricultural economists have estimated that a three- to fourfold increase in the price of kangaroo meat would make "farming" kangaroos economically attractive. At present, a typical sheep property might have six thousand sheep and three thousand kangaroos. The grazier raises the sheep deliberately and the kangaroos inadvertently. Paradoxically, instead of selling the kangaroos, he gives them away to the local kangaroo shooter and is glad to have them gone. If the price of kangaroo products were to rise, it would not take long for graziers to take an interest. If the price rose enough, such interest could overtake that in sheep. Graziers would soon realize that numbers of kangaroos can best be increased by decreasing the number of sheep, to the benefit of their own land and the whole ecology of the shrublands.

Two important assumptions behind the proposal are that the populations of kangaroos are sufficiently resilient to sustain regular harvest and that the meat and leather are good enough products to support a big increase in price. More than one hundred years of experience show that kangaroos can sustain a substantial annual harvest. Data collected by John Robertshaw and Robert Harden at the University of New England, from records from all the "kangaroo states," led them to conclude that direct killing by man has had little overall effect on the numbers and distribution of the large kangaroos. As an example, 20 million bounty payments were made on kangaroos in Queensland between 1877 and 1906, when such payments ceased. This was an average of 660,000 bounties per annum. Pest destruction and commercial harvesting continued unabated, yet between 1954 and 1987, in the same state, an annual average of 730,000 kangaroos was harvested. Similar examples could be given for other states. Quite apart from that long experience with "harvesting," for more than a decade now the populations have been the subject of extensive and regular aerial surveys which provide a good index of changes in abundance. Over that period we have seen good seasons and we have seen droughts, with kangaroo numbers rising and falling in response to the availability of grass. Throughout that decade, more than two million kangaroos have been harvested each year (about 10 percent the total population), but there has been no indication of any long-term downward trend in numbers.

What about the value of the meat and leather? The leather has always enjoyed spirited demand because of its strength and suppleness. It is in the meat, however, that there is potential for a big increase in price. Perhaps because of the traditional perception of kangaroos as pests and because it has always been a traditional food item among aborigines, white Australians have had a dim view of eating kangaroo. This is, however, changing rapidly. In most states, the sale of kangaroo meat for human consumption is prohibited, even though consignments are regularly inspected, passed fit, and exported overseas. The prohibition on local sale probably exists because beef and lamb producers fear competition. In South Australia, however, the sale is legal, and more than 150 restaurants list kangaroo dishes. What began as a novelty item a few years ago is rapidly becoming established as a routine item on menus. Well-selected steaks, cooked properly, are as tender as the tenderest beef, distinct in flavor, yet not so different as to be off-putting. Kangaroo meat also makes excellent roast, kebabs, and stew, and kangaroo-tail soup is a delicacy. Not only will the flavor and texture support a higher price for kangaroo meat, but it has less than 2 percent fat and most of that is polyunsaturated. No other red meat comes close, from a nutritional point of view. It really is the red meat you *can* eat.

Other Australian states are now reviewing the regulations. I think a three- to fourfold price increase is a conservative goal. When demand rises, so should price when supply is limited. And please note that my proposal does not advocate any increase in the proportion of kangaroos presently taken.

But how can a conservationist argue for an industry based on the killing of wildlife? In this case, I believe, very easily, because there are compelling arguments for the proposal, even apart from those related to halting the spread of deserts.

For one thing, to conserve abundant and widespread populations of kangaroos, the maintenance of habitat is mandatory. The highest densities of red, eastern gray, and western gray kangaroos all occur in the sheep rangelands. If this area continues to be pushed toward desert, it will not be to the long-term benefit of kangaroos.

There are also important animal welfare arguments. Traditional opponents of the kangaroo industry always express concern about cruelty, yet what I am proposing is a harvest of free-range animals, which live as free, wild creatures—unhindered, unhampered, and not owned by any individual. The ones who are shot, at night, in the head, will not even hear the bullet that kills them. To quote Dr. David Butcher of the Royal Society for the Prevention of Cruelty to Animals, "Paddock slaughter of animals unaware of danger is an advance over the techniques currently used with domestic animals."[3] In the same paper, Butcher listed nine other positive animal-welfare benefits that would accompany the scenario I have outlined.

Some advantages flow directly from the prediction that, under this proposal, the current pest status of kangaroos will be altered forever. No longer would the rural community tolerate illegal shooting of kangaroos by weekend or casual and inexpert shooters. No longer would graziers conduct illegal "drives" in which kangaroos are rounded up and shot by inexperienced shooters, often in daylight and often with shotguns. No longer would waterholes be poisoned to reduce kangaroo numbers. Kangaroos would simply be much too valuable. The size of the present annual illegal kill is completely unknown. Some estimates make it as large as the legal kill, which is three million or so, but nobody knows. What *is* known is that, if the illegal killing stops, the killing of kangaroos will become the province of only the skilled professional. The number of animals that die in inexpert hands would fall dramatically, probably to near zero. In my view, all considerations argue strongly in favor of the proposal.

Valerius Geist recently made a case against giving dead wildlife value claiming that it was to the detriment of wildlife.[4] That may be true in many, even most, cases. It is not true for kangaroos. None of Geist's examples even came close to paralleling the situation for kangaroos. Australia allows no legal sport shooting of any of the kangaroo species or any other native mammal, and I am not proposing active husbandry or farming in the conventional sense. Furthermore, there are few other animals for which there is such an extensive base of population data, plus one hundred years harvesting experience to draw upon. Generalizations such as Geist's are always useful, but each case must, in the end, be judged on its own merits.

In my opinion, the manifold conservation and animal welfare merits of the case for a careful, controlled harvest of free-ranging kangaroos warrant thoughtful consideration by conservationists, graziers, agricultural economists, land managers, and the community in general.

NOTES

1. D. Lunney and G. C. Grigg, eds., *Kangaroo Harvesting and the Conservation of Arid Lands*, proceedings of a Royal Zoological Society (NSW) Symposium May 14, 1988, Sydney, New South Wales, Australia. *Australian Zoologist* (Special Edition) 24 (1998).
2. G. C. Grigg, "Are Kangaroos Really Under Threat?" *Australian Natural History* 21, no. 4 (1984): 123–29.
3. David Butcher, in *Kangaroo Harvesting and the Conservation of Arid Lands*.
4. V. Geist, "How Markets in Wildlife Meats and Parts, and the Sale of Hunting Privileges, Jeopardize Wildlife Conservation," *Conservation Biology* 2, no. 1(1988): 15–26.

REFERENCES

Geist, V. "How Markets in Wildlife Meats and Parts, and the Sale of Hunting Privileges, Jeopardize Wildlife Conservation." *Conservation Biology* 2, no. 1 (1988): 15–26.
Grigg, G. C. "Are Kangaroos Really Under Threat?" *Australian Natural History* 21, no. 4 (1984): 123–29.
Lunney, D., and G. C. Grigg, eds. *Kangaroo Harvesting and the Conservation of Arid Lands*. Proceedings of a Royal Zoological Society (NSW) Symposium May 14, 1988, Sydney, New South Wales, Australia. *Australian Zoologist* (Special Edition) 24 (1998). 70 pp. (This publication, containing twenty papers *relevant* to the kangaroo-harvesting proposal, is available from the Royal Zoological Society of New South Wales, P.O. Box 20, Mosman 2088, Australia.)

17

Tourism as a Sustained Use of Wildlife:
A Case Study of Madre de Dios, Southeastern Peru

Martha J. Groom, Robert D. Podolsky, and Charles A. Munn

Nature tourism is a means of using wildlife to benefit human populations that is nonconsumptive, which distinguishes it from many of the other uses discussed in this volume. Whereas the success of harvesting wild animals may depend on their population growth rates or behavior in captivity, tourism brings a different list of concerns, including the impact of tourists on natural communities, the effects of foreign culture and capital, and the challenge of educating the public about problems facing natural systems. Perhaps the largest distinction between nature tourism and other uses of wildlife, however, is the potential size of the industry created. Whereas most consumptive uses lead to smaller commercial or subsistence economies, a well-managed and promoted tourism program can easily rival other major forms of national income, particularly in developing countries.

Tourism is a major world industry, and international tourism is its largest and fastest growing segment.[1] Many argue that investment in international tourism is risky because it depends on the vagaries of worldwide economic patterns and is sensitive to unpredictable political acts such as terrorism. However, even in countries like Peru and Zimbabwe, where terrorist incidents have resulted in large setbacks, these have proved temporary, and tourism has remained a steadily growing industry.[2] International tourism currently generates more than $40 billion per year worldwide (excluding airfares), the fastest-growing portion going to goods and services that support adventure tourism.[3] Thus, from a purely economic standpoint, investment in services supporting adventure tourism is likely to pay off increasingly for governments and private businesses willing and able to take the risks of investment.

From *Neotropical Wildlife Use and Conservation*, edited by John G. Robinson and Kent H. Redford. Reprinted by permission. Copyright © 1991 University of Chicago Press. All rights reserved.

Linking tourism to conservation, however, requires that economic gain by investors be coupled with at least three additional goals. First, a tourism/conservation program ideally should extend the economic benefits of development to a broad base of the local human population through employment, compensation fees, or the development of social services. Such an approach in Kenya's Amboseli National Park has demonstrated that local people are more likely to protect lands and wildlife when they have an economic incentive to do so.[4] Second, for tourism on reserve lands, the goals of park management must be furthered by economic gains and not counteracted by tourist activity. This involves not only supporting research on human impacts, but also creating a mechanism for visitors excited by their experience to contribute to the park following their visit, either through contributions or word-of-mouth advertising. Finally, the program ultimately should accommodate visitors of a wide range of economic status, so that access to wildlife is not restricted to the rich or foreign.

Problems in the Distribution of Income

The economic effects of tourism can be felt at four levels: the private investor, the national economy, the local economy, and, if on reserve lands, the reserve budget. Benefits are rarely well distributed, however, with revenues usually decreasing rapidly in the order listed. Creating a more equitable distribution of income among these levels is one of the main challenges to a tourism/conservation program.

Although tourism infrastructure often needs only modest capital input and can sometimes capitalize on government-supported development of transportation, entrepreneurs from developing countries usually lack the capital for initiating a tourist business. Often, control of development rests in the hands of wealthy foreign investors who do not understand or appreciate local problems and who place profits above conservation and the support of local economies. Revenues from tourism are moved out of the host country to avoid devaluation associated with the high inflation rates characteristic of developing countries. Although the exportation of profits is common to many industries, tourism usually allows a greater fraction of revenues to remain in the host country.

Of foreign capital brought into the host country, most is normally spent in the larger cities for transportation, lodging, food, and supplies.[5] The share that actually goes into economies near the tourist site can be minimal.[6] As mentioned, local people rarely have resources or experience to invest, and most therefore end up with low-paying jobs.[7] On the other hand, because it is a labor-intensive industry, tourism can provide jobs for large numbers of nonskilled workers, which makes it especially important in areas with high unemployment.[8] If properly practiced, the industry can also stimulate local economies through increased local demands for transportation, lodging, food, materials, and nature interpretation.[9] Thus, even a relatively small share of tourism revenues can provide an extremely strong boost to the local economy.

What is the Potential of Tourism as a Conservation Strategy in Developing Countries?

Meeting conservationist goals has had mixed success in east Africa, where the economic potential of watching, rather than hunting, large and attractive mammal populations has been recognized for many years. For example, tourist interest in a single—albeit spectacular—species, the mountain gorilla, is estimated to bring in $4 million per year to Rwanda, making gorilla tourism alone the country's second largest industry.[10] The small portion of revenues received by local people has played a major role in reversing negative attitudes toward wildlife. Whereas poaching was a major problem for gorillas at the start of the project, no gorillas were known to have been killed by poachers for several years, and gorilla populations showed their first size increase in 1986.[11] Newly habituated gorilla troops in Zaire have provided less expensive opportunities to see the gorilla, broadening access to new socioeconomic classes and enabling better management of the entire gorilla population in Rwanda and Zaire.

Tourism development in the neotropics has had a slower start. Since the explosion of nature tourism in Africa during the last decade, however, tourists have begun looking toward the Amazon as a destination with romantic appeal and smaller crowds. Informal interviews with tourists coming to see the cultural and natural sites of Peru indicate that, of those planning to visit the Amazon, about 70 percent are primarily interested in seeing animals and are less interested in indigenous cultures or Amazonian panoramas.[12] Unfortunately, the neotropical species that are exciting to see (e.g., giant river otter, woolly monkey, spectacle bear, and macaw and parrot species) are less familiar to the average tourist than are the large mammals of east Africa. Many tourists therefore travel to areas like Iquitos, which are well advertised but, because of hunting and land conversion, are depauperate in animal species and primary forest. Interviews with the same tourists leaving Iquitos give the impression that many are happy with their experience, but probably not excited enough to spread the appeal of a visit by word of mouth.[13] This also means that most feel no commitment to contributing financially to the preservation of the area.

In this chapter we evaluate how tourism development in the Madre de Dios region of southeastern Peru is achieving a link between private enterprise and land and wildlife protection. We focus on the magnitude and distribution of revenues coming into the region, and emphasize the degree to which growth in the tourism industry can aid development and conservation in the area. Finally, we discuss current and future limitations to tourism development in southeastern Peru and potential problems with using Madre de Dios as a model for the development of nature tourism in other countries.

Case Study: Tourism in Madre de Dios, Peru

The Madre de Dios region contains some of the most diverse and spectacular cloud and lowland tropical rainforest in the world, which is largely unmolested by hunting or commercial activity. Forests of this region boast an astounding diversity of species (with more than fifteen thousand plants, two hundred mammals, one thousand birds). Some of the only large and easily observable populations of highly endangered species are found in this area, including those of the giant river otter, jaguar, woolly monkey, spider monkey, white-lipped peccary, tapir, black caiman, and several species of macaw, parrot, toucan, and curassow. These characteristics have made the region of prime importance to conservationists and other scientists, and suggest its potential attraction to tourists wishing to see and contribute to the protection of one of the most pristine areas in the entire Amazon basin.

History of Development and Use

One of the major departments in Peru, Madre de Dios is situated at the Brazilian and Bolivian borders, just east of the department of Cuzco. It is sparsely populated, with fewer than 0.40 people per square kilometer, 39 percent of whom live in the region's capital, Puerto Maldonado.[14] Development of the region has been slow. Aside from minor rubber, Brazil nut, and hardwood extraction during the early part of this century, little commercial use was made of the area before the 1940s, when gold extraction began near Puerto Maldonado.[15] Gold is now the major industry of Madre de Dios, followed by rubber, Brazil nuts, tourism, and lumber. A recent surge in the government-supported development of cattle ranching has accelerated forest conversion in Puerto Maldonado and the surrounding lands to the north of the city.[16] Colonists along roads leading into the area have also cut some forest for small-scale agriculture.

In 1973, 1.5 million hectares (18 percent) of Madre de Dios was set aside, along with some highlands of the department of Cuzco, as the Manu National Park. The park encompasses the entire drainage of the Manu River, and spans paramo at 3,650 meters to lowland tropical rain forest at 300 meters. In 1977, with the addition of a reserved zone and a cultural zone near native and colonist settlements, the park was designated a UNESCO Man and the Biosphere Reserve, and the total area now included in the Reserve is over 1.88 million hectares.[17] Although rubber was extracted briefly at the turn of the century and mahogany trees were removed from along a portion of the Manu River in the 1960s, little hunting or other exploitation occurred within the park boundaries, leaving Manu Park in a near-pristine state. As in all areas of Madre de Dios where natural communities have not been disturbed, the diversity of species is high, and most species are tame and easily observable.

Four smaller reserves in Madre de Dios, the Tambopata Reserve Zone (5,500 hectares), Lake Sandoval (less than 2,000 hectares), Cuzco-Amazonico Reserve (10,000 hectares), and the Pampas del Heath National Sanctuary (107,000 hectares), are all located near Puerto Maldonado; the first three serve as the main destinations for tour groups. All reserves are in a semipristine state and include potential attractions not found elsewhere in Madre de Dios, such as the maned wolf in the Pampas del Heath and the grade four to five rapids on the Tambopata River. At present, only the Heath Sanctuary is officially protected, and none of the four is actually protected by guards. In February 1990, the Peruvian government established a much larger reserve (1.5 million hectares) that encompasses most of the drainage of the Tambopata and Heath rivers above and between the four existing reserves.

Little organized tourism existed in any part of the Madre de Dios region before 1975.[18] For many years the lack of transportation had restricted tourist flow into the region. Even now, of the two major roads from Cuzco, the nearest major city, one (to Puerto Maldonado) is too long to transport tourists. The other (to Shintuya, a small town on the Alto Madre de Dios River) is used frequently by tour companies operating in Manu. Large commercial jets began flights regularly into the Maldonado airport only after its dirt airstrip was paved in 1983. In addition, small twin-engine charter flights have come to Manu since 1983 by using a hard gravel airstrip at Boca Manu, a small settlement at the mouth of the Manu River. Air transportation thus has increased tourist access to the natural attractions around Puerto Maldonado and Manu, although ground transport for tourists through the area still depends chiefly on river transportation.

Tourism Near Puerto Maldonado

Although more than sixty thousand people now fly to Puerto Maldonado each year, little infrastructure exists to tap the potential nature-oriented tourism market. Only three lodges and relatively few independent guides regularly bring tourists to sites near Maldonado. One reason is that, at present, tourism is not perceived by the local and national governments or by most potential investors to be a profitable industry in the southern rain forest. Furthermore, most local people lack the expertise and resources to start such a business. As we will argue, figures on current revenue suggest that tourism is a viable industry for the region, but that it has not yet developed fully in size nor in the equitability of profit distribution.

The largest share of the tourism market in Puerto Maldonado is held by three tourist lodges, Explorer's Inn, Cuzco-Amazonico, and Tambo Lodge, each of which have about seventy beds. Explorer's Inn was established in 1975, originally as a hunting lodge for a largely European and North American clientele. When the government outlawed trophy hunting in 1975, the inn converted to nature tourism. Cuzco-Amazonico, which started in 1977, has from the start emphasized low-intensity jungle wildlife viewing with an antiquated but very widespread "dangers of the Amazon" approach. While both lodges offer a comfortable stay in intact forest, they do not offer an intact fauna. A history of hunting at the lodges, as well as continued subsistence hunting nearby, has depleted some of the more spectacular wildlife, such as large monkeys, cats, otters, caiman, and curassows.[19] Both lodges, however, offer day trips to nearby oxbow lakes on the Madre de Dios and Tambopata Rivers, where more wildlife can be dependably seen. A two-night stay is about $150 at either lodge, and both are usually full during the tourist season (June to September). Both Cuzco-Amazonico and Explorer's Inn are owned by wealthy Peruvians of European extraction, and the clientele remains mostly North American and European. A third lodge, Tambo Lodge, recently was opened by a Peruvian from Cuzco formerly employed by Cuzco-Amazonico. It is similar in goals and services to Cuzco-Amazonico, but less fancy and less expensive.

Other than the three lodges, about five independent tour guides based in Puerto Maldonado will bring people to oxbow lakes and other areas to see wildlife for relatively modest fees. Several guides are mentioned in the *South American Handbook*, and most advertise from the hotels in Puerto Maldonado and try to attract customers as they arrive at the airport. As with the lodges, some guides have moved from marketing consumptive (hunting) to nonconsumptive (photo or adventure) tours in the last few years. One guide, who now grosses $8,000 per year, led hunting safaris successfully for more than five years before learning he could make much greater profits by guiding people to see live animals. Most guides take people for one- or two-day trips to close attractions ($15 to $25 per day), particularly to Lake Sandoval, only forty-five minutes by boat from Puerto Maldonado. A few guides will run longer trips to particular sites ($20 to $30 per day), while rafting outfitters specialize in trips down the Tambopata rapids or to what is perhaps the largest macaw mineral lick in the world, attracting hundreds of macaws daily.

Where does the money go? Negative perceptions of the profitability of tourism in Madre de Dios seem unwarranted given that over $1.27 million were paid for tourist services in Puerto Maldonado in 1987 (see Table 1). This amount does not include the increase in revenues to Peru as a whole from tourists who come to Peru principally to see wildlife in Madre de Dios (particularly naturalist/bird tours), but also visit the more traditional cultural attractions of Cuzco and Macchupicchu. The average tourist spends $1,050 per visit for lodging, food, transportation, and gifts in the major cities of Peru.[20] Thus, the 6,520 tourists to Puerto Maldonado contributed about $6.8 million to the Peruvian economy above their payment to the tourist companies in Madre de Dios.

Explorer's Inn has grossed over $350,000 per year in recent years, although profits were temporarily affected by one severe but isolated terrorist incident in Cuzco in 1986.[21] The larger Cuzco-Amazonico grosses more: $432,000 in 1987 and $770,000 in 1988. Independent guides earned about $35,000 in 1987, while special-interest tours (e.g., river rafting) grossed $88,200. The combined gain of the airlines and local hotels and restaurants was at least $310,000. By gross revenue, nature tourism is thus the fourth largest industry in Madre de Dios, comparable to or above rubber and Brazil nuts.[22]

Tourism has been profitable to business owners, but the benefits are not gen-

Table 1. 1987 Revenues of tourist companies and independent tour guides in Puerto Maldonado, Madre de Dios, and Cuzco, Peru

Company	Number tourists	Average stay/days	Price/ day(US$)	Total gross revenues (US$)
Cuzco-Amazonico Lodge[a]	3,000	3	48	432,000
Explorer's Inn[a]	2,550	3	50	382,500
Tambo Lodge[a]	350	3	25	26,250
Independent guides[a]	450	3	25	33,750
Independent guides (short trips)[b]	100	1	15	1,500
River rafting[b]	70	14	90	88,200
Hotel/food in Puerto Maldonado[b]	550	1	30	16,500
AeroPeru/Faucett r/t Cuzco—Puerto Maldonado[b]	6,520	-	45	293,400
Total gross earnings				1,274,100

[a] Sources: Lodge owners, employees, *South American Handbook* 1989, and information supplied to ACSS.

[b] Figures estimated by average costs of services and estimates of number of participants supplied to ACSS.

erally widespread. The tourist lodges and some independent operators enjoy profit margins as high as 30 to 45 percent. Discounting revenues to the airlines and specialized tours, and assuming a 35 percent profit margin for the lodges and a 25 percent profit margin for all other tour operators, Puerto Maldonado tourism in 1987 paid out approximately $575,550 in salaries and for the purchase of food and local materials. These estimates are conservative, as more people may be visiting the forest than are traveling with guides known to us, and tour companies may underrepresent tourist volume for tax purposes. In any case, nearly all of the $1.4 million or more generated by tourism in Madre de Dios went to Peruvians. Tours packaged through U.S. or European companies are marked up nearly 100 percent in cost,[23] indicating that if Peruvian companies marketed directly at prices closer to those charged by foreign companies, revenue to Peru could increase.

Potential for growth. Now that transportation problems have eased with the addition of a major airport, tourism development in Maldonado is limited mostly by a lack of infrastructure. Maldonado receives less than one-tenth of the tourism volume of Iquitos,[24] despite the fact that forest clearing and hunting have decimated wild populations of large birds and mammals in Iquitos. Iquitos has the lure of being located on the banks of the Amazon and still retains some reputation as a site where wildlife can be seen. We believe that the creation of new lodges, wider advertising, and better nature interpretation in the Madre de Dios area would help to improve and spread the area's reputation and ultimately lead to an increase in tourist revenues. This process appears to have already begun: Cuzco/Amazonico received five thousand visitors in 1988, up 67 percent from 1987.

The new Tambopata/Candamo reserved zone established in February 1990 will likely foster the development of new tourist markets such as river rafting and giant otter or macaw watching. Interest in river rafting is evidenced by the popularity of day-trip rafting in the Urubamba River basin near Cuzco (3,500 tourists per year[25]). Currently, the Tambopata rapids are not well advertised, and this sector of the market will almost certainly grow once the area gains a reputation. The large size of the proposed reserve makes it likely that several new lodges could be built and operated without adverse impact on the forest and its wildlife.

Tourism in the Manu Biosphere Reserve

The value of Manu Park as an educational and recreational area has been repeatedly emphasized in statements of the park's goals and purposes.[26] Nevertheless, tourism has been introduced to the lowland forests of Manu with a great deal of caution. The first tourists to the lowlands were naturalists visiting the Cocha Cashu Biological Station in the early 1980s. In 1985, more than one hundred people visited the park in a single year for the first time. Current policy restricts the number of tourists to 500 per year, although that figure is likely to be raised pending a current reevaluation of human impact.[27] All tourism is now restricted to the reserved zone of the park to avoid any adverse impact on the original protected area. Tourism has been controlled tightly by the park to avoid compromising the other goals of park administration, including the conservation of flora and fauna and concern for the well-being and traditions of several groups of indigenous peoples.

Three organized tour companies (Expediciones Manu, Mayuc, and Manu Nature Tours) currently have permission to bring tour groups by canoe into the park. In 1987, the first tourist lodge in the reserved zone (Manu Lodge) was built near an oxbow lake to accommodate tourists brought in by Manu Nature Tours. Other groups are restricted to camping on sand beaches exposed along the Manu River from May through August. The park has designated two beaches adjacent to oxbow lakes as camping and recreation areas, and only allows camping on other beaches as long as tourists do not camp on beaches with colonies of nesting birds.[28]

While only 315 tourists entered the park in 1987, tourist companies and independent guides grossed about $172,225, or $547 per tourist from those few visitors (see Table 2). Company tours are fairly expensive ($75 to $90 per day) and bring in mostly European and North American clients, although much lower rates have been offered for Peruvian nationals and students. Independent guides are less expensive ($30 per day) and therefore accommodate the majority of foreign backpackers and nationals.

Table 2. 1987 Revenues of tourist companies and independent tour guides in Manu Biosphere Reserve, Madre de Dios, and Cuzco, Peru

Company	Number tourists	Average stay/days	Price/ day(US$)	Total gross revenues (US$)
Manu Nature Tours[a]	38	7	90	23,940
Manu Lodge[a]	32	7	90	20,160
Expediciones Manu[a]	180	6	75	81,000
Mayuc	25	6	65	8,125
Independent guides[a]	40	5	30	6,000
Amazonia Lodge[a]	150	2	30	9,000
Erica Lodge[a]	50	1	20	1,000
Hotel/Food in Puerto Maldonado[b]	100	1	30	3,000
AeroSur r/t Cuzco Boca - Manu[b]	100	-	200	20,000
Total gross earnings				172,225

[a]Source: Owner or guide of tour group and information from employees
[b]Figures estimated from known costs and numbers of tourists

Although visitation to the park's cultural zone is not restricted, the two tourist lodges located there (Erica and Amazonia Lodges) receive fewer than two hundred tourists each year. Both are close to human settlements, and while wildlife is still plentiful when compared to the Iquitos region, it is generally less observable than along the Manu River. Both lodges appear to lack the advertising and infrastructure necessary to maintain a healthy business. Most clients, in fact, are Expediciones Manu and Manu Nature Tours groups stopping for one night en route to the reserved zone of the park.

Access to the park is still one of the main limitations to tourism in Manu. The cultural zone lies a day's journey by car from Cuzco on a dirt road that is closed by landslides for parts of the year and is too narrow to accommodate more than one-way traffic (switched on alternate days). The reserved zone of the park lies an additional six to twelve hours away by river. Tourists generally reach the reserved zone by boat from the river ports on the Alto Madre de Dios River. A more comfortable (but more expensive) option is to charter a small plane from Cuzco to the mouth of the Manu River, less than a day by boat from all points in the reserved zone. Only relatively wealthy groups with Manu Nature Tours and Expediciones Manu charter a plane directly to the area.

Where does the money go? Manu Biosphere Reserve presently accommodates many fewer tourists than Puerto Maldonado and only grosses about one-seventh of the revenues (see Tables 1 and 2). Still, Manu's reputation for spectacular wildlife attracts wealthier clients, who stay longer and spend more per day. Thus, Manu earns about three times as much per tourist as Maldonado ($547 versus $195), demonstrating that the two areas cater to different economic classes.

Table 3 gives the estimated distribution of revenues to private companies. Due to an explicit emphasis on the quality of a tourist experience, Manu Nature Tours employs a greater number of people per tourist (currently about 1:1) and has a lower profit margin than any other tour company in Madre de Dios, choosing to reinvest profits to improvements in services. The other tour operations in Manu also have lower profit margins than those in Puerto Maldonado; in general, revenues are more evenly distributed among employees. Both Expediciones Manu and Manu Nature Tours use mostly Peruvian guides and employees. Nearly all food and gasoline must be brought directly from Cuzco, reducing somewhat the money going into local economies. Most employees, however, are from the area, and boats and construction materials for Manu Lodge are purchased locally.

The park itself receives tourism-generated revenue directly from two sources (park fees and concessions) and indirectly from postvisit donations to private conservation groups (see Table 4). The only park fee is a minimal per-visit charge (approximately $10 per foreign tourist and $1 per national tourist) collected from all tourists for entering the reserved zone, which generates less than $3,000 per year. Manu Lodge pays a concession fee of 5 percent of its gross revenues (about $1,000) for its access to ten hectares of reserve land. Indirect tourism-generated support for the park is collected mainly by Conservation Association of the Southern Rainforest (ACSS), a Peruvian group formed in 1984. In their fourth year, ACSS collected more than $13,000 in donations just from tourists who had visited Manu.[29] This money has been fed directly into park projects, including purchases of equipment, food, and medical care and supplies for park guards, payment for environmental education programs in local schools, applied research on the status of native groups in Manu, and other programs not supported by the park budget. Postvisit contributions thus play an essential role in furthering conservation goals in the park.

Potential for growth. If the cap on visitor numbers is raised in the next year, only infrastructure development will limit tourism growth in Manu. Manu Lodge is well under capacity due to ongoing construction and lack of publicity, but when running at full capacity the lodge should gross a minimum of $200,000 per year. The success of the lodge will indicate if enough of a market exists to develop lodges on other oxbow lakes in the reserved zone. Expediciones Manu currently runs near capacity and could increase their business with more boats and guides. Growth in beach camping could increase modestly without negative impact to wildlife as long as park regulations and guidelines for avoiding disturbing beach-nesting birds are followed.[30] Adding two more such lodges and continuing beach camping at its current level, nature tourism could gross $992,250 per year in Manu alone (excluding revenues to airlines; see Table 5). If each lodge employed local people and used local supplies, perhaps as much as 60 percent of this sum could stay within the region. Currently only about 30 percent of these revenues go into communities immediately adjacent to the Manu Biosphere Reserve.

Table 3. Estimated Distribution of Gross Revenues of
Tour Companies in Madre de Dios (1987)

Company	Gross revenues[a]	Number of employees[b]	Profit margin(%)[c]	Local expend. (US$)[d,e]
Manu Biosphere Reserve				
Manu Nature Tours and Manu Lodge	41,100	15	5	11,910
Expediciones Manu	81,000	15	30	17,000
Mayuc	8,125	6	25	?
Independent guides	6,000	6	15	4,000
Amazonia Lodge	9,000	6	15	9,000
Erica Lodge	1,000	3	15	1,000
Puerto Maldonado				
Cuzco Amazonico	432,000	20	35	84,240
Explorer's Inn	382,500	20	35	74,590
Tambo Lodge	26,250	12	25	7,875
Independent guides	35,250	15	25	26,000

[a] Figures from Tables 1 and 2.

[b] Source: Tour companies and ACSS.

[c] Percentage of gross revenues taken as profits; source: tour companies and ACSS.

[d] Estimated amount spent on salaries and local materials; based on estimates reported by Manu Nature Tours and Manu Lodge and on costs to independent guides.

[e] Information not available for Mayuc Tours.

As with other public lands, the potential of using tourist revenue to offset management costs is an attractive and, we believe, reasonable option for Manu Biosphere Reserve. The current entry fee easily could be raised, but maintaining a two-tiered rate structure is necessary so as not to exclude lower-income Peruvian nationals. Currently only Manu Lodge is paying a concession fee to the government, but the park is now considering charging Expediciones Manu and independent guides a concession fee for their use of the park.[31]

The success of postvisit donations to the park relies on the quality of the tourist experience in Manu, and tourist satisfaction seems to be substantially higher than in Iquitos or in other parts of Madre de Dios. Currently, only Manu Nature Tours groups regularly are informed of the possibility of making contributions to ACSS. It is likely that increasing the direct solicitation of donations, which is not currently done, will substantially raise this source of revenue.

Simple calculations show that tourism (as we project it) could pay for the operating budget of the park, currently set by the regional government (CORDEMAD) at $20,000. With only a modest increase in tourist volume,

Table 4. Current and Projected Revenues to Manu Biosphere Reserve from Tourism Receipts

Income source	Number of Tourists		Total gross profits (US$)
	Foreign	National	
1987 earnings			
Park fees[a]	280	35	2,835
Concession fee (ML)			1,000
Donations from ACSS[b]			13,000
CORDEMAD budget[c]			20,000
Total 1987 earnings			36,835
Projected earnings[d]			
Park fees[a]	1280	320	13,120
Concession fees[e]			43,260
Donations from ACSS[b]			50,000
Total projected earnings			106,380

[a] Assumes fee of $10 per visit to foreign tourists and $1 per visit to national tourists. Numbers of tourists projected from average proportions of foreign and national tourists visiting Manu in 1988.
[b] Source: D. Ricalde, President, ACSS.
[c] Source: A. Cuentas, Director of Parks, CORDEMAD.
[d] Source: Calculations based on projections made in Table 5.
[e] Assumes 5 percent concession fee is assessed to gross revenues of all lodges and organized tours (including Amazonia and Erica Lodges).

$50,000 per year could be raised through a combination of collection of use fees and assessing a 5 percent gross revenue charge on all lodges and organized tour operations (see Table 4). An additional $50,000 could easily be raised through postvisit contributions. This income would be enough to quintuple the current operating budget of the park without the need for input from CORDEMAD.

Concerns With Tourism Development in Madre de Dios

Economic Considerations

Although proving itself worth the risks of investment, nature tourism is not yet an economically important benefit to most of the people in Madre de Dios. A large share of the money brought into the region does not remain there. Most profits from Puerto Maldonado tourism currently return to the two lodge owners

Table 5. Projected Revenues to Tourist Companies in Manu Biosphere Reserve

Company	Number tourists	Average stay/days	Price/day(US$)	Total gross revenues (US$)
Manu Nature Tours	200	7	90	126,000
Manu Lodge	350	7	90	220,500
Two additional lodges (ML)	700	7	90	441,000
Expediciones Manu	250	6	75	112,500
Mayuc	150	6	65	48,750
Independent guides	150	5	30	22,500
Amazonia Lodge	250	2	30	15,000
Erica Lodge	250	1	20	5,000
AeroSur r/t Cusco-Boca Manu[a]	900	-	200	180,000
Total gross earnings				1,172,250

[a] assumes 70 percent of all visitors to MNT, ML, and lodges, and 50 percent of visitors with Expediciones Manu will fly into Boca Manu.

---

in Lima, while little is being reinvested in infrastructure or local employment. The average employee to tourist ratio (1:3), for example, is low relative to similar tour operations in Kenya (2:1).[32] Explorer's Inn, and to a lesser extent, Cuzco/Amazonico Lodge use nonlocal or foreign managers and guides, while local opportunities are limited mostly to lower-paid positions as cooks, waiters, construction workers, boat drivers, and groundskeepers.

This situation results from a combination of inattention to local problems on the part of some tour companies and a shortage of local guides trained both in nature interpretation and in English. This is unfortunate, because nonlocal guides in Madre de Dios know much less about the forest. Tourists are given a minimum amount of information on only a few plants and animals, and leave with the impression that there is little to see. Local people know a tremendous amount about their surroundings, but have not been trained to deliver that knowledge in a systematic way. Some tourist companies are beginning to recognize that investment in service staff and in the education of local guides is linked with the health of their own business: Local perception of the industry as an economic benefit is increased, and a high-quality experience greatly adds to word-of-mouth advertising among satisfied clients.

Tour operations in Madre de Dios do use local materials, and to some extent locally grown food, but local merchants still do not feel sufficiently patronized by tour companies. The supply of many local goods is erratic, and therefore greater awareness of the problem by tourist operators will not necessarily reduce the problem. It is difficult to envision how local suppliers can improve their dependability, since even basic supplies such as sugar and rice can be difficult to obtain in Cuzco. Patronage of local suppliers can also adversely affect availability of goods to local people. Unable to compete with the inflated prices paid by lodges, local people may suffer shortages for even basic goods.[33] This type of inflation, driven not only by tourism but even more by the gold-mining industry, is already a concern of local people in Madre de Dios. Thus, the influence of a new industry can have many unexpected disruptive effects on the social structure of the local community.[34]

Economic benefits to the park at present come mainly through donations by independent conservation groups such as ACSS. This mechanism for indirect funding of the park through independent organizations is one great conservation success of tourism development in Manu Park. Such donations, originating mostly from satisfied customers in Manu, also support conservation projects throughout the region.

Effects on Wildlife and Indigenous People

Although nature tourism encourages the economic valuation of maintaining animal populations and intact forest, it clearly has negative side effects.[35] In Manu, many animals (with the exceptions of giant river otters and beach-nesting birds) appear largely unaffected by people, although several species are sensitive to boat traffic and may experience some alterations in their ranging behavior or interruptions in foraging as a result.[36] A large volume of otherwise innocuous tourists can disturb wildlife directly through noise and overuse of critical areas (e.g., nesting areas, watering holes), and indirectly through habitat degradation (e.g., pollution, alteration through trail cutting). Lodges outside the Manu Biosphere Reserve have suffered tourist-related reductions in wildlife and habitat degradation. In the past, the two large Maldonado lodges, for example, have sold arrows with macaw feathers, and one of them also sold necklaces with monkey hands to tourists. Vines and trees along trails at one of the lodges are heavily scarred or irreparably damaged by cutting to demonstrate their latex or water-holding properties.

In the Manu Biosphere Reserve, the currently small volume of tourists has not harmed the forest or wildlife at the most heavily used areas along the river or oxbow lakes. However, incidents have been reported of independent guides bow hunting, digging up turtle nests, disturbing beach-nesting birds, and chasing giant otters, swimming jaguars, and tapir so that all members of a tour group see them.[37] Disturbance to nesting birds and turtles has been reduced by confining tourist activity to a few beaches, but problems with littering and hunting intensified in 1988 with an increase in independent tour groups, which are poorly organized and unregulated.[38] A proposal under consideration would have all tour guides entering Manu Park take a short course in park regulations and basic ecology, as is required of guides in the Galapagos.[39]

A particularly sensitive problem in Manu Park is how to handle interactions between tourists and native peoples. At present the park includes four contacted native groups, two of which (Machiguenga and Piro) are in settlements accessible by river. Several tours in the early 1980s visited a Machiguenga settlement, but such contact is now forbidden under park policy.[40] People of that community and of a Piro/Machiguenga settlement in the cultural zone, however, have expressed interest in tourist visits and the potential income from selling ceramics and other handicrafts.[41] The park is cautious about encouraging such interactions for fear of introducing outside diseases and of negatively affecting cultural traditions through participation in Western-style market activities. The introduction of indigenous peoples to diseases and cash economies has been a thorny issue throughout Latin America.[42]

It should be added that these concerns are not limited to indigenous peoples, but can create unforeseen stresses in any local community. In addition to the problems already mentioned, jealousies or power asymmetries from inequitable distribution of profits, greater movements between communities, and stresses from interacting with foreign tourists can cause a variety of social problems.[43] Finding ways to limit or mitigate social stresses that arise from tourism activity challenges not only national, regional, and local governments, but also tour companies. Here, the advice of sociologists and anthropologists must be sought and incorporated into tourism development.[44]

Nature Tourism in Developing Countries in Latin America

The economic success of nature tourism will depend on several factors: (1) the attractiveness of the natural area and its wildlife and the degree to which the wildlife is easily seen, (2) the ease of access to the area and the comfort of the accommodations, and (3) the quality of nature interpretation and other guided services.[45] In addition, the success of nature tourism in fostering a sustainable industry and promoting habitat and wildlife conservation depends on the compatibility of tourism and wildlife conservation, and on whether tourism is perceived as a benefit at local and national levels.

Increasing the Attractiveness of Latin American Sites

One reason for the huge success of nature tourism in east Africa is that east African plains mammals are both extremely attractive to the public and easy to observe. Manu Biosphere Reserve offers a wide diversity of species that, while not as visible or spectacular as their east African counterparts, are nevertheless plentiful and diverse. Indeed, few tropical forest sites in Latin America will be able to equal Manu in its attractiveness to tourists. Other ecosystems in Latin America, such as the Brazilian pantanal, the Venezuelan llanos, and similar open, wildlife-rich habitats have excellent visibility of attractive wildlife and are thus also likely to be successful nature-tourism centers. In addition, unique sites, such as the Galapagos, are highly diverse and undisturbed.

Ultimately, it is possible that only a few sites across Latin America will be able to support large-scale nature tourism. Yet skillful marketing could increase the attractiveness of many places. Individual Amazonian species, while not as well known as elephants and lions, could be publicized to increase their value. Thresher has estimated the worth of a single wild male lion to be $500,000[46] (which, given inflation and even greater interest in tourism in east Africa now, is a conservative value). We estimate that one jaguar, baited in order to be seen dependably by tourists, would increase revenues coming to Madre de Dios by at least a comparable value. Equally spectacular and rare sights such as spectacle bears, giant otters, harpy eagle nests, or salt licks that attract hundreds of macaws and parrots, could have nearly equal "worth" to the nature tourism industry of Latin America.

Infrastructure Improvements to Sustain Tourism to Natural Areas

A successful tour operation usually requires a lodge that is sufficiently comfortable to attract a wide variety of tourists. Good infrastructure support, consisting of (1) paved, reliable roads, (2) telephone or radio contact for logistics and emergencies, and (3) fail-safe coordination of travel, housing, and meals is also crucial to the success of any tourist venture.[47] Many people would never consider going to Manu because it involves a relatively arduous trip, although they will go to Puerto Maldonado or Iquitos because they are accessible by air. The costs of infrastructure development, however, can be high, although they are often lower than for alternative industries (e.g., large-scale agriculture) that require machinery and other imported goods.

The capital necessary to begin a tourism project in the rainforest is $ 10,000 to $50,000, sums that are sufficiently large to pose an immediate barrier to local entrepreneurs.[48] To overcome this barrier, economic support and guidance from governments and organizations cognizant of tourism's strengths and weaknesses is necessary.[49] Unfortunately, governmental subsidies are rarely targeted for tourism development (in Peru these are chiefly limited to tax incentives[50]), although tourism is probably a more sound investment than some businesses that are currently heavily subsidized (e.g., cattle ranching, some agriculture). Thus, an important first step is for national governments to expand loans, grants, and other support for tourism development. Nongovernmental organizations (NGOs) can provide support to tourism projects, while stipulating that they be planned carefully to enhance local participation, conservation, and educational goals. Finally, foreign NGOs can contribute by playing a greater role in advising national projects or in training local entrepreneurs to avoid the financial, environmental, and sociological pitfalls of nature tourism.[51]

Environmental Protection

Creating the infrastructure to promote tourism can create problems for protected areas. Immigrants, attracted by employment opportunities or the opening of access roads, may abuse lands surrounding and within a park.[52] At the present time, the majority of immigrants to southeastern Peru come not for tourism but for gold mining and government-supported projects such as cattle ranching and agriculture, which all promote land conversion. Madre de Dios is not an area with intense land pressures or severe conflicts between local needs and park preservation, as in Africa or Asia, where local people cut "protected" forest for firewood or are resentful of the competition of large wild animals with their cattle at water holes.[53] As such problems become more significant in Latin America, tourism may offer one of the most sustainable uses of land by requiring a minimal amount of forest conversion.

Littering, disturbance to wildlife, uncontrolled hunting, and species introductions can result from tourism programs that are underregulated or oversubscribed.[54] For example, the Galapagos—considered a model of nature tourism planning—is currently visited by more people than is legally permitted and is considered safe for the ecology of the islands.[55] Tourists and guides in the Galapagos have not followed or enforced park rules designed to protect the wildlife and flora of the islands. One direct way of limiting environmental degradation due to tourism is by improving the training of guides. Guides knowledgeable of park regulations and of the impact of tourists on ecosystems can help to determine whether nature-tourism projects were constructive or destructive. Training of guides should be supported by tour companies and by governmental and nongovernmental programs.

Improving Local Perceptions and Participation in Ecotourism

Exporting tourism revenues from a local area to the capital city or abroad eliminates the opportunity to build local support for land and wildlife preservation. It is vital that tour operators recognize the need to foster good relations and to promote economic and social cooperation with local communities to protect the park or other natural area upon which their success depends. It may be necessary for local or national governments to impose quotas for local hiring or to encourage some other means of profit sharing with local communities. Whenever possible, park policies should encourage and facilitate the participation or leadership of local and indigenous people in the tourism industry, particularly by subsidizing infrastructure development.

The degree to which local people can be assimilated into tour operations is often limited by their knowledge of tourist wants and needs, natural history, or foreign languages. Overcoming this problem may require novel approaches, such as using interpreters to communicate the knowledge of local or indigenous people to tourists. Many guides could learn English or other foreign languages under the support of government programs or grants from nongovernmental conservation organizations or by the tourist companies themselves. Tour companies should be willing to finance some education of their guides, and the use of protected areas should be limited to groups that are guided by registered people who meet certain standards of knowledge.

Local communities should be shown both how wildlife tourism benefits them economically and how tourism is dependent on good conservation practices. The indirect benefits of ecotourism must be highlighted, including the incidental expenditures by tourists in local towns, watershed protection, and other benefits of the conservation of animal and plant resources. Any indirect revenues that go to the local community must be shown to depend on intact forest and animal populations. At this time, slide shows about the benefits of conservation and appropriate development options for the southern rainforest are being given by members of ACSS in Puerto Maldonado, Cuzco, and elsewhere in southeastern Peru.

Initially, emphasis should be placed on pursuing nature tourism on a small scale in order to achieve the dual development and conservation goals described in this chapter. The successes of the Madre de Dios example come in part from the low number of tourists visiting the area and the slow development of the industry. Generally speaking, the more slowly and carefully tourist operations begin, the more equitably profits are distributed, and the greater the input of local food, labor, and materials, the more likely these projects are to succeed financially and foster conservation and community development.[56]

Conclusions

Nature tourism is a promising means of achieving sustainable development while preserving natural areas in Latin America. Examples of successful tourist operations that are managed by local people are increasing,[57] and each provides particular lessons. In Madre de Dios these lessons come mostly from the successes and shortcomings of tourism in the Madre de Dios Biosphere Reserve and from the demonstration of how accessible areas (the lodges in Puerto Maldonado) can be highly profitable, even when not managed optimally. The slow introduction of tourism to Manu, the careful monitoring of tourism's effects on the habitat, wildlife, and people in and surrounding the reserve, and the provision of alternatives to large infrastructure investments (such as permitting beach camping and providing trails to oxbow lakes), have all contributed to a greater level of participation by local people than in most nature-tourism projects. As tourism to Manu Biosphere Reserve continues to grow, we hope it can augment the benefits to local people as well as maintain its current low level of impact to the reserve, and that it will serve as a model for the development of healthy nature-tourism projects throughout Latin America. As with most sustainable-development projects, nature tourism will require guidance and support from local people, governments, and nongovernmental agencies to achieve its goals.

Notes

1. R. T. Devane, unpublished data.
2. Ibid.
3. Ibid.; L. Alpine, "Trends in Special Interest Travel," *Specialty Travel Index* 13 (1986): 83–84.
4. D. Western, "Amboseli National Park: Human Values and the Conservation of a Savanna Ecosystem," in *Proceedings, World Congress on National Parks and Protected Areas*, ed. J. A. McNeely and K. R. Miller (Bali, Indonesia: IUCN, 1982), pp. 93–100.
5. E. De Kadt, "Social Planning for Tourism in Developing Countries," *Annals of Tourism Research* 6 (1979): 36–48.
6. A. W. Weber, "Socioeconomic Factors in the Conservation of Afromontane Forest Reserves," in *Primate Conservation in a Tropical Rainforest*, ed. C. W. Marsh and R. A. Mittermeier (New York: Alan R. Liss, 1987), pp. 205–29.
7. C. L. Jenkins and B. M. Henry, "Government Involvement in Tourism in Developing Countries," *Annals of Tourism Research* 9, no. 4 (1982): 499–521; H. R. Mishra, "Balancing Human Need and Conservation in Nepal's Royal Chitwan Park," *Ambio* 11 (1982): 246–51.
8. J. V. Beekhuis, "Tourism in the Caribbean: Impacts on the Economic, Social, and Natural Environments," *Ambio* 10 (1981): 325–31; E. Cohen, "Jungle Guides in Northern Thailand: The Dynamics of a Marginal Role," *Sociology Review* 30 (1982): 234–66.
9. See, for example, C. Saglio, "Tourism for Discovery: A Project in Lower Casamance, Senegal," in *Tourism: Passport to Development?* ed. E. DeKadt (Washington, D.C.: World Bank, 1979), pp. 321–35.
10. A. W. Weber, personal communication.
11. Weber, "Socioeconomic Factors in the Conservation of Afromontane Forest Reserves."
12. Charles Munn, unpublished data.
13. Ibid.
14. *Statesman's Yearbook* (New York: St. Martin's Press, 1987).
15. M. Rios et al., *Plan Maestro del Parque Nacional de Manu, Sistema Nacional de Unidades de Conservacion* (Lima, Peru: Ministerio de Agricultura y Alimentacion, 1986).
16. Ibid.
17. Ibid.
18. G. Ruiz, *Fudamentos y Programa de Manejo Para Uso Publico del Parque Nacional del Manu* (master's thesis, Universidad Nacional Agraria, La Molina, Lima, Peru, 1979).
19. Charles Munn, personal communication; K. Renton, former guide at Explorer's Inn, personal communication.
20. According to the Ministry of Industry, Tourism, and Integration (MITI).
21. M. Gunter, personal communication.
22. According to MITI.
23. B. Gomez, personal communication.
24. According to MITI.

25. According to the Association of Canoeing Companies of Cusco (ASEC).

26. Ruiz, *Fudamentos y Programa de Manejo Para Uso Publico del Parque Nacional del Manu*; Rios et al., *Plan Maestro del Parque Nacional de Manu, Sistema Nacional de Unidades de Conservacion.*

27. L. Yallico and G. Ruiz, personal communication.

28. See M. J. Groom, "Recommendaciones Sobre el Control de los Efectos Negativos de Turismo Contra la Anidacion de Aves de la Orilla," (Lima, Peru: Technical report to the Director General of Forests and Wildlife, Ministerio de Agricultura, 1986); and M. J. Groom, "Management of Ecotourism in Manu National Park: Controlling Negative Effects on Beachnesting Birds and Other Riverine Animals," in *Proceedings of First International Symposium on Ecotourism and Resource Conservation*, ed. J. Kuslev and J. Andrews (Washington, D.C.: Association of Wetland Managers, 1990).

29. D. Ricalde, president of ACSS, personal communication.

30. Groom, "Recommendaciones Sobre el Control de los Efectos Negativos de Turismo Contra la Anidacion de Aves de la Orilla."

31. A. Cuentas, personal communication.

32. D. Western, "Tourist Capacity in East African Parks," *Industry Environment* 9 (1986): 14–16.

33. N. Myers, "The Tourist as an Agent for Development and Wildlife Conservation: The Case of Kenya," *International Journal of Social Economics* 2, no.1 (1975): 26–42; Mishra, "Balancing Human Need and Conservation in Nepal's Royal Chitwan Park"; Weber, "Socioeconomic Factors in the Conservation of Afromontane Forest Reserves."

34. De Kadt, "Social Planning for Tourism in Developing Countries"; V. Smith, "Anthropology and Tourism: A Science-Industry Evaluation," *Annals of Tourism Research* 7, no. 1 (1980): 13–33; Jenkins and Henry, "Government Involvement in Tourism in Developing Countries."

35. J. J. Pigram, "Environmental Implications of Tourism Development," *Annals of Tourism Research* 7 (1980): 554–83.

36. See Groom, "Management of Ecotourism in Manu National Park."

37. Ibid.

38. Ibid.

39. See R. S. De Groot, "Tourism and Conservation in the Galapagos Islands," *Biological Conservation* 26 (1983): 291–300.

40. G. Ruiz, personal communication.

41. H. Kaplan, personal communication.

42. Smith, "Anthropology and Tourism"; G. Halffter, "The Mapimi Biosphere Reserve: Local Participation in Conservation and Development," *Ambio* 10, no. 2/3 (1981): 93–96.

43. De Kadt, "Social Planning for Tourism in Developing Countries"; Smith, "Anthropology and Tourism"; Jenkins and Henry, "Government Involvement in Tourism in Developing Countries."

44. Smith, "Anthropology and Tourism."

45. F. F. Ferrario, "Tourism Potential and Resource Assessment," in *Tourism Planning and Development Issues*, ed. D. E. Hawkins, E. L. Shafer, and J. M. Rovelstad (Washington, D.C.: George Washington University Press, 1980): 311–20.

46. P. Thresher, "The Present Value of an Amboseli Lion," *World Animal Review* 40 (1981): 30–33.

47. P. H. Gray, "Wanderlust Tourism: Problems of Infrastructure," *Annals of Tourism Research* 8, no. 2 (1980): 285–90.

48. Gray, "Wanderlust Tourism"; C. L. Jenkins, "The Use of Investment Incentives for Tourism Projects in Developing Countries," *Tourism Management* 3, no. 2 (1982): 91–97; Weber, "Socioeconomic Factors in the Conservation of Afromontane Forest Reserves."

49. C. L. Jenkins and B. M. Henry, "Government Involvement in Tourism in Developing Countries," *Annals of Tourism Research* 9, no. 4 (1982): 499–521.

50. See *International Tourism Reports*, Peru, 1982, pp. 21–36.

51. Jenkins and Henry, "Government Involvement in Tourism in Developing Countries."

52. Western, "Tourist Capacity in East African Parks."

53. See D. Western and W. Henry, "Economics and Conservation in Third World National Parks," *Bioscience* 29 (1979): 414–18; and Western, "Tourist Capacity in East African Parks."

54. H. J. Nolan, "Tourist Attractions and Recreation Resources Providing for Natural and Human Resources," in *Tourism Planning and Development Issues*, pp. 277–82; Beekhuis, "Tourism in the Caribbean."

55. De Groot, "Tourism and Conservation in the Galapagos Islands."

56. Smith, "Anthropology and Tourism"; C. L. Jenkins and B. M. Henry, "The Effects of Scale in Tourism Projects in Developing Countries," *Annals of Tourism Research* 9, no. 2 (1982): 229–49.

57. Saglio, "Tourism for Discovery."

References

Alpine, L. "Trends in Special Interest Travel." *Specialty Travel Index* 13 (1986): 83–84.

Beekhuis, J. V. "Tourism in the Caribbean: Impacts on the Economic, Social, and Natural Environments." *Ambio* 10 (1981): 325–31.

De Groot, R. S. "Tourism and Conservation in the Galapagos Islands." *Biological Conservation* 26 (1979): 291–300.

De Kadt, E. "Social Planning for Tourism in Developing Countries." *Annals of Tourism Research* 6 (1979): 36–48.

Ferrario, F. F. "Tourism Potential and Resource Assessment." In *Tourism Planning and Development Issues*, edited by D. E. Hawkins, E. L. Shafer, and J. M. Rovelstad. Washington, D.C.: George Washington University, 1980.

Gray, P. H. "Wanderlust Tourism: Problems of Infrastructure." *Annals of Tourism Research* 8, no. 2 (1980): 285–90.

Groom, M. J. *Recommendaciones sobre el control de los efectos negativos de turismo contra la anidacion de aves de la orilla*. Technical report to the Director General of Forests and Wildlife, Ministerio de Agricultura, Lima, Peru, 1986.

———. "Management of Ecotourism in Manu National Park: Controlling Negative Effects on Beachnesting Birds and other Riverine Animals." In *Proceedings of First International Symposium on Ecotourism and Resource Conservation*, edited by J. Kuslev and J. Andrews. Washington, D.C.: Association of Wetland Managers, 1990.

Halffter, G. "The Mapimi Biosphere Reserve: Local Participation in Conservation and Development." *Ambio* 10, no. 2–3 (1981): 93–96.

International Tourism Reports. Peru, 1982.

Jenkins, C. L. "The Use of Investment Incentives for Tourism Projects in Developing Countries." *Tourism Management* 3, no. 2 (1982): 91–97

———. "The Effects of Scale in Tourism Projects in developing Countries." *Annals of Tourism Research* 9, no. 2 (1982) 229–49.

Jenkins, C. L., and B. M. Henry. "Government Involvement in Tourism in Developing Countries." *Annals of Tourism Research* 9, no. 4 (1982): 499–521.

Mishra, H. R. "Balancing Human Need and Conservation in Nepal's Royal Chitwan Park." *Ambio* 11 (1982): 246–51.

Myers, N. "The Tourist as an Agent for Development and Wildlife Conservation: The Case of Kenya." *International Journal of Social Economics* 2, no. 1 (1975): 26–42.

Nolan, H. J. "Tourist Attractions and Recreation Resources Providing for Natural and Human Resources." In *Tourism Planning and Development Issues*, edited by D. E. Hawkins, E. L. Shafer, and J. M. Rovelstad. Washington, D.C.: George Washington University, 1979.

Pigram, J. J. "Environmental Implications of Tourism Development." *Annals of Tourism Research* 7 (1980): 554–83.

Rios, M., C. I. Ponce, A. Tovar, P. G. Vasquez, and M. Dourojeanni. *Plan Maestro del Parque Nacional de Manu, Sistema Nacional de Unidades de Conservacion*. Lima: Ministerio de Agricultura y Alimentacion, 1986.

Ruiz, G. "Fudamentos y programa de manejo para uso publico del Parque Nacional del Manu." Master's thesis, Universidad Nacional Agraria, La Molina, Lima, Peru, 1979.

Saglio, C. "Tourism for Discovery: A Project in Lower Casamance, Senegal." In *Tourism: Passport to Development?*, edited by E. Dekadt. Washington, D.C.: World Bank, 1979.

Smith, V. "Anthropology and Tourism: A Science-Industry Evaluation." *Annals of Tourism Research* 7, no. 1 (1980): 13–33.

South American Handbook. Bath, England: Mendip Press, Trade and Travel Publications, 1989.

Statesman's Yearbook. New York: St. Martin's Press, 1987.

Thresher, P. "The Present Value of an Amboseli Lion." *World Animal Review* 40 (1981): 30–33.

Weber, A. W. "Socioeconomic Factors in the Conservation of Afromontane Forest Reserves." In *Primate Conservation in a Tropical Rainforest*, edited by C. W. Marsh and R. A. Mittermeier. New York: Alan R. Liss, 1987.

Western, D. "Amboseli National Park: Human Values and the Conservation of a Savanna Ecosystem." In *Proceedings, World Congress on National Parks and Protected Areas*, edited by J. A. McNeely and K. R. Miller. Bali, Indonesia: IUCN, 1982.

———. "Tourist Capacity in East African Parks." *Industry Environment* 9 (1986): 14–16.

Western, D., and W. Henry. "Economics and Conservation in Third World National Parks." *Bioscience* 29 (1979): 414–18.

18

Going Going Gone!

Jim Mason

At noon in Macon, Missouri, Lolli Brothers's stuffy, windowless sales arena is packed with people. Under a sea of cowboy hats and camouflage caps, all eyes are watching the bidding on Watusi cattle, an African breed with improbably huge horns. The auctioneer's pitch blasts from the speaker system like machine-gun fire, and after several bursts, it stops. A pregnant cow in the caged-in ring below is sold for $5,800. Gates open, handlers shout, and she trots from the ring, snorting and switching the air with her horn span.

Soon another auctioneer takes the microphone and stirs up bidding on a run of miniature cattle, then zebu, a dewlapped, humped type of domestic cattle common in Africa and Asia. I get up and meander about—out to the stalls and pens where people are huddled in groups, muttering. The usual air of a country fair is missing today; the mood here is down.

Only a couple of hours earlier, a man had been fatally gored in the Lolli Brothers auction ring. As hundreds watched that morning, a water buffalo cow had attacked her owner, thrown him across the ring, and pinned him against a fence. The cow then charged and injured another man when he tried to rescue the owner. Observers said the cow had reacted in defense of her three-month-old calf, who was also up for sale.

Now, outside the arena, braced by September's clearest air and sun, it somehow seems surreal that by evening this auction will be offering fainting goats—a breed with an inherited muscular defect called *myotonia congenita*, which causes the animal to stiffen and fall after a hand clap or some sudden

---

This article originally appeared in *Audubon* (July/August 1993): 78–83. Reprinted by permission of the author. Copyright © 1993 Jim Mason. All rights reserved.

stress. The sale is sanctioned by the International Fainting Goat Association, whose logo shows a goat flat on its back with its legs straight in the air.

By week's end, Lolli Brothers auctioneers had sold ostriches, rheas, moose, elk, various kinds of deer, kangaroos, llamas, camels, zebras, miniature horses and donkeys, primates, cats, birds, and reptiles—ten thousand animals in all. And those were the live ones. Some one thousand head and full-body taxidermy mounts also changed hands during the six-day Fall Exotic Animal and Bird Sale, which was touted as "the largest sale of its kind in the world."

Lolli Brothers is the epicenter of a broad belt of exotic animal auctions that stretches from Ohio to Texas. These sales have been "proliferating like mad" in recent years, according to Richard Crawford, an acting assistant deputy administrator for the U.S. Department of Agriculture's Animal and Plant Health Inspection Service. Together with dozens of magazines such as *Rare Breeds Journal* and *Animal Finders' Guide*, the auction circuit helps feed a market in rare and unusual animals that now exceeds $100 million a year, according to industry leaders' estimates.

Some of them say it's nothing new, that people have owned unusual animals and birds for generations. "We're not the new kid on the block," says Pat Hoctor, publisher of *Animal Finders' Guide*. "We've been around a while." Others, including Maureen Neidhardt, editor of *Rare Breeds Journal*, say the current boom started roughly twenty years ago, when zoos began selling their excess animals on the private market, mostly in North America. Through brokers and dealers, the animals have gone to private collectors, breeders, trainers, drive-through zoos, and hunting preserves.

Now, however, zoos claim to be choking off the supply. They are responding to reports—including a January 1990 story on CBS's *60 Minutes* and a 1992 *Audubon* article—that animals from publicly supported zoos were ending up in "canned hunts" on private preserves and being shot like fish in a barrel. But thousands of animals and birds usually found in zoos, including endangered species, are already in private hands and are caught up in the flood of commerce. Enough, apparently, to breed an industry that feeds on the public's fascination with exotic animals—rare animals, beautiful birds, ugly reptiles; the cute and the cuddly, the dangerous, the wild, and the weird.

The recent boom in business has stirred up a chorus of complaints. With exotics readily available, critics say, many fall into the hands of careless, naive owners and suffer or die of neglect. Some escape and hurt people. Escaped exotics can also harm native wildlife by preying on it, competing for habitat, spreading disease, and polluting the gene pool.

For some critics, the often freewheeling trade raises deeper questions about our dealings with animals: Should wild species be selectively bred to make docile, cute pets? Should we propagate miniature animals and other mutations for our amusement? Should endangered species be farmed so that everyone can own a last piece of the wild?

Wind of these criticisms probably led to some of the gloom in the crowd at Macon last September. But most of it came from the deadly goring that had occurred that morning. The incident brought still more grief for exotic-industry insiders—more bad press, more unpleasant visibility.

You could see the tension in some of the signs posted about the Lolli Brothers compound. "No cameras or recording devices," said one. Another was a laundry list of required government permits, disease tests, and health papers, as well as details on cage sizes and materials. Several fliers called for a meeting to form a national organization to protect and defend the rights of breeders, dealers, and owners of exotic animals—or "alternative livestock," as they prefer to call them.

At an earlier fall auction in Ohio, I watched the bidding on a run of bottle-raised black-bear cubs, several lynx, and a six-month-old female coyote—"leash-broke and raised with children," according to the auctioneer. Behind me, a couple had a small spotted cat on a leash; I asked them about it. The animal was a two-month-old male leopard kitten; "you just raise him like a dog," they told me. Another couple nearby had three tiny monkeys—all in diapers. A crowd of about 100 milled about, waiting for the sale of macaques, capuchins, squirrel monkeys, tamarins, grivets, and other primates to begin.

These are the sales that make pet experts like Michael W. Fox nervous. The veterinarian, who is an author and a nationally syndicated columnist on pets and wildlife, says that nondomestic animals offer more trouble than companionship. "They are not biologically adapted to live with humans," he says. "Some of these animals can be dangerous." If they escape, as many do, or are released by frustrated owners, Fox says, some species can spread diseases to domestic livestock and native wildlife that can be "quite devastating."

Ohio's public health veterinarian, Kathleen Smith, has examples on file. "I hear the horror stories," she says. "People are getting a big kick out of owning something unusual. It's fine when it's young, but it grows up, and they are probably not ready for what they have to deal with." She pulls a file and tells of a Carroll County, Ohio, man who kept a wolf at his weekend farm. Half-starved, the wolf escaped in May 1989 and stalked and attacked a boy who had gotten off a school bus. Smith also notes a pet cougar's 1990 attack on a two-year-old Ohio boy and then on his grandmother when she tried to rescue him. "You can hear these from just about any state public health official in the county," she says.

In Missouri, David Erickson, assistant chief of the wildlife division of the state Department of Conservation, cites the disease risks. He tells about a free-ranging group of Japanese macaques that were finally rounded up last fall near Branson—a town famed for its country-music shows and hordes of tourists. An Asian species that can thrive in temperate regions, the monkeys had been roaming back and forth for years between the Ozark forests and Wilderness Safari, a local tourist attraction. The animals tested were found to be infected with simian herpes—a virus that Erickson says is "fairly harmless to them but fatal to humans."

Luckily, the feral macaque population was recaptured before it could menace tourists or native wildlife. "These monkeys can forage and virtually eliminate native birds," Erickson says. "They can get right up to the treetops and raid nests."

Erickson points to the diseases that can be transmitted by the introduction of any nonnative wildlife. Missouri allows neither the possession of skunks nor the import of raccoons because of concerns about the spread of rabies, he says. And he notes there is a "very strong suspicion" that the movement of captive raccoons from southeastern states was responsible for recent rabies outbreaks in the Northeast.

With a German accent that purrs and lilts like a brogue, Canadian zoologist Valerius Geist distills the list of problems caused by runaway exotic species: interbreeding, competition, transmission of disease. The University of Calgary professor has been an outspoken critic of North American elk ranching for the Asian horn market, noting that a European subspecies of elk—called red deer—were escaping from some ranches, hybridizing native American elk and possibly spreading bovine tuberculosis. Geist gives other examples of how exotics can harm native wildlife. The mouflon—a wild species originally from Iran that is believed to be the ancestor of domestic sheep—presents native sheep with all three major threats cited by Geist. Oregon has held up reintroduction of native bighorn sheep for fear that escaped mouflons would ruin their chances of survival.

Geist also mentions Asian sika deer and African aoudads, or Barbary sheep—both favorites with exotics fanciers. The aoudad, he says, "has been liberated in the southern states, it now may be shot on sight in Arizona, and it will almost certainly outcompete the bighorn sheep very rapidly once they meet." Although sika deer cannot interbreed with American species, "they are very tough competitors," Geist says, "and they outdo white-tailed deer very quickly and efficiently." He goes on, talking of Asian sheep, European moose, European foxes, and the kinds of threats they could pose to North American wildlife.

Such talk makes people like Debbie Kolwyck angry. Kolwyck, who raises primates, elk, deer, and various other exotics at her home near Kansas City, is president of the Missouri Animal Association. She says she understands the concern about dangerous animals. "We're not comfortable with somebody with a tiger in town. It's like waiting for an accident to happen." But she believes exotics owners have a constitutional private-property right to keep their animals if they maintain them properly. "We have more control over caging our exotics than people do with Jersey bulls," she says.

Pat Hoctor says as much, but with a great deal more indignation. His editorials in *Animal Finders' Guide* lash out at "do-gooders," government officials, humane societies, zoos, environmentalists, and the rest of the motley pack of critics that is hounding his industry. In California, Wyoming, and Arizona, for example, such criticism has led to bans or severe restrictions on the possession of species whose presence could harm native wildlife or people. Louisiana, New Hampshire, Wisconsin, and South Carolina have significantly tightened their controls in recent years. Not far behind are Missouri, Ohio, Indiana, and Colorado, where new laws or regulations currently are being proposed.

Feeling surrounded by the baying critics, Hoctor, who raises and sells exotics near Prairie Creek, Indiana, insists that his industry is "very, very responsible as a whole" in the management of its animals. "The problem is," he says, "when a cougar bites a kid in California, you read about it in New York. It's a media problem, actually."

Hoctor doesn't think much of state fish and game officials either. "It's very easy to understand why fish and game departments don't like us," he says. "What good are we to them? They don't know a damned thing about us, so they feel very inadequate regulating us. And where does their money come from? From selling licenses to shoot animals."

For a $500 stud fee you can have your very own "zorse" or "zonie," according to an advertisement in the September-October 1992 issue of *Rare Breeds Journal*. These offspring of zebras and horses or ponies are becoming trendy, as are "zedonks," hybrids of zebras and donkeys. For $700 you can buy one of Hoctor's "liger" cubs—a hybrid of a male lion and a female tiger—says an ad in the *Animal Finders' Guide* of November 1, 1992. (Ligers are "the nicest animals we've ever had in the way of cats," Hoctor says.) Another advertisement in that issue lists lions, a cheetah, and leopards for sale "to approved and licensed premises"; also "eland, kudu, impala, oryx," and other animals "directly from Africa." Yet another ad lists several white tigers, a snow leopard, and a clouded leopard ("hand raised") for sale from the Tanganyika Wildlife Company in Wichita, Kansas.

In the December 1, 1992, *Animal Finders' Guide*, one ad offers wolf cubs, another, wolf-dog hybrids. On another page an Indiana couple advertises "stocking stuffers"—rhea chicks, cougar cubs, and exotic sheep. In the January 1, 1993 issue a Missourian offers a six-week-old chimpanzee for $25,000. In the February 15 issue, a Floridian offers "cougars, $500; African leopards, $1,250; North Chinese leopards, $1,250–$1,500; Bengal tigers, $1,000–$1,500; snow leopards, $5,000–$7,000. All bottle raised and well adjusted."

The laws protecting these exotic animals are not nearly as stringent as you might expect. Commercial trafficking of endangered species such as chimpanzees, cheetahs, leopards, wolves, and tigers is prohibited under federal law. But you can get one straight from the wild if your reasons persuade the U.S. Fish and Wildlife Service, which can issue permits under the Endangered Species Act if the animal is needed for "scientific purposes or to enhance the propagation or survival of the species." If you register with Fish and Wildlife, you can buy an endangered animal born in captivity in the United States. To make matters more complicated, if you are trading a native endangered species across state lines, a permit is needed for every transaction, but if you are trading exotic endangered species you need none, provided that you and the person with whom you are trading are registered.

You need not have a Fish and Wildlife permit, however, to buy from an endangered-animal breeder in your state, because the federal government has no jurisdiction over intrastate transactions. Nor do you need a permit to own a liger or some other hybrid of an endangered species, because they are deemed nonmembers of the species. As for lions, cougars, zebras and their hybrids, and other nonendangered animals, it is perfectly legal to own one for a pet as long as you comply with state and local laws.

Welcome to the confusing array of laws that are supposed to keep rare wildlife in its native habitat and out of the auction ring. This crazy quilt of state statutes and local ordinances is so diverse and rapidly changing that an attempt to sort it out just might fry the circuits of an IBM mainframe. Bear in mind, too, that even where the laws are stringent and clear, enforcement often is inadequate.

With such a complex legal tangle, it's no wonder Pat Hoctor, Debbie Kolwyck, and their industry feel unfairly regulated. And it's no wonder that government authorities feel overwhelmed—or that so many animals, individuals as well as entire species, fall through the gaps.

There are two sets of legal protections: international and domestic. The flow of rare animals from foreign habitats into the United States is controlled primarily by the federal Endangered Species Act and the 118-nation Convention on International Trade in Endangered Species (CITES), which are supposed to protect rare creatures from being destroyed by commercial demands. The problem is that the laws allow exceptions, and these create most of the confusion and the regulatory burdens.

Domestic protection of endangered animals also suffers because of what some see as the bungled implementation of the Endangered Species Act. Critics zero in on the captive-bred wildlife registration system, by which the Fish and Wildlife Service allows ownership and domestic trade in endangered animals if they were born in captivity in the United States and are used for propagation, education, and purposes that enhance conservation. This system is supposed to keep captive-bred animals out of the stream of commerce.

But does it? "Something is very wrong" with the captive-bred registration system, according to Donald Bruning, curator and chairman of the Department of Ornithology at the Wildlife Conservation Society (formerly the New York Zoological Society). "Because of the way the system works—or doesn't work—anybody or his brother has been able to get a permit. So you have all kinds of people who can legally move animals back and forth. They breed any number they want, they sell them to whomever they want, they take them to shoot them, use the skins, do all kinds of things with them, which, to me, are very inappropriate for endangered species."

Even the Fish and Wildlife Service acknowledges problems with the captive-bred registration system, though in guarded, official language. "Yes, there's a lot of concern," says Kenneth Stansell, deputy division chief of the Office of Management Authority. "We have had ten years or so to look at the system, and we now realize that we need to take another look at it." His office is currently drafting new regulations.

Bruning, who has been consulted in the draft process, thinks the new regulations are probably on hold until after Congress reauthorizes funding for the Endangered Species Act, which is expected later this year. Meanwhile, the registration system's wide gates remain open.

I asked the Fish and Wildlife Service how many captive-bred endangered animals are now in the United States. They have figures on pheasants and some birds but "no reliable estimates" for chimps, tigers, leopards, and the like. This concerns zoo professionals like Michael Hutchins, director of conservation and science for the American Association of Zoological Parks and Aquariums (AAZPA), who urges that endangered animals be bred only within programs organized or recognized by AAZPA zoos.

"The question is, why are so many [captive-bred] permits being given out?" asks Hutchins. The Fish and Wildlife Service says that as of 1990 there was a total of 850 individuals, zoos, and companies registered to trade in captive-bred endangered animals. But Hutchins estimates that maybe twice that number are actually doing captive breeding—and only 160 of them are AAZPA-accredited institutions. Though Hutchins won't point fingers, most of the country's captive breeders of endangered species clearly are the people who would like to be known as the alternative-livestock industry.

Maureen Neidhardt of *Rare Breeds Journal*, who is also an exotic-animal farmer, argues that her industry has "a very, very successful rate of preserving these animals," although she acknowledges there is "a bit of a rift" between exotics breeders and the zoo community over this issue. Conservationists say otherwise. The breeders' animals "are amateurly caged, amateurly cared for, and amateurly bred," says Sue Pressman, formerly director of captive wildlife for the Humane Society of the United States and now a private consultant.

"These quasi-zoologists will tell you that they are saving the tiger and the lion and the whatever-it-is in their backyard," says Pressman. "But if you talk to the zoo people they will tell you these are genetically dead animals." She maintains that no legitimate public zoo will buy anything from these exotic breeders because their animals are often inbred, hybridized, or otherwise not genetically fit for bona fide conservation programs. Such programs carefully organize all breeding, according to Hutchins, ensuring close monitoring of the gene pool of a captive population of endangered animals. The long-range aim is to keep the population genetically healthy so the animals will have a better chance of surviving when it becomes possible to return some to the wild.

How did the captive breeding of endangered animals get so out of control? Most experts in and out of government say it happened because there simply hasn't been much concern about captive-bred animals. The real emphasis has been on protecting wild-caught animals, on reducing the numbers being taken from natural habitat. Neither government officials nor conservationists have regarded captive-bred populations as very important in the grand scheme of conservation. The burden of closely regulating them, they say, would paralyze authorities and keep them from the more important work of protecting animals in the wild.

The good news is that most experts now realize that turning captive-bred animals into commodities hurts all conservation efforts. It keeps a price on the head of endangered animals, which encourages corruption and illegal trade. It also undermines educational efforts to build respect for the very animals we are trying to save in their habitat.

Perhaps we have been too single-mindedly focused on endangered species. If we truly want to save them and what's left of the natural world, we ought to consider the implications of the commoditization of nonendangered wildlife and freaks of nature. It does not help build respect for zebras and lions when those in captivity are turned into dancing acts and zorses and ligers. Nor does it help build respect for nature when one wild species after another is captive bred, bottle raised, and domesticated for the pet trade. And what about some of the other limits of domestication? What values about the living world are passed along when it is trendy and profitable to breed miniature horses and fainting goats?

Interestingly, many people I talked to felt disgust about the propagation of ligers, zedonks, fainting goats, and the like, but hardly anyone could explain what is wrong with it. It was just a gut reaction, hard to verbalize, they said. One said, "God almighty, whether you believe in God or what, these creatures are the products of millions of years of evolution. Just leave them to hell alone."

In a conversation with Pat Hoctor, I suggested that many saw the making of ligers, white tigers, wolf hybrids, and some of his industry's other new products of entrepreneurial breeding as frivolous—even offensive. Hoctor took offense. "I hope you realize you're talking to the biggest breeder of ligers in this state." He argued that the liger gave zoos and safari parks a healthier, more vital "crowd-getter" than the white tiger. Hoctor noted that white tigers are heavily inbred and have a lot of health problems, so many have to be destroyed. He dismissed the critics. "What I'm saying is, if that animal is of use to us and we are caring for it properly, what business is it of theirs?"

Contributors

CATHY SUE and ROGER ANUNSEN are the Pacific Northwest coordinators for the Fund for Animals.

VICTORIA BUTLER is a freelance writer from Washington, D.C. She has spent time in southern Africa as a journalist.

GRAEME CAUGHLEY was a chief research scientist in vertebrate biology at the Commonwealth Scientific and Industrial Research Organization and a fellow of the Australian Academy of Science.

ANDREW NEAL COHEN is an environmental scientist for the San Francisco Estuary Institute. His work focuses on environmental policy and the ecology of marine and aquatic invasions.

GORDON GRIGG is a professor of zoology in the Department of Zoology and Entomology at the University of Queensland, Australia. He is also the deputy director of the university's Centre for Conservation Biology.

MARTHA J. GROOM teaches in the Department of Zoology at the University of Washington. She has published articles in several journals, including *American Naturalist*.

RAYMOND HAMES is a professor of anthropology at the University of Nebraska. He has written many articles on the lives of native Amazonians and their adaptation to the Amazonian ecosystem.

RICHARD B. HARRIS teaches in the School of Forestry's wildlife biology program at the University of Montana.

REBECCA L. JOHNSON is an associate professor in the Department of Forest Resources at Oregon State University. Environmental and resource economics is one of her research interests.

GILSON B. KAWECHE is a member of the Board of Directors of National Parks and Wildlife Services in Zambia.

ROGER J. H. KING teaches philosophy at the University of Maine. He has published several articles on environmental ethics.

DALE LEWIS works with the Nyamaluma Wildlife Management Training and Research Center in Zambia. He has published numerous articles on African wildlife preservation.

ROBERT W. LOFTIN taught philosophy at the University of North Florida. He served as vice chairman of the Florida Audubon Society.

MICHAEL V. MARTIN teaches in the Department of Agricultural and Resource Economics at Oregon State University.

JIM MASON is interested in ethics and the treatment of animals. He has published numerous articles on this subject in such journals as *The Animals Agenda* and *E: The Environmental Magazine*.

CHARLES A. MUNN is senior conservation zoologist for the Wildlife Conservation Society. His work has appeared in *National Geographic* as well as in scientific journals.

ACKIM MWENYA is the director of National Parks and Wildlife Services in Zambia.

RODERICK P. NEUMANN is in the Department of International Relations at Florida International University. His research focuses on developmental issues in developing nations.

ROBERT D. PODOLOSKY is an assistant professor in the Department of Biology at the University of North Carolina–Chapel Hill.

R. J. PUTMAN is a member of the Behavioural and Environmental Biology Group at Manchester Metropolitan University. He has researched and published extensively on the biology and behavior of deer.

ALAN RABINOWITZ is the director of science in Asia for the Wildlife Conservation Society. He has conducted extensive research on endangered species.

RAYMOND RASKER works for the Northern Rockies Regional Office of the Wilderness Society. His papers have appeared in journals such as *Conservation Biology* and *Society and Natural Resources*.

VICTOR B. SCHEFFER is a biologist who has worked for many years in Washington State.

JAMES R. UDALL has written extensively on environmental issues. His work has appeared in *Audubon*, *National Wildlife*, *Sierra*, and *Outside*.